Bioeconomy for Beginners

Joachim Pietzsch
Editor

Bioeconomy for Beginners

Springer

Editor
Joachim Pietzsch
Frankfurt, Hessen, Germany

Illustrations by Stephan Meyer
Satz- und Grafik-Studio
Dresden, Sachsen, Germany

Drawings by Wolfgang Zettlmeier
Graphik & Text Studio
Barbing, Germany

Editorial advisor Ulrich Schurr
Forschungszentrum Jülich
Institutsleiter IBG-2 Pflanzenwissenschaften
Jülich, Germany

This book is a translation of the original German edition "Bioökonomie für Einsteiger" by Joachim Pietzsch, published by Springer-Verlag Germany in 2017. The translation was done with the help of artificial intelligence (machine translation by the service DeepL.com). Springer Nature works continuously to further the development of tools for the production of books and on the related technologies to support the authors.

ISBN 978-3-662-60389-5 ISBN 978-3-662-60390-1 (eBook)
https://doi.org/10.1007/978-3-662-60390-1

This Springer imprint is published by the registered company Springer-Verlag GmbH, DE part of Springer Nature.
The registered company address is: Heidelberger Platz 3, 14197 Berlin, Germany

Foreword

"Basically, bioeconomy is nothing new". With this remarkable sentence begins the first chapter of this book, *Bioeconomy for Beginners* (although, fortunately, not only for "beginners"). We, a small group of EU officials in the Directorate-General for Research, were well aware of this when, in September 2005 in Brussels, we presented the Knowledge-Based Bioeconomy (KBBE) Programme to the public as a new component of the 7th EU Research Programme, with a budget of two billion euros as starting baggage. We were of the opinion that the immense knowledge available today about the particularities of the so-called biological resources of plant, animal and microorganism (renewability, climate friendliness, elements for a circular economy and, above all, potential for new functions and properties) justified such a novel and provoking approach, especially when comparing non-fossil resources with fossil resources. But this approach was initially "limited" to research activities.

We would not have imagined at that time that, in 2017, just 12 years later, almost 60 states, international organisations and regions worldwide would have adopted this new, old economic concept in the form of national programmes, strategies, action plans and road maps going beyond research and technology!

The number of scientific and nonscientific statements (and everything in between), books, essays and reports on the bioeconomy can hardly be counted anymore; specialised web portals are useful for dealing with that. But a compendium, *Bioeconomy for Beginners*, does not yet exist, at least not in German-speaking countries, as the editor rightly states in his preface. He adds that the main criterion for this output should be intelligibility in language and content. Nevertheless, with all due respect to this qualification, this book is not a so-called popular scientific work. Fortunately, the complexity and density of the bioeconomy, as we know it today due to our enormous sum of new knowledge about the biological resources on which it is based, require more than linguistic comprehension and understanding for a successful entry into this "economy of life", namely, intellectual sincerity and honesty, scientific seriousness, freedom from ideologies, genuine knowledge and true authentic competence and, above all, openness to other newly generated flows of knowledge, for example, in the nano-field and information field or, recently, in the context of the digital revolution. In my view, these criteria have been fully met, especially in the exciting remarks on the future of the bioeconomy, including in the context of the current discussions on Sustainable Development Goals ("SDGs") and how to achieve them, the circular economy and the further development of global climate protection according to COP 21 and 22.

From the point of view of the "fathers" or "founders" of the European bioeconomy, I would like to take this opportunity to state to all authors and beginners: we have never imagined that, with this old and new form of economy, we would be offering a *silver bullet.* We only wanted to make a contribution towards ensuring that, with the help of and in harmony with nature, economic actions may continue to enable the billions of inhabitants of our planet to live a sustainable and decent life upon it. This desire and concern are very skilfully and convincingly expressed in many contributions to this book, and I would like to express my sincere thanks for that.

Christian Patermann
Bonn, Germany

March 2017

Preface

"Bioeconomy" has become a frequently used buzzword in specialist circles in politics, business and science over the past decade. Many talk about it, but it is often unclear as to what is meant when people speak of the bioeconomy. After all, experts from very different industries and disciplines are at work in service of its realisation. In the general public's perception, the term simply does not exist. "Hardly anyone knows the term bioeconomy. Yet it stands for the most ambitious economic project of the future". In fact, "bioeconomy" is far more than merely a buzzword for insiders. Rather, the word denotes a concept that must never go out of fashion if mankind is interested in long-term survival on this Earth. It is about the necessary transition from the age of fossil fuels, which began about 200 years ago, into a worldwide economic system based on renewable raw materials (and renewable energies).

The purpose of this book is to present the fundamentals of the concept of bioeconomy. Without losing sight of its possible diversity, it sees the realisation of this concept as a threefold challenge: a scientific one, an economic one and an ecological one. By explaining these three challenges from the not necessarily consistent perspectives of highly qualified authors, it offers an integral introduction to bioeconomy, a well-founded introduction to a dynamic field of research and practice that will raise more questions than it answers, and yet one that fills a gap. So far, there has been no generally understandable introduction to the field of bioeconomy.

The age of fossil fuels, the peak of which we are currently living through, will one day have been only a brief epoch in the history of human development. The reason why this statement is likely to be true is described in the introductory ▶ Chap. 1 through classification of the baseline conditions of a knowledge-based bioeconomy historically and geographically.

Bioeconomy is based on the energetic and material use of biomass, on the one hand, and the application of biological systems, on the other. Where this biomass comes from is explained in ▶ Chap. 2. It describes the provision of biomass from the fields of agriculture, forestry, fishery and waste management.

Biomass, whose use is at stake, must primarily benefit the nutrition of the growing world population. ▶ Chapter 3 therefore outlines what bioeconomy means for the food and feed sectors and highlights essential elements of nutrition in the context of bioeconomy.

▶ Chapter 4 presents the path from biomass to those platform chemicals that also form the basis of the petroleum-based economy. It shows how triglycerides, sugar, starch and nonedible lignocellulose can be processed into platform chemicals in biorefineries for the production of fuel and chemicals.

Biotechnology plays a key role in the bioeconomy. Accordingly, ▶ Chap. 5 first introduces the current importance of biotechnology as a production process and then describes the perspectives of synthetic biology.

From an economic-political perspective, the path to a bioeconomy represents a controlled transformation. ▶ Chapter 6 discusses the possibility of a transformation of the world production system towards a knowledge-based bioeconomy.

From a business management point of view, the successful transition to a bioeconomy requires the integration of a wide variety of industries and disciplines that have had little to do with each other up to now, i.e. the formation of new value creation networks. The opportunities and challenges associated with this will be discussed in ▶ Chap. 7.

Idealistic motives alone will contribute little to the success of bioeconomic products. Rather, these products must be able to compete with fossil-based products in terms of their manufacturing costs and sales price. The emerging bioeconomy must survive this competition on the market in order to find customer acceptance. Its prospects in this regard and what means will be necessary for it to become successful are described in ▶ Chap. 8.

In addition, bioeconomy must match verifiable sustainability criteria, for the prefix "bio" alone does not meet the claim of being ecological. Therefore, ▶ Chap. 9 investigates, starting from the United Nations' *sustainable development goals*, the conditions for a sustainable bioeconomy.

In official announcements, bioeconomy is often seen as a key to unlimited economic and consumption growth. But is this the core of its self-conception, and does it correspond to the goal of a transformation to sustainability? Shouldn't bioeconomy rather be oriented towards sufficiency strategies? This question is addressed in ► Chap. 10 from a philosophical perspective.

Some years ago, my friend Richard Gallagher, then editor of *The Scientist*, now president and editor in chief of *Annual Reviews*, made me aware of the importance of bioeconomy. My first thanks go to him at this point. A chance encounter with Merlet Behncke-Braunbeck from Springer Verlag gave me the impulse to develop the concept of this book. With untiring enthusiasm and competence, she supported me in implementing this concept step by step. Without the inspiring preliminary talks with "my" authors and their committed, knowledgeable and reliable work, realization of this book would have been unthinkable. I would like to thank them very much for this, in particular, Professor Ulrich Schurr, who was my scientific advisor. I would like to thank Carola Lerch for her meticulous and precise project management, as well as everyone who was involved in the production of this book on behalf of the publisher. I would especially like to thank my wife, Ellen Scheibe, for her support and her loving understanding during my not always easy but always exciting work on this book.

Joachim Pietzsch
Frankfurt, Germany

September 2016

Contents

Contributors

Chad M. Baum
University of Bonn, Institute for Food- and Resource Economics (ILR)
Bonn, Germany
ch.baum@ilr.uni-bonn.de

Silvan Berg
University of Bonn, Institute for Food- and Resource Economics (ILR)
Bonn, Germany
s.berg@ilr.uni-bonn.de

Michael Bott
Forschungszentrum Jülich, Institute of Bio- and Geosciences (IBG-1: Biotechnology)
Jülich, Germany
m.bott@fz-juelich.de

Klaus-Rainer Bräutigam
Karlsruhe Institute of Technology, Institute for Technology Assessment and Systems Analysis (ITAS)
Karlsruhe, Germany
klaus-rainer.braeutigam@partner.kit.edu

Stefanie Bröring
University of Bonn, Institute for Food- and Resource Economics (ILR)
Bonn, Germany
s.boering@ilr.uni-bonn.de

Olivier K. Butkowski
University of Bonn, Institute for Food- and Resource Economics (ILR)
Bonn, Germany
o.butkowski@ilr.uni-bonn.de

Uwe Fritsche
International Institute for Sustainability Analysis and Strategy (IINAS)
Darmstadt, Germany
uf@iinas.org

Armin Grunwald
Karlsruhe Institute of Technology, Institute for Technology Assessment and Systems Analysis (ITAS)
Karlsruhe, Germany
armin.grunwald@kit.edu

Manfred Kircher
Kircher Advice in Bioeconomy (KADIB)
Frankfurt, Germany
kircher@kadib.de

Iris Lewandowski
University of Hohenheim, Institute of Crop Science, Department of Biobased Products and Energy Crops
Stuttgart, Germany
Iris_Lewandowski@uni-hohenheim.de

Melvin Lippe
Johann Heinrich von Thünen Institute, Federal Research Institute for Rural Areas, Forestry and Fisheries
Braunschweig, Germany
melvin.lippe@thuenen.de

Jan Marienhagen
Forschungszentrum Jülich, Institute of Bio- and Geosciences (IBG-1: Biotechnology)
Jülich, Germany
j.marienhagen@fz-juelich.de

Jochen Michels
DECHEMA – Society für Chemical Engineering and Biotechnology
Frankfurt, Germany
michels@dechema.de

Joachim Pietzsch
Wissenswort, Frankfurt, Germany
j.pietzsch@wissenswort.com

Nina Preschitschek
University of Bonn, Institute for Food- and Resource Economics (ILR)
Bonn, Germany
n.preschitschek@ilr.uni-bonn.de

Johannes Pucher
German Federal Institute for Risk Assessment
Berlin, Germany
Johannes.Pucher@bfr.bund.de

Andreas Pyka
University of Hohenheim, Institute of Economics, Department of Innovation Economics
Stuttgart, Germany
a.pyka@uni-hohenheim.de

Christine Rösch
Karlsruhe Institute of Technology, Institute for Technology Assessment and Systems Analysis (ITAS)
Karlsruhe, Germany
christine.roesch@kit.edu

Ulrich Schurr
Forschungszentrum Jülich, Institute of Bio- and Geosciences (IBG-2: Plant Sciences)
Jülich, Germany
u.schurr@fz-juelich.de

Rüdiger Unseld
University of Freiburg, Faculty of Environment and Natural Resources
Freiburg im Breisgau, Germany
ruediger.unseld@waldbau.uni-freiburg.de

Introduction

Joachim Pietzsch and Ulrich Schurr

Old beech in a former grazed forest on the Schönberg south of Freiburg im Breisgau. (© N. Reif)

J. Pietzsch (ed.), *Bioeconomy for Beginners*, https://doi.org/10.1007/978-3-662-60390-1_1

1.1 The Replacement of the Original Bioeconomy

Basically, bioeconomy is nothing new. For thousands of years, mankind covered its needs for food, materials, consumer goods and energy through renewable raw materials and renewable sources. The muscle power of humans and farm animals, eventually reinforced by mechanical aids, formed the basis of their economic activity, the primary fuel of which was wood. In addition, there was wind and water for the mills, wind for the sailing ships and, above all, the rays of the sun. Almost all of the energy available on earth comes from these. Even if plants absorb only a part of it and less than 1% is used in the process of photosynthesis, solar energy generates many billions of tons of biomass in the sea and on land every year. Less than a tenth of these plants are eaten by animals, which, in turn, provide a small part of the food for carnivores and people who draw their energy from them. This energy and the heat generated by burning wood, peat and other biomass drove the economies of pre-industrial times: Until about 1780, all societies on this earth were bio-economies. But even then, humankind changed the landscape and adapted it to its needs. It created a cultural landscape that, to the furthest extent possible, no longer resembled the natural landscape as it would have developed without human intervention. Even then, humankind "overused" natural resources – with relevant consequences, such as permanent erosion and overgrazing and disasters such as famines. Even then, the use of natural resources alone did not guarantee sustainability.

Then, the industrial revolution came about and began massively to transform the earth and its landscape. The centre of this revolution was in Europe. Encouraged by the invention of the steam engine, which was able to convert combustion heat into mechanical labour, coal – initially in England – emerged as the most important source of energy, whereas it had merely been a special energy carrier for the smelting of iron in the previous centuries. Outside of north-western Europe, however, the use of coal in production processes did not become noticeable macro-economically before the 1820s. This had far-reaching consequences: "Coal sets steam engines in motion, and steam engines move spindles and pumps, ships and railways. The era of fossil fuels that began in the third decade of the 19th century was therefore not only one of an unprecedented production of goods, but also an era of networking, speed, national integration and facilitated imperial control" (Osterhammel 2009). Industrialization accelerated economic development and – coupled with scientific progress in medicine and the natural sciences – laid the foundation for the fastest known development of a species in Earth's history.

The rise of coal as an energy source took place only gradually, but, around 1890, coal had overtaken biomass as the most important energy source worldwide. Between 1850 and 1914, global coal production increased 16-fold to around 1300 million tonnes per year. 43% of these were mined in the USA, followed by Great Britain, with a share of 25%, and Germany, with a share of 15% (Osterhammel 2009). Hard coal supplied not only energy, but also raw materials for the manufacture of new products, which primarily originated from the coal tar that resulted from her coking. This made it possible to manufacture synthetic dyes and medicines, which, in the second half of the nineteenth century, led to the emergence and rise of the chemical industry.

At the same time, crude oil began its career as a second fossil fuel. Its first commercial spring was developed in 1859 in the US state of Pennsylvania. The decisive impetus for the birth of this industry came from the development of a process for refining gasoline in the 1890s and from worldwide automobilization resulting from the introduction of combustion engines in the twentieth century. Only then did it find a broad basis through the discovery of large oil deposits in Russia, the USA, Mexico, Iran, Arabia, and other countries, creating the basis for worldwide industrialization and the global integration of human economic activity. Due to its greater yield and flexibility, after World War II, crude oil replaced coal as the primary source of raw materials for chemical production. This means that, today, around 90% of the basic chemicals used in all chemical value chains are created from crude oil and petroleum gas. In addition to the manufacture of pharmaceuticals, paints, lacquers and detergents, the production of plastics and fibres is of particular importance. Until about 1965, coal remained the predominant fossil fuel. Only then did crude oil displace it from the top position, while natural gas established itself in third place (McNeill and Engelke 2013).

Today, the primary energy consumption of humankind has reached dimensions that can only be represented in unimaginably high units of numbers (▫ Fig. 1.1).

Expressed in petroleum equivalents, this consumption is around 14 Gt per year, which corresponds to the energy produced by burning 14 billion tonnes of oil and can be converted into around 580 EJ. Fossil fuels account for almost 80% thereof. It is estimated that this share will fall to just over 70% by 2040, despite the persistently steep rise in demand for energy (▫ Fig. 1.2). In terms of continents and countries, energy consumption is very unevenly distributed. At the beginning of our century, for example, the average consumption of a North American was 70 times higher than that of an inhabitant of East African Mozambique (McNeill and Engelke 2013). This reflects the asynchronous and asymmetrical course of the fossil age in terms of economic geography and power politics that has existed from its very beginning. As early as "around 1910 or 1920, the world fell into a minority of those who had achieved access to fossil energy stores and established the infrastructure necessary for their use, and the majority of those who had to get along with traditional energy sources under growing pressure of scarcity" (Osterhammel 2009). As a result, the fact that, "for a century after 1850, high energy consumption was limited mainly to Europe and North America and, to a lesser extent, to Japan ... is probably one of the most important reasons for the political and economic

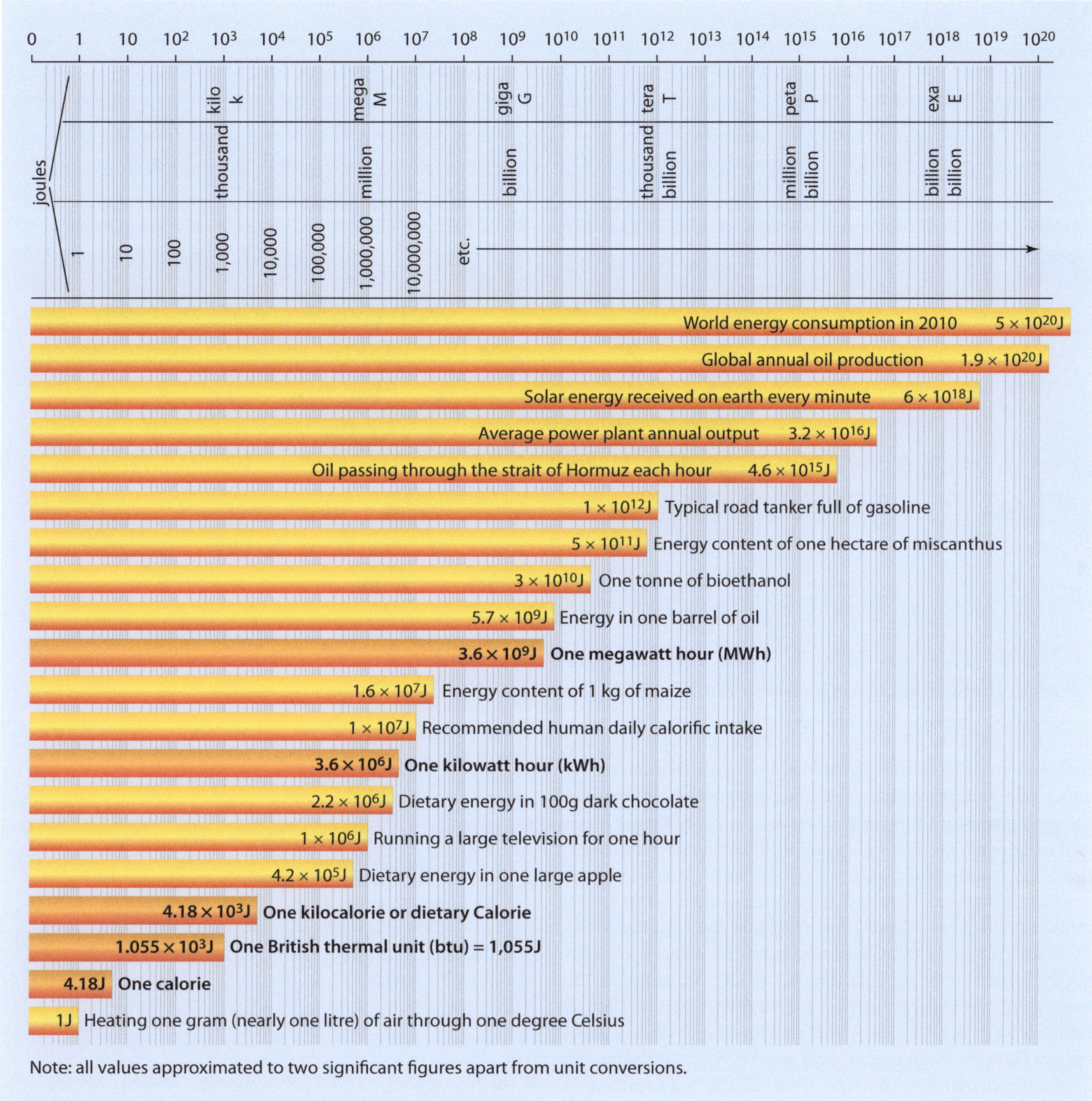

Fig. 1.1 A scale of our energy consumption. (Source: Davis et al. 2014)

dominance of these regions in the international system" (McNeill and Engelke 2013). On the other hand, since 1960, these countries in particular have been confronted with the demands of the Organization of Petroleum Exporting Countries (OPEC), a cartel whose power they felt during the two oil crises in 1973 and 1979, because their prosperity depended on provision with crude oil. This was also evident in Germany. "While the Federal Republic of Germany, with its coal, was almost energy self-sufficient at the beginning of the 1960s, two decades later, it had become dependent on imports by a share of 61%" (Rödder 2015).

1.2 The Ambivalence of the Anthropocene

Undoubtedly, access to fossil energy resources is primarily responsible for the progress to which humankind owes its immense prosperity today, at least in large parts of the world. Let us not forget that, by current standards, we must regard almost all people as having been poor before the beginning of the industrial revolution. They were at the mercy of failed harvests and epidemics without the ability to exert much control, they lived relatively short lives on average, and they existed mostly within a narrowly limited horizon and sphere of

1

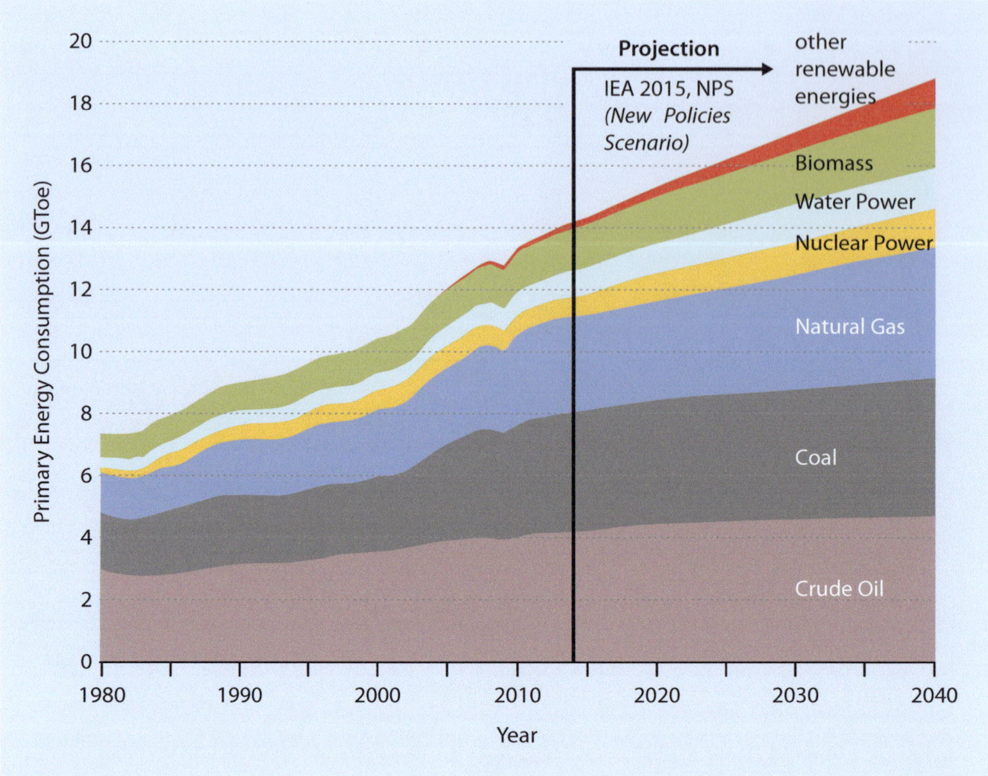

Fig. 1.2 Development and projection of total global energy consumption. (BGR 2015)

influence. The fossil age has catapulted us into the comfort of a modern age whose equipment requires the sufficient and inexpensive availability of energy. Let us also not forget, however, that this was and is only possible at the expense incurred by enormous overexploitation. Fossil fuels are nothing more than the geological storage form of biomass. They contain the accumulated energy from roughly 500 million years of photosynthesis. This sounds reassuring, but it is not: "The fossil fuels consumed between 1950 and 2010 corresponded to 50 to 150 million years of stored sunshine" (McNeill and Engelke 2013).

Fossil reserves are therefore not unlimited: we are approaching the peak of their exploitation (Fig. 1.3). According to the findings of the Federal Institute for Geosciences and Natural Resources (BGR), "there are still enormous amounts of fossil energy that can, from a geological point of view, also cover an increasing demand for energy" (BGR 2015), but crude oil is "the only non-renewable energy resource for which rising demand can probably no longer be met in the coming decades." Overall, more crude oil has already been consumed worldwide than is currently reported in conventional reserves (180 vs. 171 billion t).

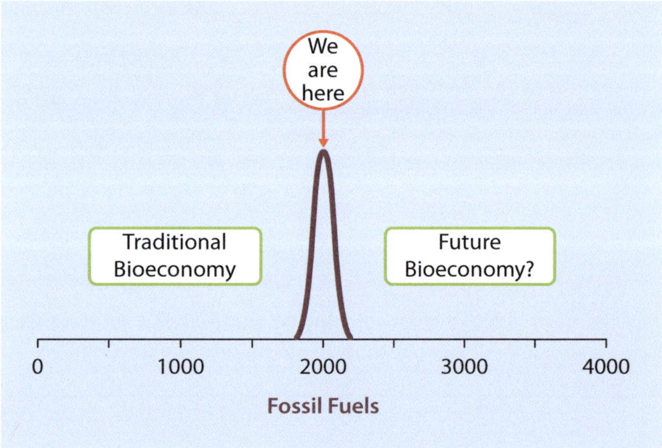

Fig. 1.3 Today's economy and prosperity are based on fossil resources deposited during geological periods. Fossil coal and energy resources are finite. We are therefore currently at the peak of the fossil age, which will one day represent only a brief epoch in human history (Schurr 2015)

Therefore, although fossil fuels will not become scarce in the near future and will still be available at least through the twenty-first century (beyond that in the case of coal), it would be better to use them more and more sparingly. Their use has increased the prosperity of mankind, but it has also had a considerable impact on the environment. These include local, massive interventions in landscapes such as open-cast mines, oil and gas fields, where concentrated fossil fuels occur – in the industrialized countries themselves, but also, for example, in the destruction of the Niger Delta and the rainforests of Ecuador. Even today, despite much stricter safety standards, oil spills still occur along the transport routes, and, on a global scale, air pollution, smog, acid rain and climate change result from the use of raw fossil materials. In an ambivalent dynamic, the use of raw fossil materials has opened up an epoch to which can be attributed its own geological dimension: the Anthropocene, in which humankind, through profound interventions, has become the determining force in shaping and changing the earth. First proposed in 1873 by the Italian geologist Antonio Stoppani, the term became popular only around the year 2000, thanks to the atmo-

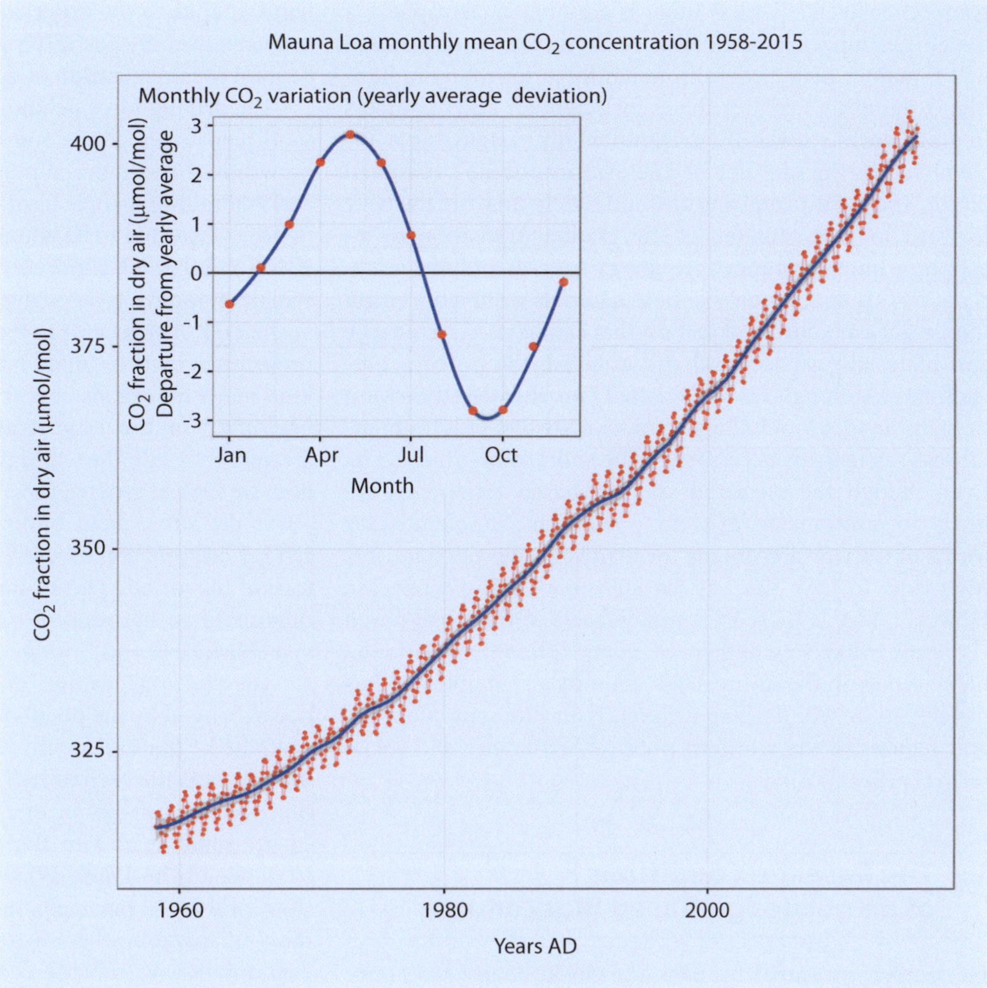

■ **Fig. 1.4** Development of CO_2-content of the atmosphere in ppm since the beginning of its measurement in 1958 at the Mauna Loa Observatory in Hawaii. The sawtooth pattern of the measurement results is seasonal. During the summer months, the atmosphere contains less carbon dioxide, because more of it is stored in the leaves of trees and bushes. (© Delorme 2015, CC BY-SA 4.0, ► https://commons.wikimedia.org/wiki/File:Mauna_Loa_CO2_monthly_mean_concentration.svg)

spheric chemist Paul Crutzen, who was awarded a Nobel Prize in Chemistry in 1995 for his contribution to the discovery of the hole in the ozone layer: "It seems appropriate to assign the term 'Anthropocene' to the present, in many ways human-dominated, geological epoch, supplementing the Holocene – the warm period of the past 10-12 millennia. The Anthropocene could be said to have started in the latter part of the eighteenth century, when analyses of air trapped in polar ice showed the beginning of growing global concentrations of carbon dioxide and methane. This date also happens to coincide with James Watt's design of the steam engine in 1784" (Crutzen 2002).

The anthropogenically induced increase in carbon dioxide and other greenhouse gases would thus be a characteristic feature of the Anthropocene, and climate change was one of its culmination points. Before the beginning of the Industrial Revolution, the concentration of carbon dioxide in the atmosphere was about 280 ppm; today, it is about 385 ppm. By far, the largest portion of this increase was caused by the combustion of raw fossil materials. Since the beginning of its targeted measurement in 1958 alone, carbon dioxide concentration has risen by around 80 ppm (■ Fig. 1.4).

The greenhouse effect, which is mainly due to the increase in carbon dioxide emissions, has, in all probability, caused the average warming of the earth's atmosphere near the surface by 0.8 °C compared with the pre-industrial era in the last decades of the twentieth century. According to NASA, the first decade of the twenty-first century was warmer than any previous decade documented. Many scientists are convinced that the consequences of this warming are already being felt, e.g., through the melting of glaciers: "The potential risks of climate change are numerous, but none is more alarming than the upheavals in the global water balance. Atmospheric warming is likely to change many of the planet's ecosystems, alter precipitation patterns, cause more frequent and extreme weather events, raise sea levels and flood coasts, adversely affect biodiversity, promote the spread of infectious diseases, cause more heat-related deaths and much more" (McNeill and Engelke 2013). For good reason, international climate policy has therefore set itself the goal of strictly limiting global warming in relation to the temperature at the beginning of industrialization – to a maximum of 2 °C, if possible, or even 1.5 °C, as agreed upon at the world climate conference in Paris in December 2015. But this goal can only be achieved if 70% of all available reserves of coal and a third of the oil and gas remain underground and are never used (IPCC 2014).

To counter global warming with a sustainable containment policy is therefore one of mankind's most urgent tasks in the Anthropocene, which Crutzen defines even more comprehensively: "Unless there is a global catastrophe – a meteorite impact, a world war or a pandemic – mankind will remain a major environmental force for many millennia. A daunting task lies ahead for scientists and engineers to guide society towards environmentally sustainable management during the era of the Anthropocene" (Crutzen 2002). The most promising and ultimately unavoidable way forward in this situation is the gradual transition to an economy based on renewable energy sources and raw materials. To this end, renewable raw materials will be of central importance as sources of carbon that can be used for energy and material purposes and will establish the modern bioeconomy. Although resource scarcity justifies this transition only in the medium to long term, climate and environmental reasons make it necessary in the short to medium term. Even though the oil-based and bio-based economies are likely to continue to exist in parallel as complementary forms of the world economy for many decades to come, it is high time to pave the way for sustainable bio-economies. However, just as there exist many facets of the fossil-based economy today, so will there be many different facets to the bioeconomy in the future. Both embody a principle that has developed or will develop differently in different countries and regions of the world in political, economic and social terms (▶ Sect. 1.4).

1.3 The Baseline Conditions of a Knowledge-Based Bioeconomy

Of course, you can't turn back the clock: Future bioeconomies will have almost nothing in common with the preindustrial forms of bioeconomy. On the one hand, today, we are faced with a completely different starting position and with challenges that were unknown in the past. On the other hand, thanks to the experience and knowledge that we have acquired over the past 250 years, we are able to find solutions beyond the imagination of our ancestors.

The driving force behind the ever-increasing demand for food, consumer goods and energy is the continuing rapid growth in the number of people on this planet. Between 800 and 900 million people lived on Earth in 1780 – by 2030, this figure is expected to be about ten times higher, at 8.3 billion. If the number of children per woman worldwide were to remain at today's level on average, then the world population at the end of this century would be 26 billion, according to projections by the United Nations (◘ Fig. 1.5). It took mankind many thousands of years to reach the first billion. By the beginning of the nineteenth century, this milestone had arrived. By 1930, the world population had doubled to 2 billion. In 1960, it crossed the 3 billion mark. In 1975, 1987 and 1999, the subsequent billion marks were reached, and by 2011, 7 billion people were living on earth. "In the entire history of life on our planet, no primate and possibly no other mammal has reproduced so frenetically and secured its survival" (McNeill and Engelke 2013).

The starting position of the new bioeconomy has also been changed by the phenomenon of globalisation. While its roots lie in the nineteenth century, its shoots did not really unfold until the second half of the twentieth century, when they reached full bloom after the world political turn of 1989. In the same year, Tim Berners-Lee invented the universal language of the *World Wide Web*. Globalisation means that the world is, in principle, open to all people, and that they meet in markets where a transboundary worldwide supply encounters a worldwide demand, be it in the exchange of labour or capital, commercial goods or digitally mediated information. This global networking offers great opportunities for economic development, increased prosperity and knowledge. However, it also harbours major risks and makes

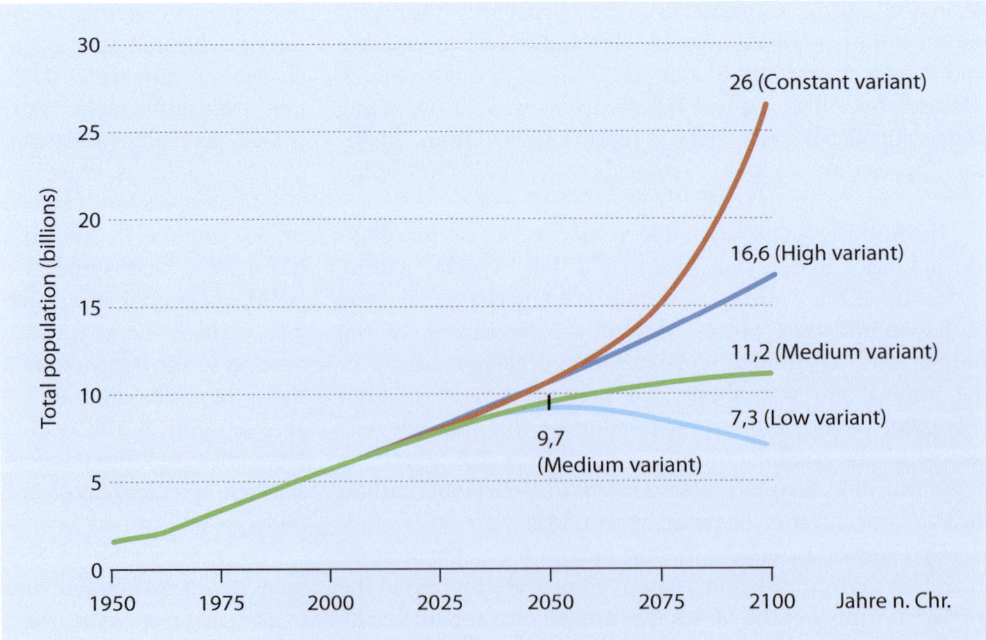

◘ **Fig. 1.5** Even small differences in the average number of children per woman will have a significant impact on the future development of the world population. If the current average of 2.5 children per woman remains constant, 26 billion people will be living on this earth by the end of this century. If the average were to fall to 2 children per woman, it would only be 11.2 billion, and if it were to fall to 1.5 children per woman, it would be as low as 7.3 billion. (© Stiftung Weltbevölkerung; Source: United Nations, World Population Prospects: The 2015 Revision)

the world more vulnerable than it used to be, as the emergence of the global financial crisis in 2008, for example, showed. In its context it also became clear to what extent the prices of raw materials and agricultural products depend on the mechanisms of speculation on the financial markets.

As anachronistic as it may sound, the realization of new approaches to bioeconomy is complicated by the fact that – in contrast to the bioeconomy of our forefathers – we live today mostly in democratically constituted states, in which change cannot be imposed from above, but the political shaping rather emanates from the people, and therefore has to be promoted for approval in participatory processes. To successfully embark on the path to a bioeconomy therefore requires an intensive discourse so as to balance often diverging interests.

At the multinational level, too, the complexity of political decision-making and shaping options is enormous. However, shaped by the terrible experiences of the first half of the twentieth century, institutions and procedures have been established whose significance for global peace- and consensus-building must not be underestimated, even if their mills often seem to grind so slowly. This also applies, in particular, to environmental policy, which was first put on the agenda and made internationally acceptable by the United Nations. In the summer of 1972, the first World Environment Conference was convened in Stockholm, providing the impetus for the establishment of the United Nations Environment Programme (UNEP). This conference is regarded as the first milestone on the road to global sustainable development. The concept of *sustainable development* was coined in the report *Our common future*, which was published in 1987 by the UN World Commission on Environment and Development, founded in 1983 and chaired by Norwegian Prime Minister Gro Harlem Brundtlandt. It formed the basis of the 1992 World Environment Conference in Rio de Janeiro and the Agenda 21 adopted there, which has since come to be regarded as a compass for sustainability and emphasises the triad of economic, ecological and social factors in the concept of sustainability.

That environmental policy conferences can go beyond declarations of intent and have a lasting positive impact is demonstrated by the Montreal Protocol, signed in 1987. It was a quick political reaction to the discovery two years earlier of a huge thinning of the stratospheric ozone layer over the South Pole. The ozone-depleting influence of chlorofluorocarbons (CFCs), which were mainly used as refrigerants and propellants in spray cans around the world, had already been experimentally indicated since the mid-1970s, but had initially been disputed or played down. The discovery of the ozone hole forced the international community to take action. In the Montreal Protocol, which came into force in 1989, the signatory states undertook to introduce a binding ban on the production of CFCs and other ozone-depleting substances, to be implemented step by step, in accordance with international law. Since then, the ozone layer has gradually begun to recover. Climate change has also only become the focus of public attention through the work of international bodies, in particular, the Intergovernmental Panel on Climate Change *(Intergovernmental Panel on Climate Change, IPCC)*, which was jointly founded in 1988 by UNEP and the World Meteorological Organization.

The Agenda 21 adopted in Rio also provides the basis for the United Nations' *sustainable development goals*, which emerged from the Millennium Goals formulated in 2000 and were adopted in autumn 2015. Many of these 17 sustainability goals are directly linked to the challenges of a sustainable bio-economy (▶ Chap. 9). This includes not only the fight against poverty, hunger and climate change and for health, but also the protection of natural resources and the responsible production and consumption of goods. Topics such as education and gender equality also play very important roles.

However tough the negotiations at international conferences dealing with pressing problems of sustainable development may be, diplomats can draw on a wealth of solution options that are scientifically sound and can be communicated to a better-educated humanity than ever before. Since the beginning of the nineteenth century, in a dynamic interaction with the exploitation of raw fossil material sources and economic and social progress, an increase and deepening of knowledge has taken place, starting in Europe and North America and later involving more and more parts of the world, giving humankind immense scope for action. It started with achievements in literacy – before the industrial revolution, very few people could read and write – and has led to digitisation and its assorted options. Mathematics and the natural sciences had already laid their exact and experimental foundations in the 17th and 18th centuries, but only began their triumphal march in the nineteenth century. The modern humanities and social sciences (Marquard 2015) emerged as a response to the challenge posed by the natural and technological sciences. They offered orientation in an increasingly confusing world and revealed its anthropologically, psychologically and sociologically conditioned ways of functioning. Supported by these, medicine, building on the findings of physics, chemistry and biology, learned to understand physiological and pathological processes better and better and to push back against illness and death. With the help of mathematical methods, economics developed models of economic action that could also be used for the analysis and planning of political processes, such as, for example, in game theory.

From the invention of electricity to the standard model of particle physics, from urea synthesis to environmentally friendly catalysis, from the theory of evolution to the invention of genetic engineering and the decoding of the human genome, from the first anaesthesia to antibiotics to magnetic resonance imaging, from railways to cars to space satellites, from calculating machines to computers to the Internet of Things: Breathtakingly wide and not nearly comprehensively to be hinted at stretches the arc of scientific knowledge that allowed humankind in a relatively short time to develop technologies with which it has profoundly changed the world – a knowledge gain from which it can benefit, together with the insights from the humanities, social sciences and economics, when it comes to determining a path into a future bioeconomy.

The revolution in knowledge has led to technologies that today open up for humankind ways out of unsustainable dead ends, for example, in the area of the generation and application of energy from renewable sources. The bioeconomies of the future will therefore be knowledge-based. Politics and society will have to deal with all fields of science and practice in order to implement a sustainable bioeconomy. An isolated view of the necessary change from only one or a few perspectives can no longer do justice to the complexity of modern social and economic systems. Bioeconomy is a necessary but not sufficient condition for the development of a sustainable economic system. In addition, there are complementary and parallel areas whose sustainability is based on non-biology-based approaches, for example, in the field of electromobility.

1.4 Starting Points for National Bioeconomic Strategies

Against this background, more than 40 countries around the world have already made plans to take greater account of bioeconomic principles or have even developed national bioeconomic strategies (◘ Fig. 1.6). All of them are committed to the sustainable use of biological systems. Moreover, those countries regard bioeconomy as a great opportunity to tackle central problems of their economic development in innovative ways. However, the definition of which areas belong to the bioeconomy and the objectives pursued along the respective bioeconomical paths are both very different. This is a consequence of the different challenges and opportunities in the individual countries (regionalisation of the bioeconomy).

◘ **Fig. 1.6** Bioeconomic strategies are becoming increasingly important worldwide, as the team from the office of the German Bioeconomy Council demonstrated here at the first Global Bioeconomy Summit, which took place in Berlin in November 2015

While some countries have formulated very broad bioeconomic strategies that not only address regional problems but also relate to global issues, others are much more specific with regard to their own country's concrete objectives based on the natural and spatial resources available there. Some countries generally regard the life sciences as the basis of the bioeconomy, and therefore attribute great importance to applications of biotechnology in the health sector. These include the USA, India, South Africa and South Korea. On the other hand, other states or communities of states deliberately exclude medical biotechnology, focusing their strategies on agricultural production, the provision of food and feed, and the replacement of raw fossil-based materials with renewable ones. This primarily includes the European Union (EU), which is one of the political pioneers of the bioeconomy. Already in 2005, its Commissioner for Research, Science and Innovation presented the concept of a knowledge-based bioeconomy, which was adopted at the Cologne Summit under German EU Presidency in 2007. In addition to national and European interests, aspects of responsibility for global problems such as food security and sustainable development played an important role from the outset. Important impetuses were subsequently provided by the Organisation for Economic Cooperation and Development (OECD) in 2009 with its strategy paper entitled *The Bioeconomy to 2030: Designing a Policy Agenda*.

Nowhere does the bioeconomy develop from scratch; it always stems from a starting point that depends on the raw material base, the economic specialisation and the level and path of development of the respective country. From an economic perspective, the different national motivations to pursue bioeconomic strategies can be divided into four categories (German Bioeconomy Council 2015a, b):

- Some countries suffer from a structural food shortage. Bioeconomy is therefore primarily seen as a way to facilitate more effective production of food and feed, which is predominantly aimed at *food security*. This is the case in *Tanzania*, for example. With almost 50 million inhabitants, the East African country is one of the poorest countries in the world. More than three quarters of the working population are employed in agriculture. The Tanzanian *National Biotechnology Policy* of 2010, which is based on the *Strategy for growth and reduction of poverty* developed 5 years earlier, therefore focuses, above all, on the agricultural sector. Its main objective is to promote food security and ensure that the country is self-sufficient and not dependent on expensive imports. The introduction of modern food technologies and the development of food adapted to local conditions and needs are central components of this strategy. An important role for biotechnology is also seen in improving health care, e.g., through vaccines, and in preserving biodiversity. Countries such as Kenya, Mozambique, Paraguay and Uganda have a comparably strong focus on food security.

- Some countries have a large and almost inexhaustible amount of natural resources, which they have not yet learned to use efficiently enough. Bioeconomy is therefore understood as a way to establish new biomass value chains that profitably link the raw materials of the primary sector with downstream sectors in order to increase the gross national product. This is the case in *Finland*, for example. Almost three quarters of the country's surface area – equivalent to about 23 million ha – is covered with forest. As a result, forestry is at the heart of the 2014 *Finnish Bioeconomy Strategy – Sustainable growth from bioeconomy* (Albrecht and Ettling 2014). Strengthening the timber market and diversifying the supply of timber products will contribute to Finland's prosperity and competitiveness. Wood-based fuels are also gaining in importance there. The value-enhancing link between biomass production and the potential of industrial biotechnology is emphasised. The Finnish expertise in the field of biotechnology shall also benefit the healthcare industry. The Finnish strategy also underlines the importance of clean water as a critical resource of the bioeconomy. Among the countries that have made a large supply of biomass the starting point for their bioeconomic strategies are Argentina, Brazil and Russia.
- Some countries are home to diverse and large industries that depend on sufficient amounts of raw materials. In view of the foreseeable shortage of raw fossil materials, they are interested in tapping new sources of raw materials for their industries through the bioeconomy. In this way, they can simultaneously make optimal use of their respective domestic knowledge and know-how. This is the case in *Germany*, for example. North Rhine-Westphalia, its most populous and industrial federal state, has based its bioeconomic strategy on a potential study based on the key questions: What industries do we have today? What resources do they need? How can we address this through bioeconomy? Nationwide, the National Research Strategy Bioeconomy of 2010 and the National Policy Strategy Bioeconomy of 2013 are ambitiously and comprehensively dedicated to bioeconomy. As its five main fields of action, the research strategy formulates: securing global food supplies; designing sustainable agricultural production; producing healthy and safe food; making industrial use of renewable raw materials; and expanding energy sources based on biomass. As early as 2009, the Bioeconomy Council was established as an independent political advisory body, which, in November 2015, organized the first Global Bioeconomy Summit in Berlin. Countries whose bioeconomic starting point of a potential shortage of raw materials is comparable to that of Germany are, for example, Japan and the USA.
- Some countries have developed so well in recent decades that they are on the verge of becoming industrial countries. They have both abundant natural resources and considerable high-tech potential. The effective technological use of biological systems in a bioeconomy is therefore primarily seen as a means to "cross the threshold." This is the case in *Malaysia*, for example. The country is rich in soil resources and biomass. It has experienced a rapid economic upswing over the past quarter century and is one of the most politically stable countries in Southeast Asia, among which it has been pioneering the formulation of a holistic bioeconomic strategy (▫ Fig. 1.7). Malaysia, as its government already stressed in its 2005 *National Biotechnology Policy*, regards biotechnology as a key driver of its future growth, hoping to become a knowledge-based economy by 2020 with its help. In this context, *Bioeconomy Transformation Programs* was published in 2012, flanked by the amended *National Biomass Strategy 2020* in 2013. It focuses on the industrial value creation that is possible with agricultural biomass (especially palm oil) and its residues, and thus concentrates on promoting technologies that generate a higher added value from the country's biological resources than in the past. Thailand, South Africa and India, for example, have comparable starting positions and motivations that they see as key to their industrialization on the way to a bioeconomy. For such emerging countries – including Peru – the bioeconomy also offers an opportunity to improve the economic situation of the rural population, and thus reduce rural exodus and the swelling of megacities.

However, the different starting points of national bioeconomic strategies should not obscure the fact that all nationally pursued goals must ultimately be oriented towards the overarching global goal of sustainability, and thus meet the UN sustainability goals. The final communiqué of the first Global Bioeconomy Summit points to five cornerstones that are essential for the establishment of a sustainable bioeconomy (El-Chichakli et al. 2016):

1. International collaborations between governments and public and private researchers to optimise resource use and knowledge sharing;
2. Definition of internationally agreed criteria for the evaluation of progress in the bioeconomy, which can be specified in national monitoring systems;
3. Stronger linkage of bioeconomy initiatives with multilateral policy processes and intergovernmental discussions, particularly the UN sustainability goals, and the follow-ups of the Paris climate and Aichi biodiversity agreements;
4. International collaboration of educators to define the knowledge, skills and competencies required for developing a sustainable bioeconomy;
5. Establishment and promotion of specifically selected research and development programmes so as to encourage global collaborations for a sustainable bioeconomy.

1

Fig. 1.7 With the technological use of biological systems, Malaysia (represented here by its capital Kuala Lumpur) is taking the leap across the threshold to becoming an industrial nation. It was the first country in Southeast Asia to formulate a dedicated bioeconomic strategy. (© Paulista/Fotolia)

References

Albrecht K, Ettling S (2014) Bioeconomy strategies across the globe. Rural 21 Vol. 48 Nr.3:10–13

Bundesanstalt für Geowissenschaften und Rohstoffe (BGR) (2015) Energiestudie 2015. Reserven, Ressourcen und Verfügbarkeit von Energierohstoffen. Hannover

Crutzen PJ (2002) Geology of mankind. Nature 415:23

Davis, SC, Hay W & Pierce J. (2014). Biomass in the energy industry: an introduction

Delorme 2015, CC BY-SA 4.0, https://commons.wikimedia.org/wiki/File:Mauna_Loa_CO2_monthly_mean_con-centration.svg

El-Chichakli B, von Braun J, Lang C, Barben D, Philip J (2016) Five cornerstones for a global bioeconomy. Nature 535:221–223

German Bioeconomy Council (2015a) Bioeconomy policy. Synopsis and analysis of strategies in the G7. Berlin

German Bioeconomy Council (2015b) Bioeconomy policy (Part II). Synopsis of national strategies around the world. Berlin

IPCC (2014) Climate change 2014: mitigation of climate change. In: Edenhofer O et al (eds) Contribution of working group III to the fifth assessment report of the IPCC, Cambridge, New York

Marquard O (2015) Über die Unvermeidlichkeit der Geisteswissenschaften. In: Zukunft braucht Herkunft. Reclam Verlag, Stuttgart, pp 169–187

McNeill JR, Engelke P (2013) Mensch und Umwelt im Zeitalter des Anthropozän. In: Iriye A Geschichte der Welt 1945 bis heute. Die globalisierte Welt. C.H. Beck Verlag, München, pp 357–534

Osterhammel J (2009) Die Verwandlung der Welt. Eine Geschichte des 19. Jahrhunderts. C.H.Beck Verlag, München

Rödder A (2015) 21.0 Eine kurze Geschichte der Gegenwart. C.H. Beck Verlag, München

Schurr U (2015) Forschen für eine nachhaltige Bioökonomie – von der Idee zur Umsetzung. Vortrag im Rahmen der Vorlesungsreihe „Weltweite Ernährungssicherheit als Herausforderung der Bioökonomie" Düsseldorf, 27 April 2015. http://download.fz-juelich.de/ibg-2/Forschung_zur_nachhaltigen_Biooekonomie_Schurr.mp4. Accessed: 22 August 2016

The Origin of Biomass

Melvin Lippe, Iris Lewandowski, Rüdiger Unseld, Johannes Pucher, and Klaus-Rainer Bräutigam

Field of maize shortly before harvest, for use in a biogas plant at the University of Hohenheim's experimental station "Unterer Lindenhof", Eningen, Baden-Württemberg. (© Oskar Eyb, Picture Archive of the University of Hohenheim)

J. Pietzsch (ed.), *Bioeconomy for Beginners*, https://doi.org/10.1007/978-3-662-60390-1_2

In the origin of biomass, a distinction is made between primary and secondary biomass flows. Primary biomass is formed by autotrophic organisms. These are plants, algae and certain bacteria. They are able to produce their mass through the use of solar energy in the process of photosynthesis. Secondary biomass is produced on the basis of the consumption of primary biomass, e.g., in animal production or as organic residues and waste streams.

Fresh and dry biomass differ in their water content. Plants are made up of carbon, nitrogen, various macro-elements, trace elements and water. Most plant organs consist of 75–85% water, with the exception of organs in which carbon or nitrogen is stored in higher concentrations, resulting in a significantly lower water content. These include, for example, grains, but also fruits and below-ground storage organs such as potatoes and beets. Even the permanent supporting structures of plants such as wood have a low water content. They are often composed of dead cells and cell walls. Living tissues, on the other hand, are largely shaped by a high internal pressure in their cells. If the available water is no longer sufficient to maintain the shape, the plant withers. Biomass is therefore characterised by a high degree of heterogeneity in its composition, made up of water, carbon and nutrients taken up from the environment.

Photosynthesis is of particular significance in the formation of biomass. It is (almost) the only process by which organisms can acquire energy. This process, which is essential for all life on Earth, also produces the oxygen that animals and aerobic microorganisms need to live. During photosynthesis, autotrophic organisms take up carbon dioxide (CO_2) and incorporate carbon (C) into their organisms (assimilation), converting light energy into chemical energy. The end products of photosynthesis are C6 sugars (hexoses). The entire process is described in the following equation:

$$6CO_2 + 6H_2O \xrightarrow[\text{Chlorophyll}]{\text{Light}} C_6H_{12}O_6 + 6O_2$$

The conversion of carbon dioxide (CO_2) and water (H_2O) into sugars and oxygen takes place in chloroplasts, which contain the green light-absorbing pigment chlorophyll. CO_2 is diffunded through the stomata and interstitial spaces of plants to their photosynthetically active cells, where it is converted into carbohydrates. These carbohydrates serve the plants both as energy and carbon sources for their metabolic processes and as building blocks for their biomass. Photosynthesis takes place exclusively in the leaves. From the leaves, carbohydrates are transported to non-photosynthetically active areas of the root, where they are either used to build plant structures, provide energy (respiration) or are stored as carbohydrates, fats or proteins (together with the macronutrient nitrogen).

2.1 Biomass from Agriculture

Melvin Lippe and Iris Lewandowski

Agriculture is the production of biomass within the framework of humankind's cultivation of land. The way in which the land is managed, and the quantity and quality of the products produced, depend on site-specific (biophysical and climatic), socio-economic and political conditions. Agricultural production systems are subdivided into plant and animal production. The biomass is then used as food or animal feed, for the production of materials (fibres, chemicals) or for energy (biogas, biofuels, solid fuels) purposes.

2.1.1 Basic Principles of Plant Production

2.1.1.1 Yield Formation and Quality

In agriculture, plant growth, and thus also yield formation, is essentially determined by the (genetic) properties of plants and by the site factors soil, nutrients, temperature, water and solar radiation, as well as by farming management (Fig. 2.1).

The physiological processes that take place within a plant during one growing season depend on its genetic information and the surrounding environmental conditions. They determine yield - the area-related production output of a crop. In the ideal case, the selection and cultivation methods of cropping plants are aimed at exploiting the production potential of a specific site as much as possible (Diepenbrock et al. 2012). Plant growth and development processes take up the largest part of its metabolic activity. Production and distribution of dry matter are the decisive internal plant processes for yield formation. In most cases, crop yield is provided by certain organs, which are formed during plant

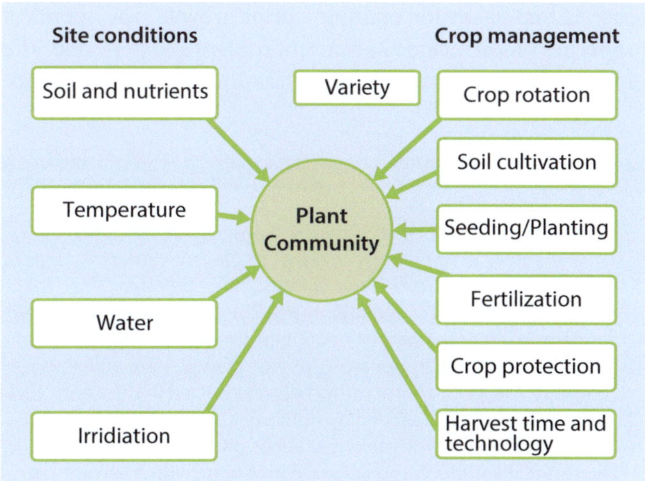

Fig. 2.1 Factors influencing plant production

development and can be further promoted by cultivation interventions and other measures. These include fruits and seeds, stems, leaves, below-ground storage organs (tubers, beets) and roots. Their share of a harvest is called the **harvest index**. The harvest index only reaches the maximum value of 1 if the growth potential of a plant is fully utilised – whereby the above-ground part of the plant is often taken as basis. This is the case for most forage crops and some energy crops (e.g., biogas maize, fast-growing tree species, miscanthus). In all other cases, the harvest index value is lower. Crop cultivation measures are generally aimed at optimising the harvest index (Diepenbrock 2014).

In addition to the yield, the quality of biomass is a central objective of plant production. The term 'quality' covers a broad spectrum of parameters that depend on the utilisation path of the respective biomass. Examples of quality parameters include: colour and taste, nutrient content of fruit and vegetables, low water content in cereals, protein composition of malting barley, and composition of lignocellulose if the whole plant is to be used for energy or material purposes.

2.1.1.2 Soil and Nutrients

Soil is formed through weathering of the Earth's crust with the participation of soil organisms (microorganisms and soil biota). It consists of minerals of different types and sizes, as well as humus formed from organic substances. It also contains water, air and various living organisms, such as bacteria, earthworms and insects. Soil offers plants root space for anchoring and supplies them with water, nutrients and oxygen.

Plant growth and yield formation are strongly influenced by the physical, chemical and biological properties of the soil. One of soil's physical properties is the depth of the upper layer accessible to plant roots. Other physical properties are texture or grain size, the proportion of air-carrying pores and ability to retain water and store or release heat. A sufficiently large root space is important for optimum plant growth. The chemical properties of soil include its nutrient content and pH value. The biological properties of soil are determined by the occurrence and activity of soil microorganisms. These organisms rely mainly on organic matter as a food source, supplied to the soil by dead plant material. Microbial activity and physical-chemical processes, in turn, release nutrients that the plants absorb via their roots.

All plants require mineral and non-mineral nutrients. The non-mineral nutrient elements carbon (C) and oxygen (O_2) are absorbed out of the atmosphere by plants' leaves. In contrast to abundant oxygen, carbon dioxide (CO_2) is present in the air in a very low concentration of only 0.03 vol.%. For this reason, the CO_2 supply of chloroplasts limits the production rate of a plant population stand while exposed to strong sunlight. The main mineral nutrient elements nitrogen, phosphorus, potassium, calcium, magnesium and sulphur, as well as the trace elements iron, manganese, zinc, copper, molybdenum, chlorine and boron, have to be taken up by plants from the soil via the roots. The better the root is able to develop, the more nutrients and water the plant can absorb from the soil to form the largest possible root surface. Rootability of a soil decreases with increasing bulk density, as well as with the occurrence of compaction zones caused, for example, by improper soil tillage.

2.1.1.3 Temperature

Temperature influences all plant growth processes. This applies, in particular, to photosynthesis, respiration and transpiration. Plants show a species-specific optimum in their activity. C_4 plants are characterized by a higher temperature optimum (above 30 °C) than C_3 plants (about 20 °C) (▶ Excursus 2.1). The lower limit for photosynthesis activity - the temperature minimum - is a few degrees below zero for plants in cold and temperate climates.

If the average annual temperature rises (up to 30 °C), the yield potential of a site also increases, provided there is sufficient water supply. The upper temperature limit is between 38 and 60 °C, depending on the plant species. Higher temperatures destroy the proteins, which results in reduced enzyme activity and damage to cell membranes, ultimately leading to the ceasing of all metabolic processes.

Excursus 2.1 C_3 and C_4 plants

C_3 plants work with the basic form of photosynthesis. Since their stomata close in hot and dry weather to prevent excessive evaporation of water, they have a lower photosynthesis performance than C4 plants under these conditions. However, they are more efficient under moderate temperature and light conditions. Most plants of middle and high latitudes are C_3 plants. Examples are wheat, rye, oats and rice. In C_3 plants, carbon dioxide is fixed in the Calvin cycle during the *RuBisCO* reaction by ribulose-1,5-bisphosphate.

C_4 plants use a different metabolic pathway, in which they first spatially prefix carbon dioxide for photosynthesis and then, like C_3-plants, use the Calvin cycle to produce carbohydrates. The name C_4 is derived from the first fixation product generated by the assimilation of carbon dioxide. In C_3 plants, this is a carbon compound with three carbon atoms (D-3 phosphoglycerate), whereas in C_4 plants it is oxalacetate, which is composed of four carbon atoms. In C_4 plants, carbon dioxide assimilation and the Calvin cycle are spatially separated from each other. By applying energy, carbon dioxide is actively enriched, which leads to a higher photo synthesis rate - especially under conditions of water shortage and the resulting narrowing of the stomata. Therefore, C_4 plants are ecophysiologically superior to C_3 plants under arid conditions. Due to the active enrichment, photorespiration - and thus the fixation of O_2 instead of CO_2 - is greatly reduced. Typical C_4 plants are grasses, including well-known crops such as maize, sugar cane and millet, but also include other species such as amaranth.

In Earth's history, C_3 plants developed first. Their key enzyme RuBisCO emerged during a period when the atmosphere was rich in carbon dioxide (CO_2) and poor in oxygen (O_2). In this environment, assimilation did not cause any problems, as there were no losses due to photorespiration.

2.1.1.4 Water

Green plants consist of 70–90% water, with water content of plant organs varying according to the type and age. Water performs various functions in a plant. For example, it transports dissolved substances and maintains hydrostatic pressure in plant cells, which keeps tissue taut. Water is also an important starting material for metabolic processes such as photosynthesis. In addition, almost all biochemical reactions take place in aqueous solution.

Water uptake and water release determine the water balance of a plant. Water is mainly absorbed via the roots. Water is lost mainly by leaf transpiration. If water release is greater than water absorption, a water deficit occurs. This can be the result of high transpiration, low water availability in the soil or inhibited water supply through the roots. The root absorbs water from the soil via the suction power of its cells. However, the water absorption capacity ends at the wilting point. There, the water content of the soil becomes so low that its water retention capacity exceeds the absorbency of the root.

The biomass production of plants depends largely on their water supply. Each plant species has a specific water consumption for mass formation. The **transpiration coefficient** describes the amount of water required by a plant to produce 1 kg of dry matter. C_4 plants like maize *(Zea mays)* and miscanthus *(Miscanthus spp.)* (◘ Fig. 2.2) have the lowest transpiration coefficient at 220–350 l/kg, and thus the most efficient water utilization of all crops. This is due, among other things, to the dense arrangement of their photosynthetically active cells and the associated lower water loss through transpiration. C_3 plants such as cereals and willows, which belong to the fast-growing tree species *Salix*, require 500–700 l of water for the production of 1 kg of biomass. The biomass productivity of a site is potentially higher the better the water supply is. These figures impressively demonstrate that the amount of water transpired is a multiple of the amount that remains as water in the biomass.

◘ **Fig. 2.2** Miscanthus is one of the crops that make the best use of water for growth. (© Kiesel, picture archive of the University of Hohenheim)

2.1.1.5 Radiation

The net photosynthesis rate increases with increasing radiation intensity up to a saturation point. However, if the radiation is very low, the respiration of carbon dioxide (CO_2) exceeds its assimilation. The radiation intensity at which the quantity of respired CO_2 is the same as the quantity of assimilated CO_2 is called the **light compensation point**. A plant absorbs only part of the incident radiation; the rest is reflected or transmitted. The absorption of radiation in plant tissue occurs selectively depending on its wavelength. Particularly within the range of infrared radiation from 0.7 to 1.1 µm, a considerable amount of radiation energy penetrates the plant stand without being absorbed. The net radiation results from the non-reflected total radiation and the long-wave reflection. The **reflection coefficient** is the ratio of reflected to absorbed radiation. It mainly depends on the angle of incidence of the light, as well as the surface condition and colour of the plant. In green plants, the reflection coefficient is between 0.1 and 0.4., and CO_2 assimilation increases as a function of the radiation and the type of photosynthesis. With the same amount of radiation, the assimilation of C_4 plants is higher than that of C_3 plants.

2.1.1.6 Farm Management

In addition to the factors determined by the natural site conditions, the anthropogenic influence on plant growth through plant cultivation measures plays an importanct role. These include the selection of suitable crops for the respective location, soil cultivation, sowing methods, fertilisation, application of plant protection agents and harvesting measures. The most important prerequisite for the success of plant cultivation measures is the selection of plant species adapted to the ecological conditions of the production site. This applies both to soil quality and the amount and distribution of precipitation, as well as the local temperature regime during the cropping period.

Soil preparation operations are carried out to loosen the soil, to incorporate harvest residues and organic and mineral fertilisers, to control weeds and to prepare the soil for sowing. The time and method of tillage must be adapted to soil conditions and the requirements of the selected crop types.

Crop rotation describes the chronological sequence of cultivated plants on a field. Individual species of cultivated plants are usually components of a crop rotation. Even though the design of crop rotations is often limited by economic constraints, the positive and/or negative effects of crop rotations need to be considered. These effects are due to the fact that the previous crop has an influence on the occurrence of weeds and pests, as well as on the soil water and nutrient supply of a cultivated area. Crop residues and their decomposition can have an effect on the subsequent cropping cycle, for example in the form of chemical interactions (allelopathy), which occur both between species and within the same species (Diepenbrock 2014). There are biological limits to crop rotation, because the cultivation of the same or related crop species in successive years favours the outbreak of plant diseases and often leads to so-called soil fatigue, and thus poorer crop yields. Nevertheless, cropping systems that do not use any rotation component are widespread, especially in subtropical and tropical regions. For example, rice is mainly produced without rotation with other crops. Here, the intensive soil preparation and submerged growth has the advantage of strong suppression of weeds and plant diseases. However, crop rotation must be planned in such a way that there is sufficient time for soil preparation between the harvest of one crop and the next. Crop types with early sowing dates like winter rape and barley can therefore not be grown after late harvesting species such as maize or sugar beet.

Fertilisation is the main field management measure aimed at improving plant nutrient uptake (e.g., by mineral or organic nitrogen fertilisers) and soil properties (e.g., by liming or organic matter uptake). The extent of fertilization is based on the amount of nutrients removed from the soil by plants. The strongest influence on yield levels is achieved through nitrogen fertilisation, because nitrogen strongly promotes growth and its supply in the soil is often a yield-limiting factor. Nitrogen can be supplied to the soil in the form of either mineral or organic fertilisation through the nitrogen fixation by legumes and from the atmosphere via rain. In addition to nitrogen, phosphorus and potassium are commonly applied as fertilisers on a regular basis. Calcium is important both as a plant nutrient and also for soil fertility. It influences soil pH, chemical reactions in the soil and thus the availability of various nutrients, and stabilizes the soil structure via its bridging function. Apart from magnesium, which is often contained in potassium fertilizers, all other nutrients are usually sufficiently present in most soil types and are only applied if there is an obvious deficiency.

Crop protection measures are used to prevent or control weeds, diseases and pest infestations during and after vegetation period. Weeds compete with crops for all factors affecting growth. They can inhibit or completely suppress the cultivated plants. This usually leads not only to a reduced yield, but also to a lower quality or undesirable condition of the harvested biomass. The same effects are caused by infestation with diseases and pests that live off of the photosynthetic products and reserve substances of the plant.

The harvesting method determines both the quality of the biomass and the extent to which it is available for later use. To keep harvest losses low, particular attention must be paid to the right harvest time and the right harvesting technique.

2.1.1.7 Climate Zones and Focus on Global Agricultural Production

The supply of biomass is determined by the combination of soil quality, amount and distribution of precipitation, temperature curve and light supply. ◘ Figure 2.3 shows the amount of biomass growth, expressed in anual increase in carbon per square metre.

The spectrum ranges from 0 to over 1200 g per year. These 1200 g correspond to 12 t carbon (C) or around 24 t biomass (dry matter) per ha per year. Tropical areas around the equator, where high precipitation meets high temperatures and a year-round vegetation period, have the highest biomass productivity. The representations in ◘ Fig. 2.4 show the current focus of global arable, grassland, pasture and forest areas (Foley et al. 2005).

In many regions of the world, biomass production is limited by a lack of water supply. Climate change is expected to increase water scarcity in agricultural production systems and, as a consequence, to increase irrigated

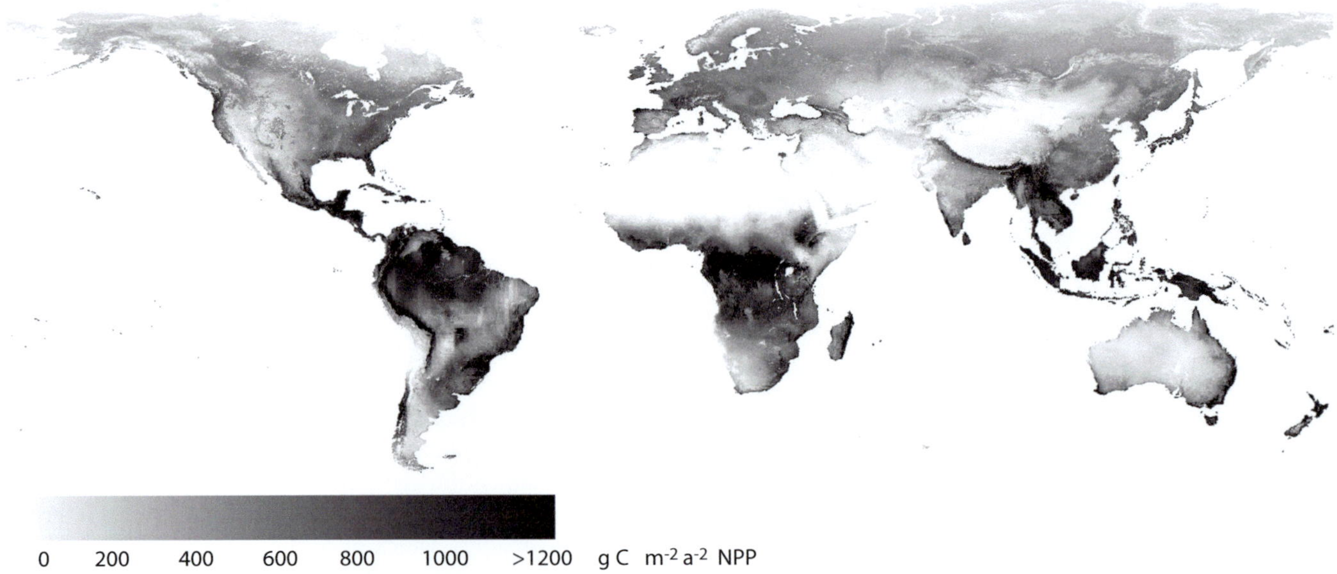

◘ **Fig. 2.3** Net primary production (NPP) of biomass, in g annual increase in carbon (C) per m^2. (Source: Imhoff et al. 2004)

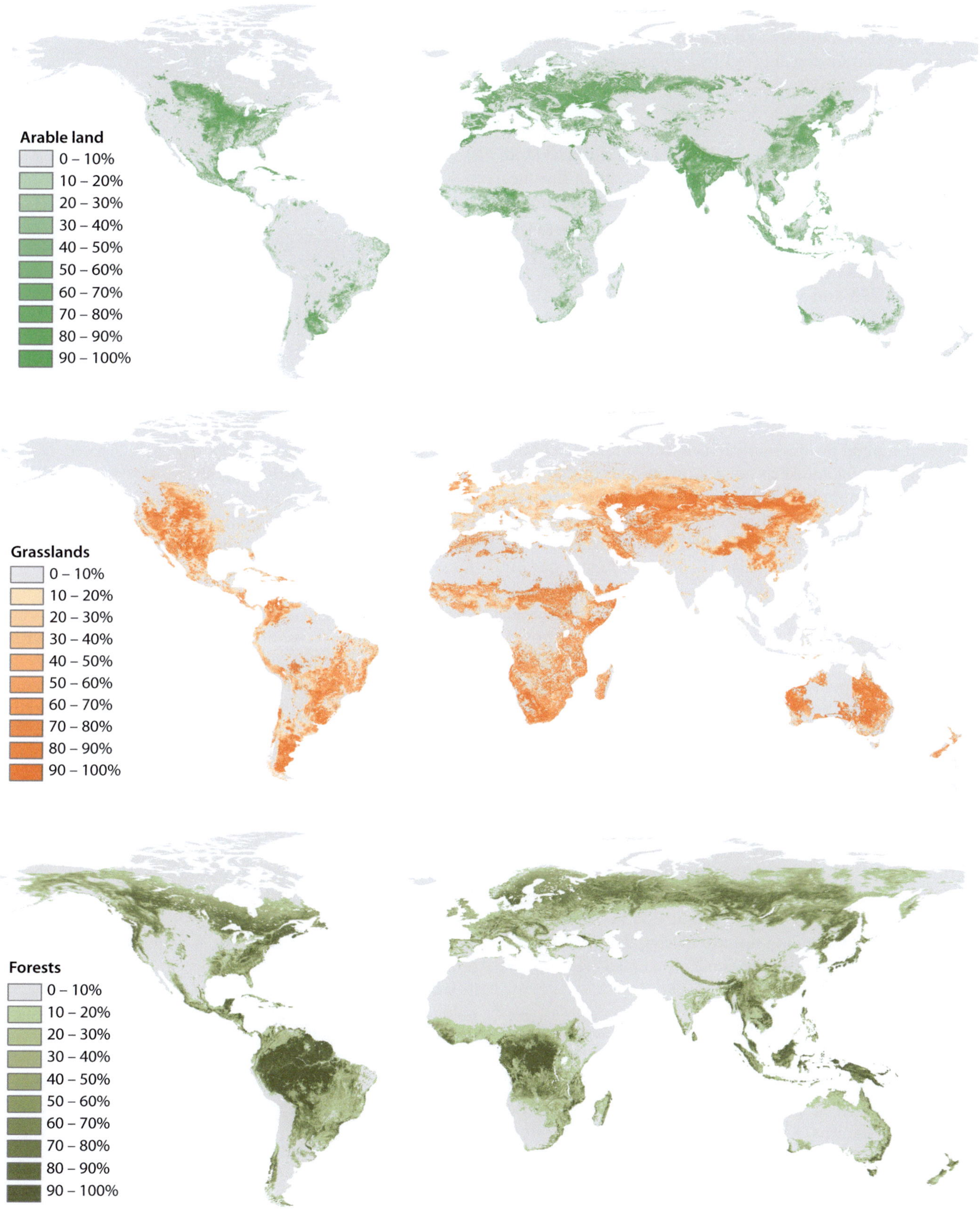

Fig. 2.4 Location of major global agricultural production systems and remaining forest areas. (Scource: Foley et al. 2005)

agriculture. For this reason, it is important to ensure that water is used sustainably and drawn from resources that can be completely replenished by rain or snow. Towards the north, productivity decreases mainly due to the combination of low temperatures and short vegetation periods. Accordingly, tundra and savannah areas are more likely to be found here. As a result, global agricultural land is concentrated at temperate latitudes, such as in Europe, the Midwestern United States, large areas of the Indian subcontinent, and China. By contrast, pasture and grassland areas, which are mainly used for livestock farming, are found more in Central Asia, sub-Saharan Africa, the west of the USA and South America (parts of Brazil, Chile and Argentina).

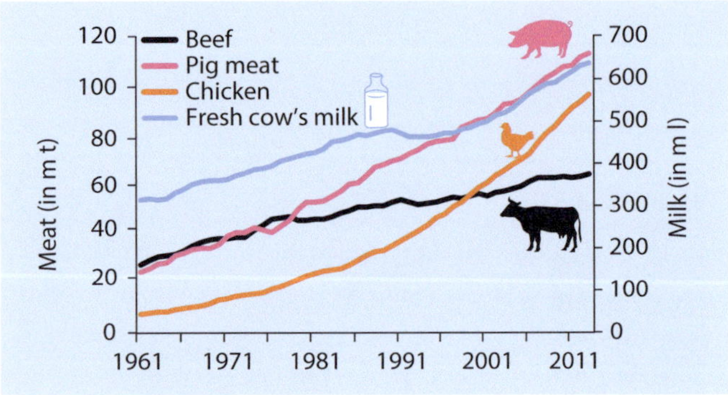

Fig. 2.5 Trends in global animal production from 1961 to 2013: meat production (in million tonnes) and fresh cow's milk (in million litres). (Source: FAOSTAT 2014)

2.1.2 Basic Principles of Livestock Production

The term 'animal production' refers to the interaction of the breeding, feeding and keeping of farm animals on the basis of anatomical and physiological conditions and under consideration of economic principles. The aim of animal production is primarily the production of food (meat, milk, eggs, honey and fish). It is also used to produce hides for leather production, wool, down and feathers, as well as raw materials (especially fats) for the chemical industry. Cattle, pigs, poultry, sheep and rabbits are kept for these purposes. Other services such as landscape management by sheep and cattle and pollination of flowers by bees can also be part of an animal production system. Whereas quantitative food security used to be regarded as the primary task of animal production, its main focus at present is on improving the quality of existing supply, insofar as this is possible, via animal breeding, husbandry and nutrition (Weiß et al. 2011).

Global annual production volumes rose sharply in the period from 1961 to 2013. The quantity of beef produced more than doubled and pig meat production increased even more than quadrupled. Poultry production increased, from about 7 to over 90 million tonnes. During the same period, annual production of fresh cow's milk increased from around 360 to over 640 million litres (Fig. 2.5).

2.1.2.1 Extensive and Intensive Livestock Systems

Animal production is a key component of global food security, because it provides important proteins, amino acids and vitamins. On the production side, live-stock farming serves either to diversify income with the aim of minimising risk (mostly in small-scale farming structures) or to maximise profits in large market-oriented farms. These different business models are reflected in the different production systems of extensive and intensive animal husbandry. The space utilisation rate, i.e., the number of animals per unit area, is the most important distinguishing criterion.

Extensive animal husbandry is characterized by large pastures with small numbers of livestock. The animals are usually kept almost exclusively on natural pastures without additional feeding and rarely or only temporarily in stables. Extensive animal husbandry systems usually involve several animal species. They often rely on traditional forms of use with a high proportion of self-sufficiency. They closely link their livestock production with agricultural farming systems by using manure and slurry as organic fertilisers to maintain soil fertility, and thus represent an important component of the local nutrient cycle. Those forms of modern organic livestock farming, in which the preservation of pastures is the main focus, can also often be regarded as extensive systems (Weiß et al. 2011).

Intensive animal husbandry is operated on small pasture areas with high stocking density or in stables with mass livestock farming and generally with a high degree of technology. It is also characterised by supplementary feeding using purchased feed. As a rule, stocking rates are over ten livestock units per hectare (► Excursus 2.2). Intensive animal husbandry systems usually include only one animal species, because they focus on productivity and profit generation (Doppler 1991).

Excursus 2.2 Livestock unit and livestock stocking

The number of livestock on a holding is expressed in livestock units (LU or SLU). It is either calculated according to feed requirement, or by taking the live weight per head of adult cattle of 500 kg as its benchmark. In tropical regions, there is the *tropical livestock unit* (*TLU*), which is based on a live weight of only 250 kg. Livestock units are used to calculate the required area of a farm holding with livestock so as to avoid overuse.

The term stocking rate means the ratio of the number of livestock to the area on which their fodder is produced (◘ Fig. 2.6). This can refer to pastures and alpine pastures, but also to meadows and fields, provided that fodder is abundantly available. The stocking rate is based on the number of livestock units per hectare (LU/ha). The pressure of grazing animals on the land is referred to as grazing pressure (Chilonda and Otte 2006).

◘ **Fig. 2.6** Exemplary representation of a low (left) and high stocking rate (right) as a ratio of the number of livestock per unit area (authors' own representation)

2.1.2.2 Feeds

Feeds in animal production systems are mostly based on primary biomass. The composition of the feed is crucial in meeting nutrient requirements for the health and performance of the animals. Especially in intensive forms of animal husbandry, feeds are specifically tailored to the farm animal species and intended use. They are often subject to state approval and control (Jeroch et al. 2008).

Feeds are scientifically assessed according to their constituents, which are classified in feed tables according to feed type and harvest. Such feed tables indicate the nutritional value of the feed. In daily practice, feeds can be subdivided according to various criteria. These include botanical characteristics, economic and market considerations, consistency and water content, constituents and nutrients, and intended use. A distinction is mainly made between forages (also called roughages), concentrated and intermediate feed forms:

- **Forages** refers to fodder produced from whole plants, either as fresh greens, ensiled or dried fodder or in the form of cobs and straw. These feeds have medium to low energy contents in dry matter and are characterised by high structural efficacy in the animal's digestive tract.
- **Concentrates** are energy- and protein-rich single-component feeds and industrially produced compound feeds. This includes all mixable components with a water content of less than 45% and an energy content of more than 7 MJ/kg dry matter. In addition to air-dried concentrated feeds with less than 12% water content, moist cereals, soda grain, molasses and dry greens are typical examples of this type of feed. Compared to forages, concentrated feeds have practically no structural value. Due to their high energy concentration, however, they can be transported over long distances. They usually have a good shelf life and are suitable for trade, especially as their use is necessary when there are not enough forages available.
- **Intermediate forms** have an energy concentration similar to that of concentrates, but have a relatively high water content of more than 45%. They are therefore handled in a similar way to forages. If they are not fed fresh, they usually need to be preserved in the form of silage. Beets, roots, tubers, brewer's grains, pressed pulp, stillage and maize by-products are often used as such kinds of feed. The structural value of these feeds lies between that of concentrates and forages (Weiß et al. 2011).

Animal production systems are also divided into the following categories according to the amount (as dry matter) of green fodder used:

- **Grassland-based pasture management systems** with less than 10 livestock units per hectare, in which more than 90% of the fodder consumed is supplied from pasture or grassland systems and via green fodder
- **Industrial or stationary systems,** which have an average stocking rate of more than 10 livestock units per hectare and in which less than 10% of the fodder is produced on the farm
- **Mixing systems,** in which more than 10% of the feed is derived from by-products or residues of agricultural production such as combs or straw

2.1.2.3 Feed Use Efficiency

Feed efficiency or feed conversion efficiency indicates how many kilograms of feed must be used to produce 1 kg of a particular animal product. The smaller their value within a production system, the more efficiently animal protein or protein is built up from the feed consumed.

Fodder use efficiency varies globally (◘ Table 2.1). Beef production in industrialised countries is about four times more efficient than in sub-Saharan Africa and three times

Table 2.1 Feed conversion efficiency of animal production systems (expressed as the reciprocal of kilograms of feed as dry matter per kilogram of animal product). (Source: Smeets et al. 2007)

region	beef	pork	mutton	chicken and eggs
North America	26	6.2	58	3.1
Western Europe	24	6.2	71	3.1
Eastern Europe	19	7.0	86	3.9
C.I.S.[a]/Baltic States	21	7.4	69	3.9
Sub-Saharan Africa	99	6.6	108	4.1
East Asia	62	6.9	66	3.6
South Asia	72	6.6	64	4.1
Global	45	6.7	79	3.6

[a]C.I.S - Commonwealth of Independent States, also known as the Russian Commonwealth

more efficient than in South Asia. Other forms of production are also subject to strong regional fluctuations. Sub-Saharan Africa is above the world average in all production systems, which can be explained by the low availability of biomass for feed use and the prevailing local climatic conditions.

2.1.3 Characterization of Agricultural Production Systems

There are a number of different pathways that can be taken to develop a bioeconomy. The pathway taken by a specific region or country depends above all on the agricultural production systems already in place and the options available to farmers.

Aricultural holdings are characterised by factors including available resources, which cultivation and livestock systems are possible in the locality, characteristics of the household and the given natural limitations of the ecosystem such as climate and soil (Ruthenberg 1980; Seré and Steinfeld 1996; Dixon et al. 2001).

Agricultural production systems can be distinguished on the basis of climatic conditions (e.g., length of day, temperature curve, amount of precipitation, relative humidity) and socio-economic factors (e.g., population density, agricultural policy, organisation of available market and farm structures) (Diepenbrock 2014).

From a global perspective, about one fifth of total agricultural production is still performed in traditional subsistence systems, mainly in the rural areas of numerous developing countries. In those regions, smallholder agricultural production systems form the backbone of local grain production and livestock farming, which primarily serve to supply the smallholder families with food. Traditional subsistence systems are often characterised by extensive farming, low fertiliser inputs and simple forms of tillage, such as hoes or animal ploughing. However, these systems are increasingly being converted to market-oriented farming approaches, partly or predominantly serving market demands.

The economic development of many emerging economies such as Brazil, Argentina and China evolved from the agricultural sector. These countries exploit their natural advantages (large areas for arable farming) for the industrialisation of this sector, but subsistence systems continue to exist in parallel. By contrast, central Europe and the USA are dominated by intensive arable and livestock farming. These systems are often based on a few cultivated plants, monocultures and intensive livestock farming. Market-oriented production systems such as those in Central Europe have increasingly become an integral part of globally connected value chains, which, in addition to maximizing revenues, are often designed to maximize profits. In addition to focusing purely on yield maximisation, other objectives such as biodiversity and the maintenance of forests and landscapes as recreational areas are also pursued to varying degrees.

2.1.3.1 Agro-Ecological Zones

Agro-ecological zones (AEZ) are geographically limited areas with similar climatic and ecological characteristics in which only certain agricultural use forms occur. AEZs are used to determine regional cultivation potentials and for land use planning. The core principle of the AEZ is the classification of farm types according to the duration of the possible growth period of cultivated plants during the year as a function of soil moisture content and potential evapotranspiration (water evaporation). The growth period is thus defined as the "number of days on which the soil moisture content is greater than half the potential evapotranspiration" (Fischer et al. 2002). From this definition, the following core AEZs are derived:

- Arid: growth period <75 days per year
- Semi-arid: growth period >75 < 180 days per year
- Sub-humid: growth period >180 < 270 days per year
- Humid: growth period >270 days per year

2.1.3.2 Types of Agricultural Farming Systems

One classification of agricultural farming types that is very frequently used throughout the world was developed by the *Food and Agricultural Organization* (FAO), a subsidiary of the United Nations. According to this classification, an agricultural farming or operating system is regarded as a "population of individual farms that share, in the broadest sense, equal or similar natural conditions, business structures, household conditions and limitations (biophysical and socio-economic)" (Dixon et al. 2001). On a global scale, agricultural farming types are assigned to the following main classes:

- Irrigated agricultural farming systems operated by mechanical or motorised pumps throughout the year, or only during dry periods;
- latitudes of the Indian subcontinent and Rice-based systems in wetlands, mainly found in tropical ntinent or East Asia;

- Rain-irrigated systems in humid areas with high resource potential, characterised by fertile soils and/or year-round water supplies;
- Rain-irrigated systems on steep slopes and in mountainous regions, which are typical for many rural regions in East Africa or the mountainous regions of Central and Southeast Asia;
- Rain irrigated systems in dry or cool regions with low resource potential, e.g., due to shallow soils;
- Dualistic systems (large commercial enterprises and small farmers); and
- Urban systems, which are often found on the outskirts of cities or metropolitan regions and primarily serve to supply the adjacent urban population.

2.1.3.3 Regional Characteristics

On the basis of the aforementioned natural vegetation zones, regional characteristics of agricultural production systems can be summarised according to their agro-ecological zones and respective farming types (◘ Table 2.2).

Crops such as maize or sorghum are, for example, characterized by very high cultivation amplitudes. Both crops can be cultivated in almost all vegetation and climate zones. Oil palms, on the other hand, can only be cultivated in areas along the equator with high annual precipitation of more than 1500 mm/m^2 and annual average temperatures of more than 25 °C. Other crops such as jatropha and agaves are well adapted to drier locations, which is advantageous in areas that have not yet been used for agricultural production.

◘ **Table 2.2** Biome and types of agricultural production system. (Source: Davis et al. 2014)

Biome/vegetation zone	Agricultural production systems	precipitation (mm•annually)	Temperature (°C)[a]	Growth period (days)[b]	Possible crops (selected)
Subtropical/temperate humid forest	Large commercial and smallholder: intensive mixed agriculture, cereals and livestock, tree crops, agroforestry[c]	1000–2500	10–30	270–365	Maize, sugar cane, soybean, sorghum, wheat, other
Temperate broad-leaved forest	Large commercial and smallholder: tree crops, forest-based livestock, large-scale cereal and vegetables, cereal/livestock, agroforestry	350–1500	−10–30	90–365	Maize, switchgrass, miscanthus, soy, wheat, rapeseed, sorghum
Temperate coniferous forests	Forestry, large commercial and smallholder: cereals/roots, forest-based livestock, agrosilvipastoral[d]	100–1500	−30–5	30–180	Miscanthus, switchgrass, wheat, potatoes, other
Temperate grassland	Large commercial and smallholder: irrigated mixed agriculture, small-scale cereal/livestock	50–1500	−10–30	0–320	Maize, sugar cane, switchgrass, miscanthus, soybean, wheat, rapeseed, sorghum
Tropical dry forests	Large commercial and smallholder: tree crops, rice, cereals/roots, agroforestry	700–2500	15–30	30–300	Maize, wheat, soy, sugar cane, rape, rice, sorghum
Tropical grassland	Large commercial and smallholder: extensive mixed cropping, cereal/livestock	500–2500	15–30	30–300	Corn, sugar cane, soybean, wheat, sorghum, jatropha
Tropical humid rainforest	Large commercial and smallholder: subsistence agriculture, livestock, tree crop, root crop, agroforestry	1500–5000	25–30	300–365	Oil palm, sugar cane, sorghum, maize
Desert	Subsistence pastoralism	0–350	10–40	0–30	Agaves, jathropha

[a]Annual average temperature based on: FAO GeoNetwork
[b]See also: Explanations AEZ above
[c]Agricultural system in combination with trees
[d]Agricultural system in combination with trees and livestock farming

Fig. 2.7 Jatropha plantation in Madagascar (© JatroSolutions GmbH, picture archive of the University of Hohenheim)

Evidence in recent years has shown, however, that producers' and smallholders' cropping decisions mainly follow economic incentives. Hence, farmers grow crops such as jatropha, which were actually intended for unused or degraded sites, on areas that were previously used for other cropping systems due to the promise of higher yields and economic returns (Fig. 2.7). In recent years, this development has led to the so-called 'food-vs-fuel' debate. This is due to the fact that locations previously used exclusively for food production are now being used, to a large extent, for renewable raw materials to meet the increasing demands of the bioenergy sector.

2.1.4 Material Flows and Biomass Yields

2.1.4.1 Forms of Agricultural Biomass Use

Renewable raw materials from agricultural production can be used for food, material and energy use. Material uses include, amongst others, textile fibres, insulating materials, cellulose, and special platform chemicals. Energy uses are the production of bioenergy and biofuels (▶ Chap. 4).

Biomass can be divided into main products, by-products and residues. The main products of crop cultivation include, for example, maize and cereal grains, sugar cane stems, rapeseed and, in the case of miscanthus and switchgrass (*Panicum virgatum*), the total above-ground biomass. Table 2.3 lists selected annual and perennial crops with various typical attributes. It shows that the components of their main harvest products (e.g., sugar or starch) are suitable for different energy and material usage pathways, which can lead to competition with food production (see also: food-vs.-fuel debate).

Table 2.4 shows a number of by-products resulting from the harvesting of crops. Of these, barley straw has the highest energy content and ethanol yield. However, the withdrawal of harvest by-products from the agricultural system for external use is limited if soil fertility is to be maintained in line with sustainable land use.

2.1.4.2 Components of Agricultural Biomass

The decision to use crop biomass either as food or feed, or for material or energy production depends not only on yield but also on its plant components and organic metabolites. The elemental composition (C, H, O, macro/micro elements) is a key criterion for energy use, because it is particularly important for the type and composition of residues and exhaust gases formed during thermochemical transformation. If biomass is to be used as a solid fuel, its water content and calorific value, its shelf life and combustion temperature are also decisive criteria (Diepenbrock 2014).

Carbohydrates, fats, oils, proteins and lignin are the most important valuable organic metabolites (▶ Chap. 4). Carbohydrates include sugar, starch and cellulose. Plant sugars

Table 2.3 Selected crop plants with information on constituents, main products, water, fertilizer and pesticide requirements

	sugar cane	maize	soy	oil palm	miscanthus
Cultivation Type	perennial	1 year	1 year	perennial	perennial
Photosynthesis type	C_4	C_4	C_3	C_3	C_4
Water requirements (mm)	high: 1500–2500	moderate: 670–800	moderate: 600	high: 2000–2500	low: > 450
Fertilizer requirement (kg/ha/annual)	N: 45–300 P: 15–50 K: as required	N: 145–200 P: 26–110 K: 25–130	N: 0–70 P: 32–155 K: 30–320	N: 114 P: 14 K: 159	N: 0–92 P: 0–13 K: 0–202
Pesticide use	yes	yes	yes	yes	no
Main harvest product	leaf, stem	grain	seed	seed	stem
Component	sugar	starch	oil	oil	lignocellulose
Form of use	food, bioenergy	food, feed, bioenergy	feed, food, bioenergy	food, bioenergy[a]	bioenergy fibre materials

[a]Oil derivatives are used in the cosmetics industry (Source: Davis et al. 2014)

Table 2.4 By-products of agricultural production resulting from the harvest of the main product, expressed as a ratio of crop residue/field crop, possible energy content and ethanol yield

Harvest residues	Harvest residue/ crop ratio	Energy content (MJ/kg dry matter)	Ethanol yield (l/kg dry matter)[a]
Maize straw	1.0	0.29	98
Barley straw	1.2	0.31	100
Rye straw	1.3	0.26	88
Rice straw	1.4	0.28	94
Wheat straw	1.3	0.29	98
Sorghum straw	1.3	0.27	91

[a]estimated (based on Kim and Dale 2004; Lal 2005)

are divided into mono-, di- and oligosaccharides (single, double and multiple sugars made up of two to ten monosaccharides). The monosaccharides glucose and fructose are two of the most important organic molecules for metabolism. They also serve as basic building blocks for all other carbohydrates and as a carbon source for other organic compounds. The disaccharide sucrose is composed of glucose and fructose. In most crops, it is the most important disaccharide for metabolism and, for example, the main sugar component of sugar cane, sugar beet and sorghum.

Starch is the most common polysaccharide and is found in many cereals, as well as in potatoes. The basic building block of starch is glucose, which is present in the form of the polymers amylose and amylopectin. Starch is a mixture of about 20–30% amylose and 70–80% amylopectin.

Cellulose is a homopolymer composed of about 1000–1400 d-glucose units. Their spatial structure corresponds to a chain with a slight fold. Cellulose is the main component of the plant cell wall.

Lignin is a high polymeric macromolecule composed of phenyl-propane molecules. These units form a three-dimensional network in plant cell walls, which is very resistant and retains its shape even after the cells die. As a renewable raw material, lignin is formed together with cellulose and hemicelluloses in lignocellulose plants such as miscanthus. These plants supply biogenic raw materials and are used to produce materials and energy in the construction, chemical, wood, paper and pulp industries.

Fats and oils are very rich in energy. Their energy content is 37 kJ/g, compared with 23 kJ/g for protein and 17 kJ/g for carbohydrates. In cultivated plants, fats are often present in the reproductive organs as triglycerides. Due to their hydrophobic properties, their extremely low oxygen content and their tight spatial packing, triglycerides are ideal storage substances that ensure maximum energy reserves in the storage organs of the plant at relatively low weight. In the context of renewable raw materials, fats and oils serve as starting materials for detergents, lubricants and fuels (biodiesel) (Diepenbrock 2014).

Proteins are biological macromolecules, built up of amino acids connected by peptide bonds, that fulfil a multitude of vital functions. They contain a large part of the nitrogen bound in plants. Proteins are of particular importance for the nitrogen supply of humans and animals. They can be used for the production of amino acids (e.g., lysine, methionine, phenylalanine), adhesives and cosmetic additives. Because of their complete biodegradability, their edibility, non-toxicity and compostability, as well as their functional diversity, they are regarded as promising raw materials for the industrial production of biopolymers (Diepenbrock 2014).

2

2.1.4.3 Crop Yields

The following tables provide an overview of the global average yields of the world's most commonly cultivated food crops (◘ Table 2.5), those used as feeds (◘ Table 2.6), those used for bioenergy (◘ Table 2.7), and those used as raw material for fibres and textiles (◘ Table 2.8). The yields are mostly given as dry matter, but for permanent grassland, fodder/sugar beet and silage maize (◘ Table 2.6) as fresh weight. Food and feed yields are given for conventional and organic farming, as the applied farm management methods differ. Conventional cultivation systems are characterised by the often high application of synthetic fertilisers and the use of chemical-synthetic pesticides, whereas organic plant cultivation does not use any of these synthetic farm inputs.

◘ **Table 2.5** Average yields (conventional/organic) of selected crops used as food/feed (in dry matter, dm)

Crop type	Main harvest product	Yield (t dm /ha per year)					
		conventional			organic		
		low	medium	high	low	medium	high
Oat	grain	2.2	4.5	6.1	2	3	4
Barley	grain						
- summer		2.2	4.9	6	2.5	3	4
- winter		5.1	6.5	9.1	2.5	4	5.5
Maize	grain	6.2	9.5	12			
Rapeseed	seed	2.2	3.7	4.7	1	2	3.5
Fodder beets[a]	turnip	67	90	114	30	45	60
Sugar beets[a]	turnip	45	67	85	25	40	50
Sunflowers	seed	1.3	2.5	4.3	2	3	3.5
Wheat	grain						
- summer		3.4	5.4	7.1	3	4	5
- winter		5.4	7.4	9.5	3	4	7

[a]Fresh weight (Source: KTBL 2015)

◘ **Table 2.6** Average yields (conventional/organic, fresh or dry weight) of forages (t/ha per year)

Forage type	Yield (t/ha per year)					
	conventional			organic		
	low	medium	high	low	medium	high
Permanent grassland[a]						
- intensive	38	56	75	30	45	55
- extensive	20	30	42	20	25	35
Maize						
Grain straw-mix[b]	12	14	16			
Silage maize[a]	40	50	60	20	30	50

[a]Fresh weight
[b]Dry matter (Source: KTBL 2015)

Table 2.7 Average yields of selected bioenergy crops (t dm/ha per year)

Bioenergy crop type	Principal harvest product	Yield (t DM/ ha per year)	Use	Main cultivation areas
Sugar cane	Stem	71 (fresh)[a]	Bioethanol, biogas, lignocellulose ethanol	Brazil, India, China
Miscanthus[c] (*x giganteus*)	Stem/whole plant	18 (Europe) 38 (Northern America)	Combustion, lignocellulose ethanol, biogas	Europe, USA
Switch grass	Whole plant	14	Combustion, lignocellulose ethanol	USA
Oil palm	fruits oil	14 2.9	Biodiesel	Indonesia, Malaysia, Nigeria
Soy	seed oil	2.9 0.44	Biodiesel	USA, China, Brazil
Jatropha[b]	fruits oil	1.25–6 0.5–2.5	Biodiesel	

[a]Fresh weight at harvest, as sugar cane stems are processed directly
[b]Variable due to irrigated/non-irrigated cultivation systems
[c]Can also be used as substrate in automotive/construction industry (Source: Davis et al. 2014)

Table 2.8 Average yields of selected fibre plants (t dm/ha per year)

Fibre plant	Fibre yield (t dm/ha per year)	Use	Main cultivation areas
Cotton	0.79	Textile industry	Australia, India, USA
Jute	0.47	Yarn, fabrics, ropes	India, Bangladesh
Flax	0.66	Textile/automotive industry	Europe, China
Sisal	1.19	Floor coverings, polishing agents	Brazil, Tanzania, Kenya
Kenaf u. a.[a]	0.77	Yarns, fabrics, floor coverings	India, Bangladesh
Abaca	1.46	Yarns, ropes, etc.	Philippines
Hemp	0.77	Textile/automotive industry	China, Europe
Coconut fibres	[b]	Combustion, automotive industry	India, Sri Lanka

[a] By-product only;
[b] jute-like by (Source: FNR 2008; FAOSTAT 2014)

2.1.4.4 Biomass Supply and Demand

In 2011, around 11.4 billion tonnes of biomass were available as dry matter globally (Fig. 2.8). Biomass from the agricultural sector accounted for the largest share (40%). Agricultural biomass originates mainly from the most common crops: maize, wheat, rice and soybeans. If the harvest by-products are added, more than half of the globally produced biomass can be attributed to an agricultural production system. One should bear in mind that harvest by-products such as straw, stems and leaves are not fully covered by statistical information. For their quantification, therefore, the harvest index is usually taken. This gives the weight ratio between the main product (e.g., grain) and the by-product (e.g., straw).

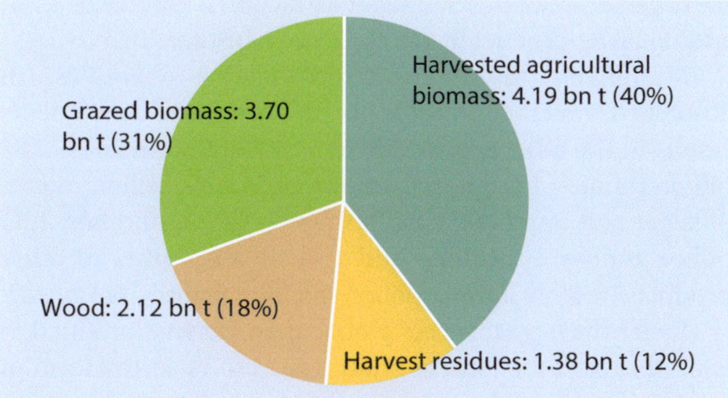

Fig. 2.8 Global biomass supply in 2011 by sources (in billion t dm, total 11.4 billion t dm). (Source: FAOSTAT 2014; nova-Institut 2015)

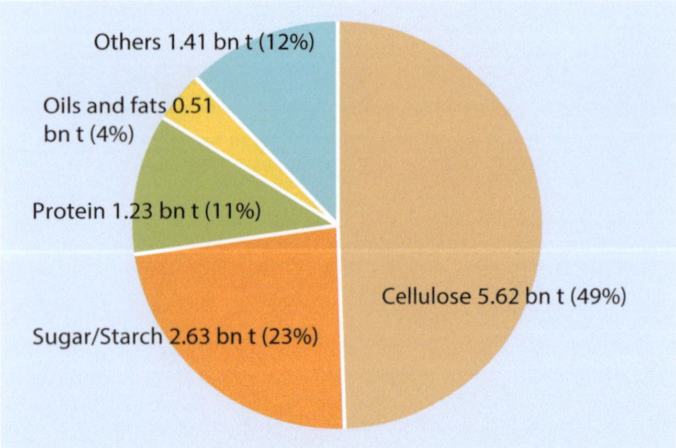

Fig. 2.9 Global biomass supply in 2011 by biomass constituents (in billion t dm, total 11.4 billion t dm). (Source: FAOSTAT 2014; nova-Institut 2015)

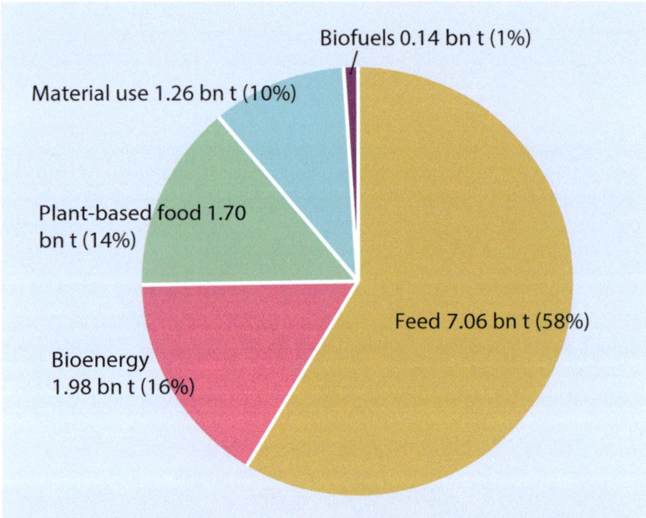

Fig. 2.10 Global biomass demand in 2011 by sectors (in billion t dm, total 12.1 bn t dm). (Source: FAOSTAT 2014; nova-Institut 2015)

In the case of sugar cane, the statistically recorded harvest includes the processable stem without the inflorescence and leaves, which are allocated to the harvest by-products. However, the bagasse remaining after sugar extraction is not treated as an additional by-product in this case (nova-Institut 2015).

Multiplying the respective production quantities (in absolute dry matter) by the mass proportions of constituents results in the following worldwide distribution in 2011: 5.62 billion tonnes of cellulose/hemicellulose, 2.63 billion tonnes of sugar and starch, 0.51 billion tonnes of oils and fats, 1.23 billion tonnes of proteins and 1.41 billion tonnes of other products (mainly lignin, rubber and fibre plants) (Fig. 2.9).

By far the largest part of global plant biomass production is used as feed for farm animals - as fresh or in the form of cereal grains or by-products of sugar beet processing (Fig. 2.10). In 2011, feed at 7.06 billion tonnes of dry matter, accounted for 58% of global biomass demand, while food accounted for only 1.7 billion tonnes, or 14%. A total of 17% of the global biomass stock was used for bioenergy and biofuels, and 10% for material use.

The complexity of global biomass flows from agricultural production can be illustrated using a Sankey diagram (Born et al. 2014). In a Sankey diagram, quantities are represented by arrow thicknesses proportional to their magnitude. Figure 2.11 uses the quotient of reference weight and energy content in exajoules (EJ), based on FAOSTAT 2010 datasets and additional literature. A direct comparison of the data from Fig. 2.11 with those from Figs. 2.8, 2.9 and 2.10 is therefore not possible.

Nevertheless, the Sankey diagram shows the relationship between the different categories of biomass, how they are related to and how they influence each other. For example, the possibility of increased recycling of residues from agricultural production is currently being intensively discussed in order to increase its potential for energy and material use. However, these residues are also necessary in order to ensure the maintenance of soil fertility. Their increased use could therefore have a negative long-term impact on the biogeochemical cycling of soil nutrients if an insufficient amount of residual material were to be returned to the field.

2.1.5 Potentials for Increasing Biomass Production

2.1.5.1 Definitions of Terms

The factors influencing plant growth, and thus yield formation (► Sect. 2.1.1), are divided into determining, limiting and reducing factors (Fig. 2.12).

The determining factors include, for example, solar radiation, local temperature gradients and the growth characteristics of a specific crop. These factors determine the potential yield to be generated at a particular location. The determination of the theoretical biomass potentials is generally based on this potential yield.

The crop yield, however, that can be achieved at a site depends on limiting factors such as the availability of water, nutrients and soil minerals. The determination of the technical biomass potential is usually based on this attainable yield. The gap between the technical and the potential yield can be closed through yield-increasing measures such as fertilisation or irrigation. Nevertheless, the attainable yield does not reflect the actual yield that is harvested at a specific site. In practice, the attainable yield is diminished by reducing factors such as weeds, pests, contaminants and diseases. Socio-economic factors, such as farmers' access to inputs (seed varieties, fertilisers, pesticides) and their level of education, also influence the size of the actual yield. Closing the gap between the actual yield and the potential yield to the greatest degree possible is an essential prerequisite for increasing biomass production (► Excursus 2.3).

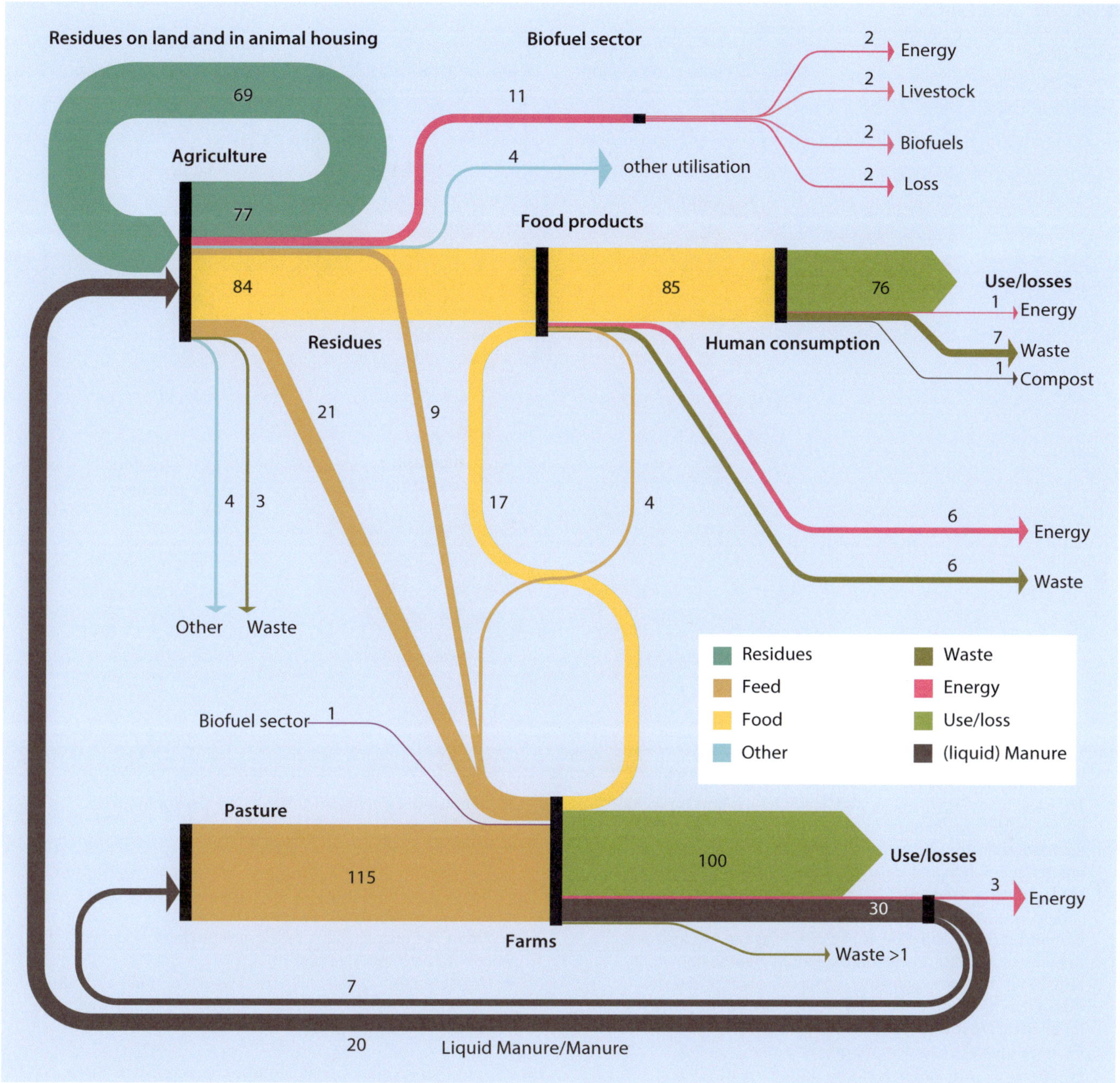

Fig. 2.11 Global biomass flows resulting from agriculture (2010) in energy equivalents. (Source: van den Born et al. 2014)

2.1.5.2 Determinants and Magnitude of Estimated Global Potentials

The potential of agriculturally produced biomass that could be available to the future bioeconomy will thus essentially depend on the quantity and quality of land available for agricultural production, the type and intensity of land management, existing production factors (especially water and nutrients) and competing demands for both land and biomass.

Land availability

According to FAO estimates, around 4.15 billion hectares of land are available worldwide for rainfed agriculture, i.e., the form of land management that uses rainwater. Of this, only less than half, 1.61 billion hectares, are currently used (FAO 2003) (Fig. 2.13). Most of the unused land is located in Latin America and sub-Saharan Africa. But, even in industrialised countries, there is still unused land that would, in principle, be suitable for agricultural production.

The extent to which the potentially usable land should actually be used for agricultural production is controversial. There are serious arguments against it. On the one hand, part of this area is home to species-rich vegetation that would be lost. This would be the case, for example, if grasslands were to

2

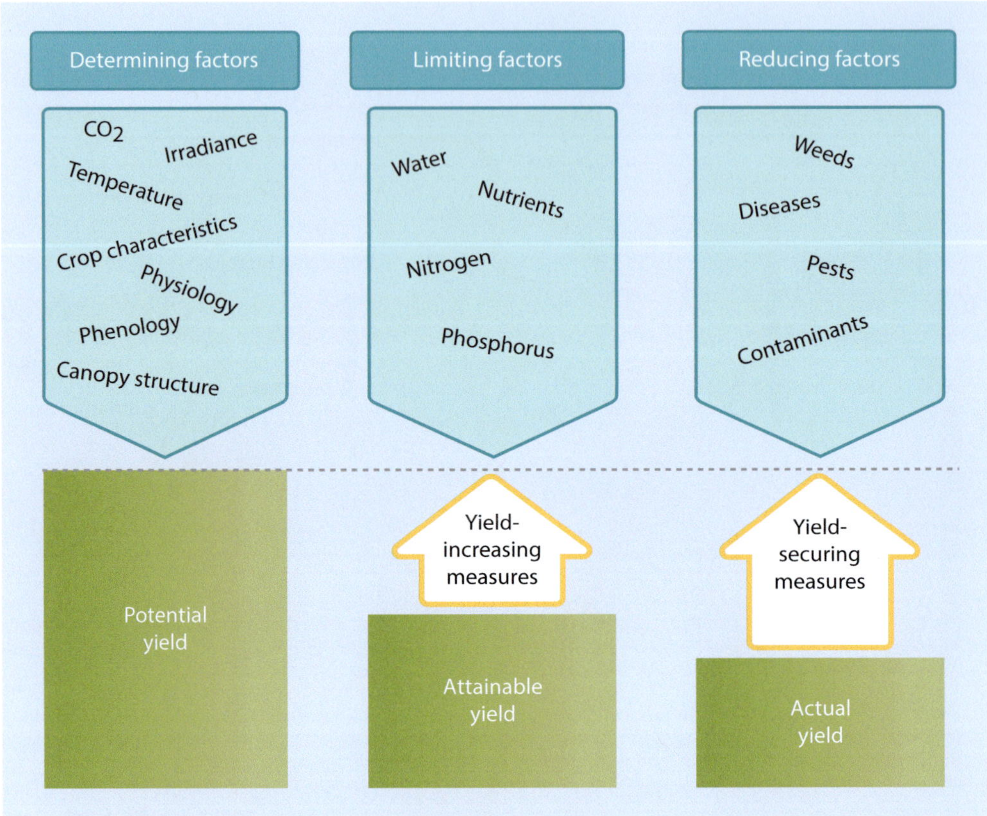

Fig. 2.12 Relationship between potential, attainable and actual yields

Excursus 2.3 In search of the ideal bioenergy plant

The ideal bioenergy plant should have a high biomass productivity per unit area in order to minimise the required cultivation area. However, in order to have a significant influence on future energy generation and to satisfy increasing demand, it must also be grown over large areas. Its physical and chemical properties should be such that the conversion of its constituents into biofuel, biogas or any other forms of energy can be carried out as easily as possible. In addition, it should meet sustainability requirements in environmental, economic and social terms. Altogether, this results in a long list of desirable key properties for an ideal energy plant (from Davis et al. 2014):

- High energy yield per unit growing area
- Low-inut, low-cost processing requirements
- Low greenhouse gas emissions and energy requirements during cultivation and processing
- Easy handling during establishment, cultivation, harvesting and storage
- Tolerant to extreme and/or varying environmental conditions
- High nutrient and water use efficiency
- Provision of additional ecosystem functions and/or co-products
- Suitable for a range of conversion processes for the production of various forms of bioenergy
- Productive on marginal soils less suited to food production
- Low-or zero-invasive potential
- Unrelated to naturally occurring weeds or main agricultural weed species so as to avoid outcrossing or mixing with the predominant gene pool

It is obvious that no plant has all of these characteristics. Perennial plants such as miscanthus and switchgrass come closest to the sum of these properties. These plants do not require annual soil cultivation measures, recycle nutrients efficiently and contribute to the formation of soil humus.

be converted for agriculture. Part of this area is also of considerable importance for the Earth's climate system. On the other hand, it is difficult to determine whether this land is truly unused or perhaps, only temporarily out of use, but occasionally used for grazing animals, for example by nomadic tribes. Where this is the case, the further question of a fair prioritisation of land use rights arises.

Additionally, a large part of the potentially available land is not used because of difficult production conditions and because the cultivation of biomass on these areas would not. From an economic point of view, land is marginal if the farmer cannot make a profit from farming because the expenditures are greater than the financial return. Economic marginality is often a consequence of biophysical factors that complicate land management. Such factors include low soil fertility, steepness of the terrain, drought and contamination of the land. In some cases, management measures can improve the soil and make marginal land accessible for cultivation.

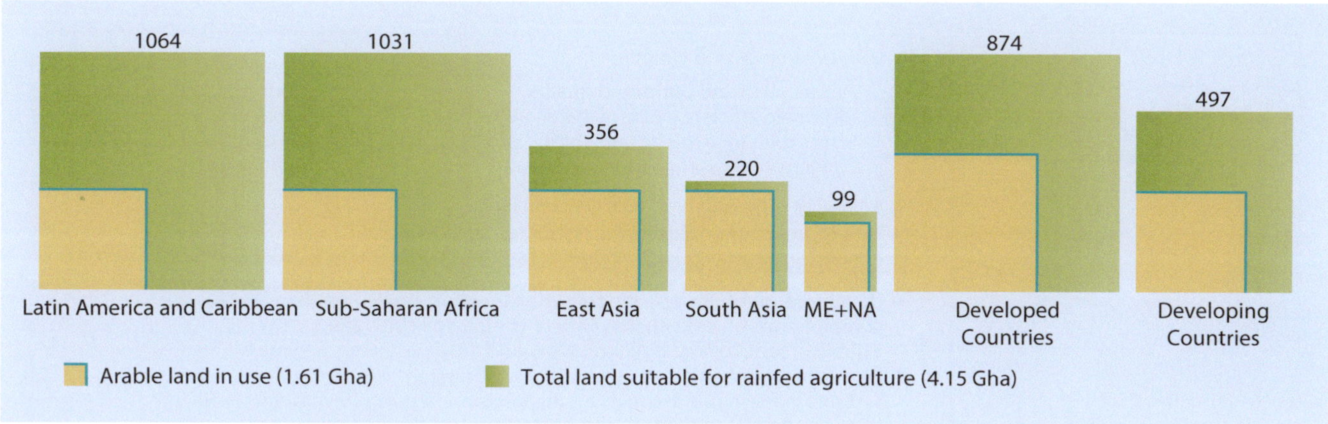

Fig. 2.13 Comparison of the global availability of land resources suitable for crop production and area currently used (million ha); ME: Middle East, NA: North Africa. (Based on FAO 2003)

The term "marginal land" is therefore used to describe different types of land that are either not used or are used at a low intensity (Fig. 2.14). This includes fallow land or land that has been abandoned, as well as land that has been degraded by soil erosion or contamination.

Type and intensity of land use

If more land cannot be made available, an intensification of land use must contribute to higher quantities and qualities of raw agricultural materials for the bioeconomy. Current productivity can, in principle, be increased by breeding plant varieties and animal breeds, improving crop management through more efficient production systems, optimising land use systems and harvesting techniques, and reducing crop losses. It is also very important to process, store and use biomass as efficiently and with as little loss as possible after harvest (Fig. 2.15).

The potential to increase yields is particularly high in countries with less well-developed agricultural sectors. In sub-Saharan Africa, for example, only 30–60% of the attainable wheat yield is actually harvested. In Europe, on the other hand, 80–90% of the attainable wheat yield is achieved. There are also major differences with regard to the causes of biomass losses. For example, in countries with poorly developed agricultural infrastructure, depending on the perishability of the product, it is estimated that 30–55% of the yield is lost to diseases or other perishable processes during transport and storage. In industrialised countries, on the other hand, losses occur mainly at the end of the food supply chain, when consumers do not use food in time and throw it away (▶ Sect. 2.4).

The total annual global food loss is estimated at 1.3 billion tonnes (Gustavsson et al. 2011). Reducing these losses would undoubtedly contribute significantly and sustainably to increasing biomass production and availability. By contrast, the sustainability of seeking to achieve an increase in productivity by increasing agricultural inputs, especially fertilizers and pesticides, is currently the subject of controversial debate.

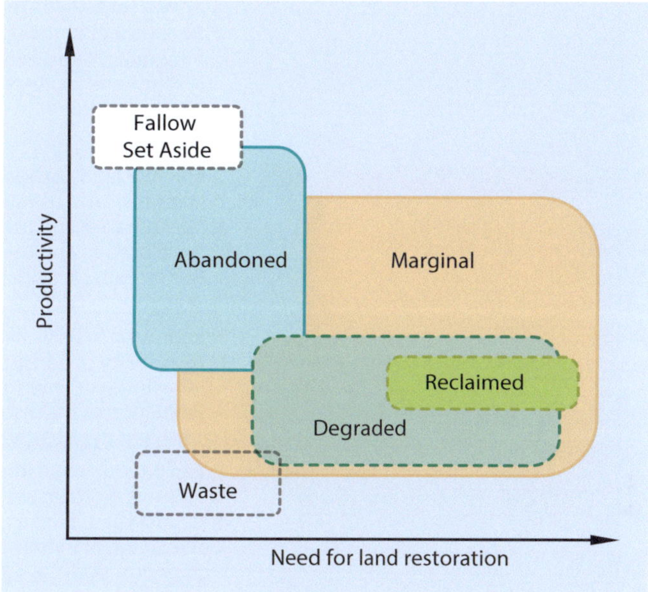

Fig. 2.14 Classification of "marginal land". (Source: Dauber et al. 2012)

In Europe, opportunities are seen for a "sustainable intensification" of agriculture in the context of the bioeconomy. The aim is to produce more yield from the same area and, at the same time, reduce negative environmental impacts and increase the contribution to ecological services (Pretty et al. 2011). In addition to breeding and using higher-yielding varieties and breeds, the instruments of sustainable intensification include methods of *precision farming* that focus on new agricultural technologies and the efficient use of production inputs.

Sustainability and increasing biomass production

There are several concepts of sustainable management in agriculture. These include "Good Agricultural Practice/*cross compliance*", "Integrated Farming", "Organic Farming", and the development of indicator & criteria systems.

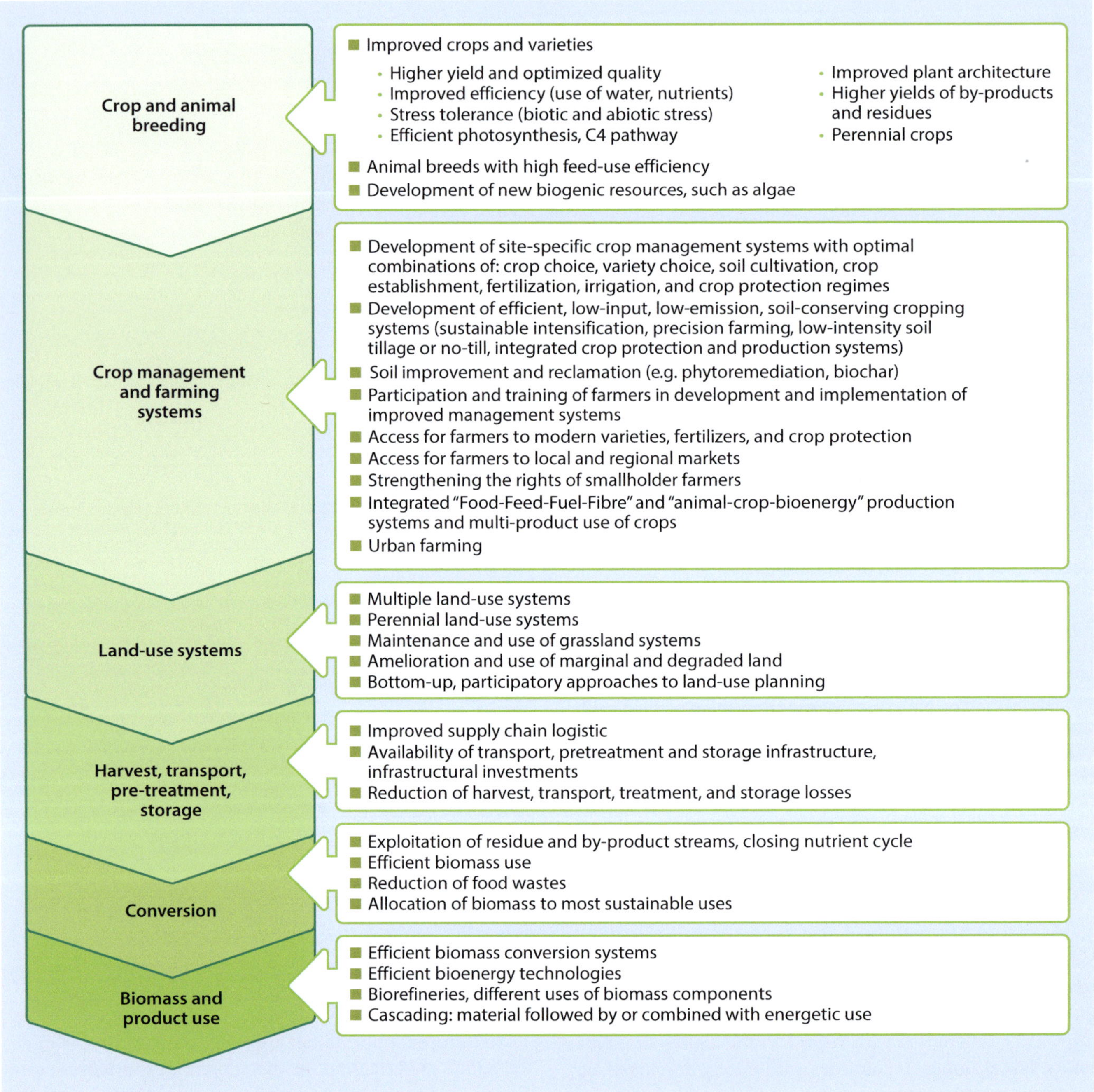

Fig. 2.15 Possibilities for increasing biomass production in agriculture; presented along the bio-based product chain. (From Lewandowski 2015)

Good Agricultural Practice (GAP) are largely defined by legislation in the Europan found in the *cross compliance* guidelines that determine how farmers must manage their land in order to receive financial subsidies. In Germany, the rules of good agricultural practice are aligned with legal regulations such as the Federal Soil Protection Act and, for the most part, are based on the methods of integrated farming.

In the **integrated farming approach,** mineral and chemical fertilizers are permitted, but should be used according to the expected benefit, i.e., they should only be used if the economic benefit is greater than the costs incurred. Here, too, the aim is to apply as few agricultural chemicals as possible, to use natural processes to control weeds and plant diseases and to use legumes such as clover for nitrogen fixation.

Organic farming defines sustainable agriculture primarily through closed material cycles. It considers the agricultural holding as a unit and dictates a number of management rules. Only organic fertilisers may be used; no mineral fertilisers and no chemical pesticides. Plant protection is largely achieved through the use of natural processes. This includes of crop rotations with many different crops and the use of natural enemies against pests.

International agreements on criteria and indicators for sustainable biomass production have been reached in several working and discussion rounds between all relevant interest groups (farmers, industry, environmental protection groups, and others) within the framework of multi-stakeholder roundtables. Such criteria have been compiled either for specific agricultural products (e.g., palm oil, in the *Roundtable on Sustainable Palm Oil*) or for agricultural production in general (e.g., the *Roundtable on Sustainable Biomaterials*). Based on these criteria, farms can apply for sustainability certificates. The criteria for sustainable biomass production are summarised below (▶ Chap. 9). In practice, it is usually not possible to fulfil all criteria at the same time, as there are often conflicting objectives (▶ Chaps. 8 and 9).

Compilation of criteria for sustainable biomass production

The criteria were developed by Roundtable on Sustainable Palm Oil (RSPO), Roundtable Responsible Soy (RTRS), Bonsucro and the Roundtable on Sustainable Biomaterials (RSB) (Source: Lewandowski 2015)

- Social criteria
  - Respect of human and labour rights
    - No child labour
    - Consultation/stakeholder involvement
    - Payment/fair salary
    - No discrimination (gender, race)
    - Freedom of association
    - Health and safety plans
    - Respect of cultural rights and local people
    - Rights of smallholders
    - Responsible community relations
    - Socio-economic development
    - Well-being
- Ecological criteria
  - Protection of biodiversity, wildlife and areas of high conservation value (HCV)
  - Environmental responsibility
  - Minimisation of waste
  - Reduction of greenhouse gas emissions
  - Efficient use of energy
  - Responsible use of fire
  - Preservation of soil fertility
  - Water resources/quality
  - Level of air pollution
  - Use of best practice/responsible agricultural practices
  - Responsible use of agrochemicals
  - Training of workers and employees
  - Responsible development of infrastructure and new areas of cultivation/plantations
  - Impact assessment prior to establishment
  - No replacement of areas with high protection value after year X
  - No establishment on fragile soils
  - Restoration of degraded lands
  - Compensation of local population, informed consent
  - Maintenance of sites with high soil carbon content
- General and economic criteria
  - Commitment to continuous improvement
  - Wise use of biotechnology
  - Climate change and greenhouse gas mitigation
  - Food security
  - Use of by-products
  - Traceability
  - Transparency
  - Legality
  - Responsible business practices
  - Respect of land-use rights

▪ Magnitude of global biomass potentials

Biomass potential analyses determine the amount of biomass that is sustainably available for material and energy use. As a rule, potential analyses assume that only the biomass that is left after satisfying the demand for food and animal feed is available for material or energy use. Currently, around 60% of agricultural biomass is used as animal feed (◘ Fig. 2.10). Feed is mostly used for meat production. However, because animals use plant biomass with relatively low efficiency (◘ Table 2.1), the demand for plant biomass has grown disproportionately due to increasing meat consumption. Therefore, biomass potential studies often report their results as a function of the expected meat consumption.

Initial biomass potential analyses identified a potential that would suggest the entire global primary energy demand could be covered by bioenergy (Smeets et al. 2007). The underlying scenario, however, was based on the assumptions that all land suitable for agricultural production is used, livestock farming is efficient and livestock is mainly kept indoors, the potential yield is achieved and bioenergy is supplied from wood and short rotational coppice, i.e., a type of biomass production that promises the highest yields. This scenario is unrealistic in practice, however, because the development of agricultural infrastructure alone would require huge investments and many years of implementation.

A biomass potential study performed in 2015 compared the global biomass supply, subdivided according to its origin, with the global biomass demand and its use

2

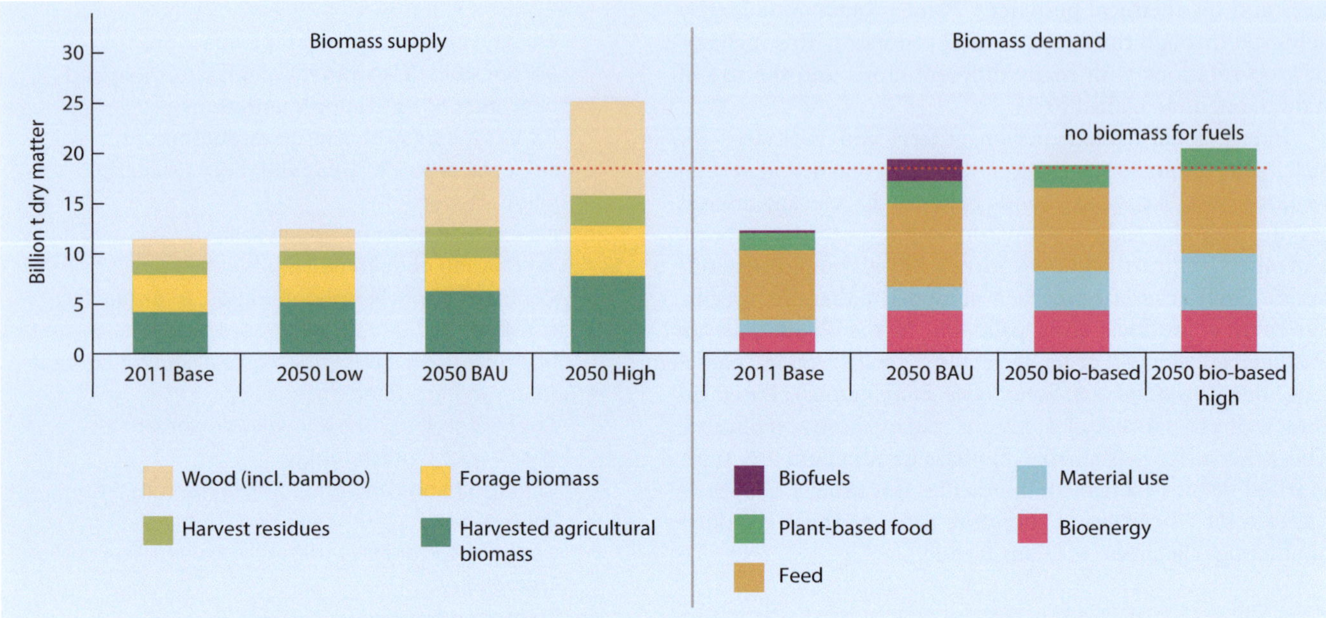

Fig. 2.16 Comparison between global biomass supply and demand scenarios by biomass sources and uses. (Source: nova-Institut 2015)

(nova-Institut 2015). This division makes sense given the varying suitability of different biomasses for different applications. For example, the cultivation of lignocellulose plants such as trees and grasses produces the highest yields of biomass per unit area. However, these can be used neither as food nor as animal feed.

This study also developed a number of varying biomass scenarios (Fig. 2.16). For example, in the most optimistic scenario, a total of 25 billion tonnes of biomass from agriculture and forestry could be available annually if there were a strong demand-driven conversion of land into arable land and intensive forestry (including plantation management). In this scenario, around 8 billion tonnes of biomass from agriculture and an additional 4.5 billion tonnes of residues or harvest by-products are assumed. The expansion of arable farming, to the detriment of grassland, would reduce the supply of pasture biomass, i.e., plant biomass growing on permanent grassland or pastures, from 3.7 to 3.0 billion tonnes. In this scenario, 20 billion tonnes of biomass, almost half of which is used for material and energy purposes. Indeed, compared with today's biomass production of around 11.5 billion tonnes, more than twice as much biomass would be available. In a business-as-usual-scenario (BAU), which merely extrapolates the current trends of productivity growth and land use, global biomass production will increase to only 18 billion tonnes by 2050 but could still meet the demand for food and feed, as well as for material and energy use. Only a cessation in the further development of the agricultural sector ("low" scenario) would lead to a bottleneck in the supply of biomass for a growing bioeconomy.

2.2 Biomass from Forestry

Rüdiger Unseld

The extent and nature of woody biomass, which forestry provides for various paths of use, is determined, on the one hand, by natural factors such as the forest site and, on the other, by anthropogenic factors. These include, for example, socio-political demands on the forest, rationalisation trends and technical innovations in the harvesting and processing of the raw material, as well as developments in the timber market. The task of silviculture as a central discipline of forestry is to control the growth and use of forest stands and individual trees as sustainably as possible. As in agriculture, there are proponents of both more intensive and more extensive use strategies in forestry.

2.2.1 The Importance of Wood as a Raw Material

In forestry circles, the use of wood as biomass for the bioeconomy is often seen less as a trend than as a tradition. This is because wood has always been the most important raw and building material for the economy (Fig. 2.17) and, accordingly, humans have a familiarity in dealing with it (Fig. 2.18).

In pre-industrial and early industrial times, wood played a key role almost everywhere: in crafts and house-building, in iron smelting and salt extraction, in textile production and processing. It was the building material of simple machines, such as cranes and mills, and the basis of construction for all means of transport, including large ships. Last but not least, it was used for heating and for preparing food. Historians therefore also refer to this period as the "Wooden Age" (Grewe 2011).

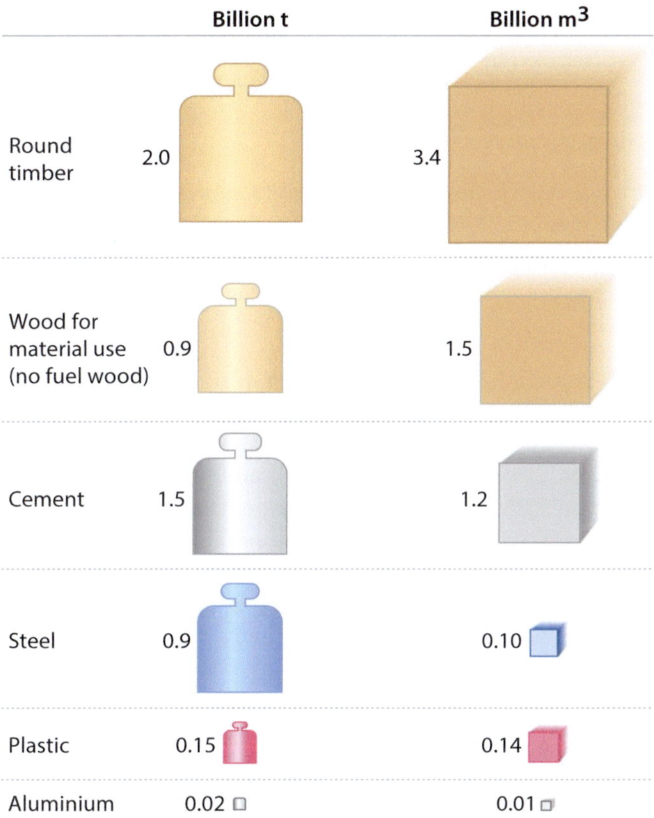

Fig. 2.17 Worldwide annual production or harvest of wood (1997) in comparison with other important raw materials and materials for building. (Source: Wegener and Zimmer 2001)

Early in human history, comprehensive knowledge as to how to process and use a wide variety of wood species was developed (Radkau 2007). Today, the use of wood extends to three main areas:

- The sawmill industry and industry for wood-based materials
- The paper industry
- Power generation (with the use of wood chips in large heating systems, new utilization lines have recently been set up)

Overall, however, "the visible contribution of forestry in the context of bioeconomy still falls short of its potential" (Hüttl 2012). The reasons for this are the "hitherto rather traditionally oriented product range of forestry," but also the "current scientific-institutional equipment in this area in relation to the affected land area." Accordingly, the future key tasks of forestry in regard to shaping a future bio-economy are:

- Development of new products (▶ Excursus 2.4)
- Increasing biomass production through intensification
- Increasing the sustainability of land use with regard to soil and water protection, erosion reduction, carbon storage and biodiversity
- Adaptation of forestry to climate change

Fig. 2.18 Water-operated saw gate in southwest Germany (l. M. Bailiff; r. R. Unseld)

Excursus 2.4 New products made of wood

For the packaging industry, as bioplastics, for the hygiene and health sector, and for the printing industry, there are many new options for the use of wood and derived products (FTP 2013).

Special hopes have been pinned on biotechnologically obtained nanocellulose (Kralisch 2014). It is not only extremely absorbent, but also extremely tear-resistant. The products that can probably be developed from it range from dressing materials to cosmetics and chip cards. Three new types of nanocellulose are currently of particular interest (Burbiel 2014): microfibrillated cellulose (MFC), nanocrystalline cellulose (NCC) and bacterial nanocellulose (BNC). However, their production has so far only been possible in pilot plants. Nevertheless, due to the good availability of raw materials for MFC and NCC, mass production at favourable prices is expected to develop in the coming years. One important advantage is likely to be the involvement of the existing timber- and paper industries, which have a strong interest in the production of new cellulose products.

Promising prospects are also offered by lignocellulose-based biorefineries (► Chap. 4). They produce fuels, energy and building blocks for organic syntheses from raw materials containing lignocellulose. The celluloses, hemicelluloses and lignin of the raw materials are used. The most important raw materials for these refineries, which have only so far been tested as pilot plants, are currently wood and agricultural residues such as cereal and corn straw. It is expected that the use of beech and poplar wood will be a focus of the raw material supply of lignocellulose biorefineries in the future (Michels 2013).

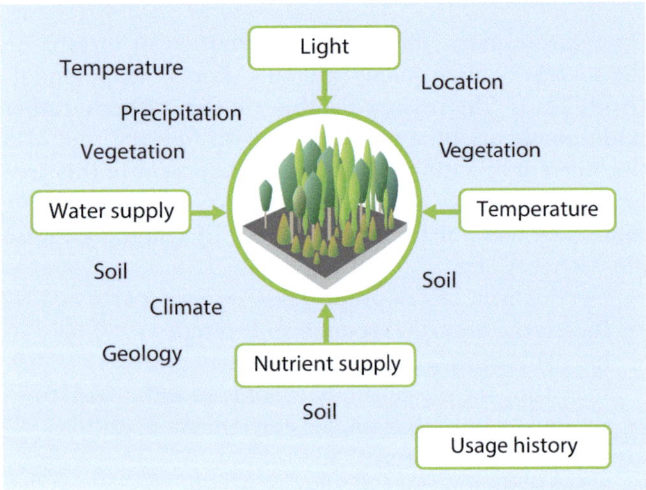

Fig. 2.19 Local factors influencing forest growth. (From English 2009)

2.2.2 Forest Sites as a Production Base

The site quality of a forest is essentially determined by the soil on which it grows and the prevailing climate there. The nutrient supply, light, temperature and water supply of a forest stand depends, above all, on these two factors (Fig. 2.19). They can be influenced and controlled, to a certain extent, through silvicultural measures like tree species selection, thinning and final-cutting.

Forestry site maps are the basis for the silvicultural planning of larger forest enterprises in Germany. Forest experts evaluate their soil properties, local climatic conditions and potential hazards, such as the risk of windthrow, during field surveys. Their analysis produces maps that are linked to text descriptions. They describe the tree species or tree species mixtures suitable for the respective forest locations. According to the definition of the German board for forest site mapping, tree species are suitable for a particular site "if the ecological requirements correspond to the site characteristics recorded, if the tree or tree population is vital and sufficiently stable with appropriate management and if it has no negative effects on the site" (AKS 2003). Thus, spruce, with a share of 25%, is still the most common tree species in Germany (BMEL 2014), and foreign tree species such as Douglas fir can also be found to be suitable for sites, even if they would not naturally occur there. However, the term "suitable for a site" must be distinguished from the term "indigenous to a site." The latter includes only those tree species whose natural distribution, even historically, corresponds to the site in question (ANL 1994).

2.2.2.1 Focus on Soil Fertility

Forestry pays particular attention to forest soil. Due to its diverse forms and complex ecological interactions, it represents much more than just an economic production factor. As humankind was unaware of this fact for so long, most forests in Central Europe were used in an unregulated manner for centuries. The forests were also of great importance as agricultural reserve areas (Grewe 2011). They were burned and ploughed if necessary. Livestock was driven into them, cattle feed was obtained, and the litter was used as fertilizer for arable land (Fig. 2.20). Such overuse led to significant losses in soil fertility, which can still be seen today in some forest soils. Forestry therefore saw the preservation of soil fertility as a central task at an early stage, and it was subsequently anchored in the forest laws of the federal states in Germany. In view of today's efforts to intensify the use of biomass from forests, these negative experiences from previous overuse should be taken into account (Meiwes et al. 2008).

In contrast to agricultural areas, forest soils are not homogenised by mechanical soil cultivation, drainage, fertiliser application and irrigation. Forests therefore often resemble a small-scale mosaic with different growing conditions for the trees (Fig. 2.21).

Long production periods allow the forest soil to develop as naturally as possible, even if this is increasingly influenced by anthropogenic acidification of the topsoil. The acid inputs

Fig. 2.20 Re-managed grazed forest. (© N. Schoof)

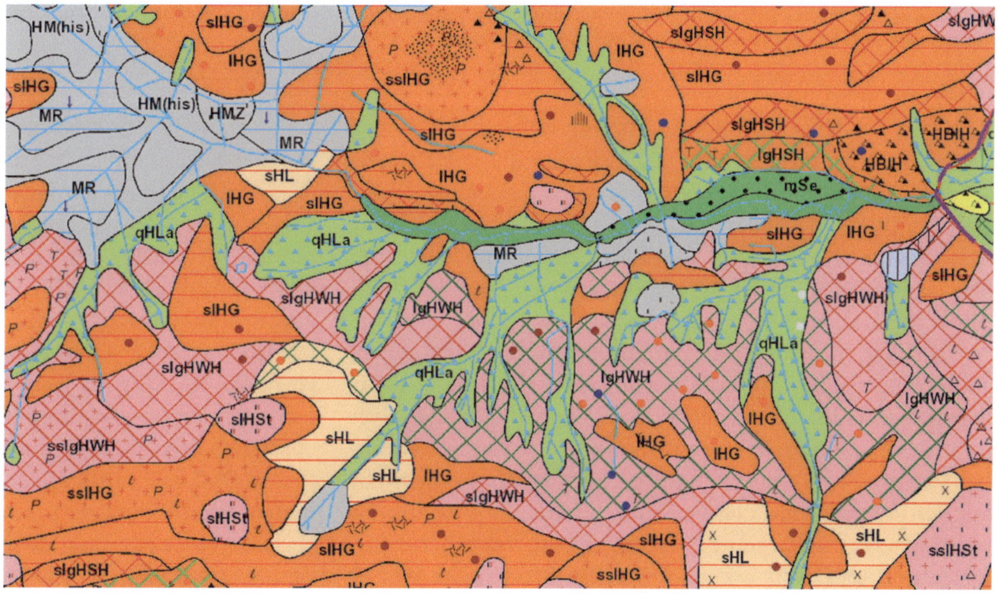

Fig. 2.21 Mosaic created by heterogeneous soil conditions on a site map (© Geoportal Baden-Württemberg 2016)

originate partly from precipitation (today mainly from its nitrogen, in former times mainly from its sulphur) and partly from the permanent cultivation of pure coniferous stands.

2.2.2.2 Consequences of Climate Change

In the past, climate was regarded as a constant site factor. Since the beginning of the 20th century, however, average global temperatures have risen noticeably (IPCC 2013). In Germany, the average temperature has risen by 1.3 °C since 1881 (DWD 2015). This is in line with the rise in European temperatures, but clearly exceeds global warming. In the last 30 years alone, the average temperature has risen by more than 0.7 °C. The precipitation trend is not so clear. In the meantime, climate change has been classified as a dynamic process that will change the growing conditions for forests in the long term. Since wood production in forests also takes place over long periods of time, the forest farmer is already faced with the question of which tree species will be suitable for a particular location in the future. For this purpose, various climate scenarios for forest locations have been calculated, from which the future suitability of different tree species for a given location can be derived.

Climate change will have an impact on the supply of certain types of wood. Spruce already has massive problems with increasingly frequent droughts. This is because spruce

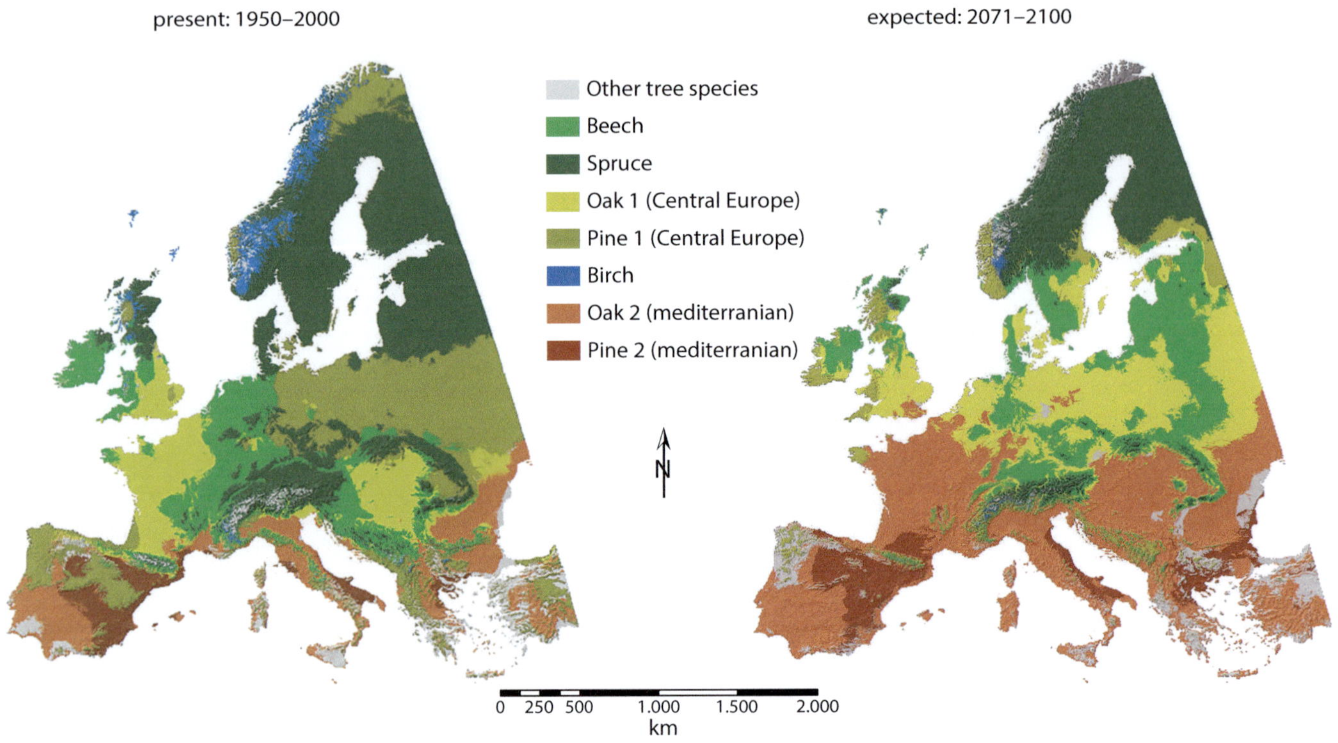

Fig. 2.22 Potential distribution of the most important tree species in Europe for the climate periods 1950–2000 and 2071–2100. (© Hanewinkel et al. 2013, with permission from Macmillan Publishers Ltd.)

was cultivated far beyond its natural range limit and is naturally more adapted to a cold climate and moderately humid soils. At the same time, spruce is the most important supplier of sawn timber, and thus the most economically important tree species in Germany. According to current forecasts, their potential cultivation areas will shrink significantly throughout Europe (Fig. 2.22).

In order to compensate for the resulting feared gap in coniferous timber supply, in Germany, the fast-growing Douglas fir is to be cultivated above all, as well as the silver fir and the grand fir. While cultivation of the native silver fir is limited by browsing, due to too high game populations, and by the preferred cultivation in submontane and montane locations, a forced introduction of the two other coniferous tree species is viewed critically through the lens of nature conservation. The current FSC certification system limits the proportion of foreign tree species to a maximum of 20% per stand (FSC 2013). Douglas fir and grand fir are sometimes even referred to as "invasive species" (Nehring et al. 2013). According to other experts however, neither tree species fulfils the main criteria of invasiveness - a displacing effect for other tree species and a threat to biodiversity (DVFF 2014; Vor et al. 2015).

There are indications that slow-growing, drought resistant tree species, such as Mediterranean oaks, can benefit from climate change in Europe and will expand to the north (Hanewinkel et al. 2013). The Central European oak types, which are extending their natural range, above all, in the northern and eastern directions, are also regarded as climate profiteers. Under current market conditions, there are fears that the reduction of coniferous sawn timber as a result of a decreasing number of suitable sites will lead to major economic losses (Hanewinkel et al. 2009). The question as to whether the expected shift towards hardwoods as a result of climate change will lead to a decline in the amount of woody biomass does not yet have a clear answer. Hardwood usually shows lower volume increases, but significantly higher wood densities, than coniferous wood. In addition, global warming is extending the vegetation period of Central European forests, which could lead to higher overall biomass growth. Accordingly, in southwestern Germany, for example, growth in beech and oak trees would be significantly better at most of the higher altitudes, whereas, at the lower altitudes, growth losses would be expected (Nothdurft et al. 2012). Although global warming can increase the yields of many tree species, it also increases the cultivation risk: dry periods or pest infestation could cancel out the additional yields (Fig. 2.23). Stable, low-risk stands consisting of different tree species are a conceivable compromise in this situation in order to achieve a balance between cultivation risk and an adequate yield (Kölling et al. 2013).

Whether increased or significantly reduced tree growth is to be expected in the future is likely to depend, above all, on one key factor, namely, the development of precipitation in regions that already have less rainfall (Sabaté et al. 2002). In

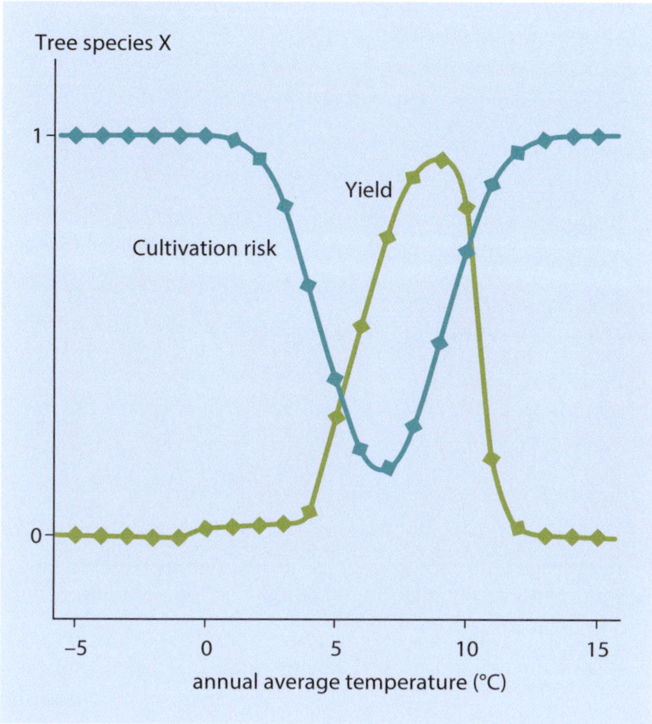

Fig. 2.23 Exemplary representation of the development of cultivation risk and yield along a gradient of the annual average temperature. (Source: Kölling et al. 2013)

comparison to temperature-controlled climate parameters such as the duration of the vegetation period or the number of hot days, statements on future precipitation levels and their annual distribution are still marked by relative uncertainty.

2.2.2.3 Wood Consumption and Forest Loss

Worldwide, forest area is shrinking by an average of around 15,000 ha every day, albeit at a declining rate in recent years (FAO 2012a). Continental focal points of the decline are South America and Africa (Fig. 2.24). The main cause is the conversion of forests into agricultural land (BPP 2010), on which palm oil and soya are cultivated. Illegal logging also accelerates the decline. Its share is estimated at 7–17% of global timber production. It also affects large areas in Eastern Europe (BMEL 2013). The information provided by nature conservation organisations is still well above these values.

Over 50% of the world's annual timber yield is used as firewood or for charcoal production (Fig. 2.25). While, in the northern hemisphere, most wood is used industrially, in Asia and Africa, it is mainly used as fuel. North America consumes the most industrial roundwood, but also produces it.

In the past decades, the demand for industrial roundwood has increased in particular. It is mainly processed into paper or cardboard (Altwegg and Meier 2008). This trend is likely to continue in the coming decades. The reasons cited for this are the growth of the world population and the

Fig. 2.24 Percentage change in forest area from 1990 to 2015 (Source: FAO 2016)

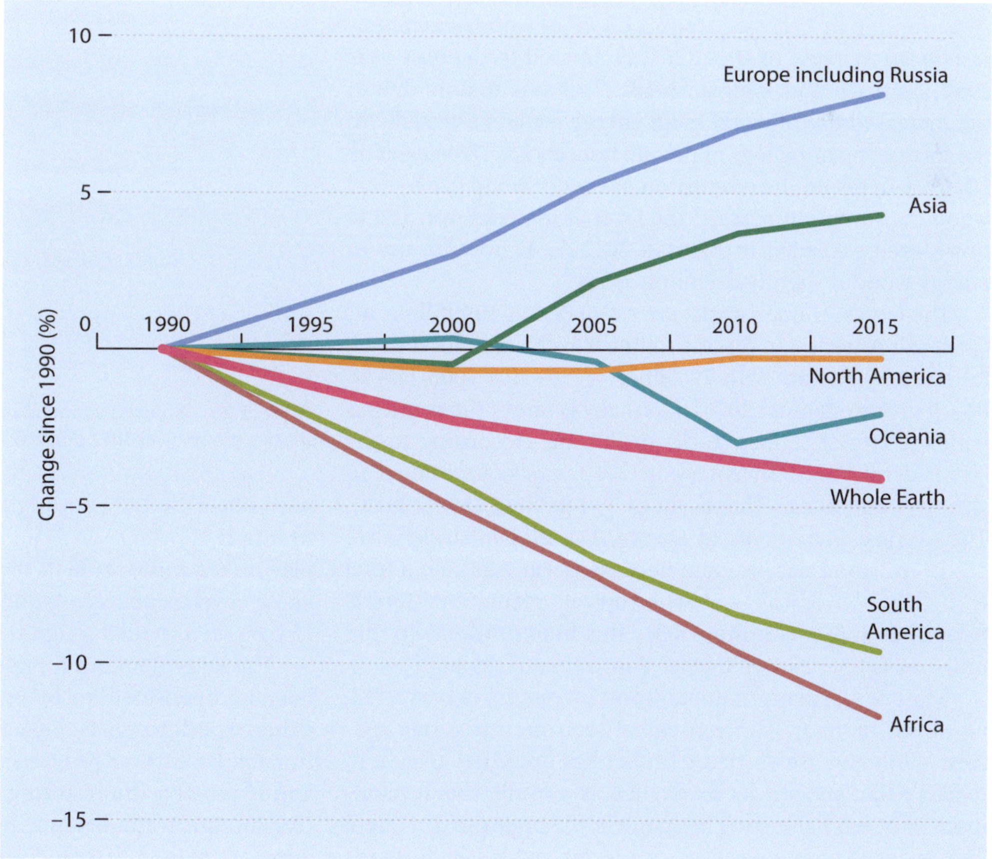

2

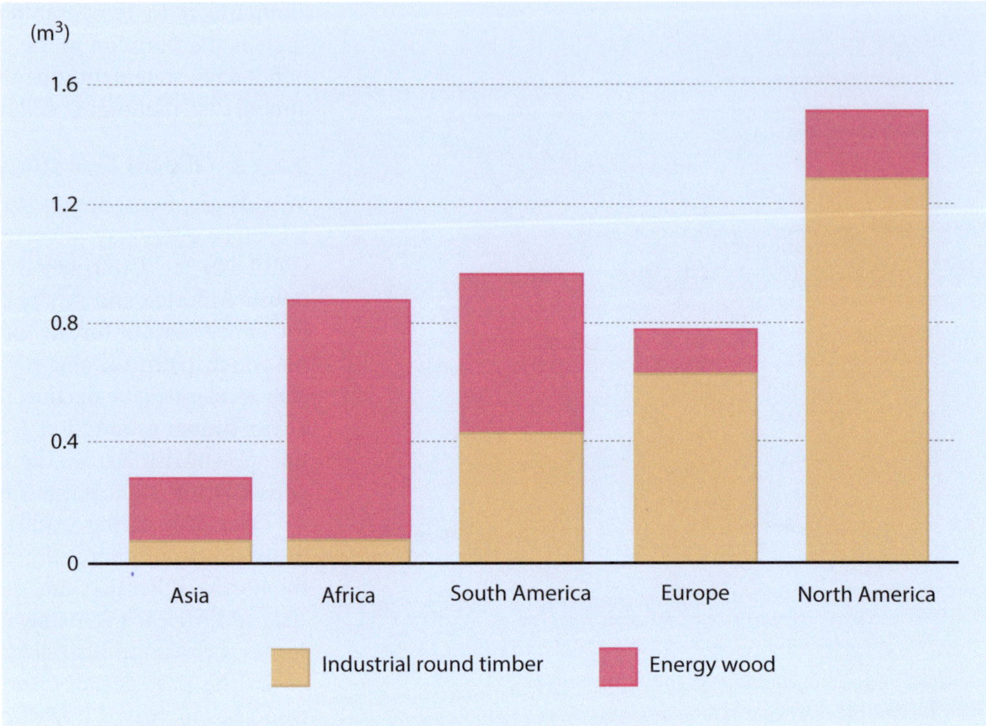

Fig. 2.25 Per capita consumption of raw wood. (Source: Altwegg and Meier 2008)

pronounced economic dynamism in countries such as China and India and the associated higher incomes (Altwegg and Meier 2008; FAO 2012a). In Germany, the annual per capita consumption in 2012 was around 1.3 m^3 of wood, which represents an increase of about 20% compared to the base year 1997 (Seintsch and Weimar 2013). This was mainly due to the increased use of wood as an energy source, especially in the form of burning logs in private households (Döring et al. 2016). As a result, the proportion of energy wood use in total wood consumption reached the level of material use, and at times even exceeded it (Mantau 2012a). At present, use of energy wood is slightly declining again.

The largest timber stocks are estimated at 109 billion m^3 for South America, in this case, almost exclusively hardwood, followed by Russia, with 79 billion m^3 (Status 2003 Jaakko Pöyry in Sommerauer 2006). Germany is one of the countries with the largest timber stocks in Europe. According to the last Federal Forest Inventory in 2012, stocks amounted to almost 4 billion m^3. That is about 340 m^3/ha (BMEL 2014). The average increments of spruce, Douglas fir and fir are 15–19 m^3 wood per year and hectare (Fig. 2.26), on a level with many tropical and subtropical plantation forests (Cossalter and Pye-Smith 2003). This high productivity per unit area is the result of intensive forest management.

Most of Germany's annual wood increment is harvested. However, there are differences between the main tree species. While the use of spruce timber has exceeded growth in recent years, and stocks have fallen as a result, the development of beech harvesting has gone in the opposite direction. The average volume per hectare and the age of beech stands have increased significantly since 2002. In contrast to the agricultural markets, the timber market has, to a large extent, been liberalised. Accordingly, the market reacts dynamically to supply shortages and changes in demand. Transport costs play only a minor role in the globalised timber market. The latest developments in wood consumption in Germany can also be seen in the foreign trade balance (Fig. 2.27).

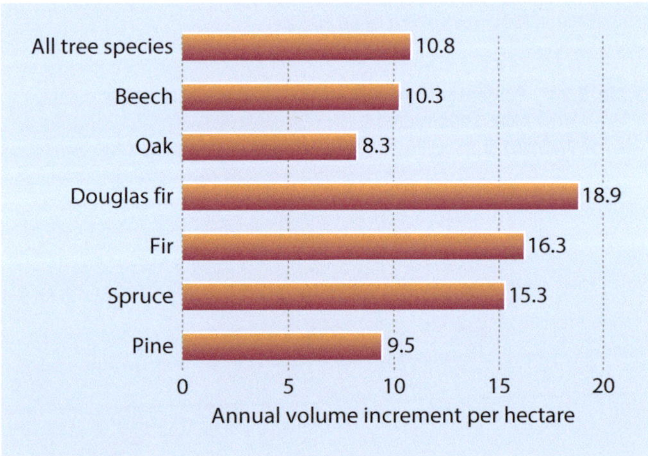

Fig. 2.26 Annual increment of stand volume per hectare by tree species group as of 2012 (Source: BMEL 2014)

The large demand for coniferous wood in Germany has been compensated for by imports since 2009. In the meantime, wood imports have more than doubled the total impact. By further processing the raw wood in the country and exporting the resulting wood products, a clearly positive foreign trade balance has been achieved, measured in euros (Weimar 2014). Solid firewood is mainly traded regionally. With the emerging pellet market, however, new trade flows have developed. The major part of the pellet

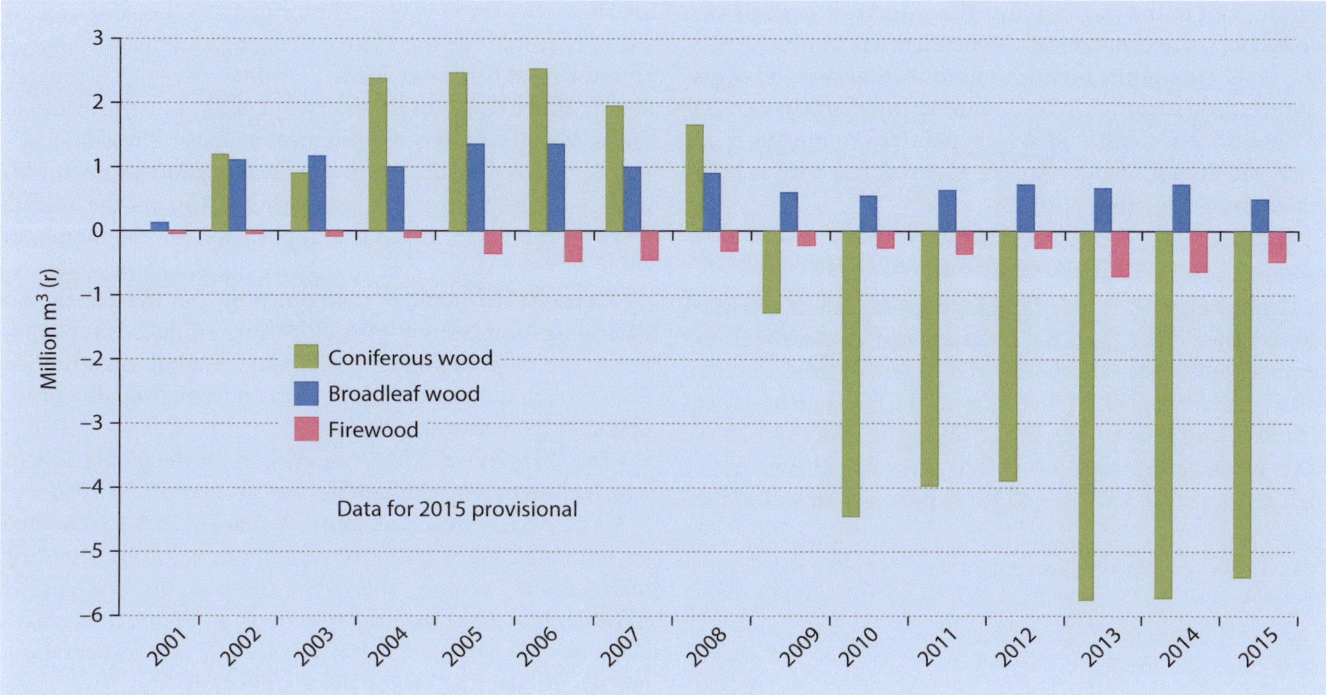

Fig. 2.27 Foreign trade balance of raw wood in Germany. (Source: Weimar 2014)

supply is covered by domestic production. With regard to the European Union, North America and Russia have, in the meantime, developed into important pellet importers (Goetzl 2015).

2.2.3 Forest Use and Forest Management Systems

2.2.3.1 Integration or Segregation?

Sustainable use of forests requires the preservation, protection and appropriate enhancement of biological diversity in forest ecosystems (MCPFE 2003). With the integration and segregation of protective measures for the forest, two different strategies are available to meet this demand. The strategy of integration assumes that wood utilization does not have a negative impact on the conservation of sufficient biodiversity. Its motto is "protect through use" or "protect and use." Through suitable forest management systems, it strives for integrative, multifunctional management over large areas, which, at the same time, permits an adequate supply of wood.

Typical integrative measures are the support of rare tree species, the enrichment of deadwood and the conservation of so-called "woodpecker tree groups" or "habitat tree groups". The strategy of segregation, on the other hand, aims to make a clearer distinction between forest areas used for nature conservation and those used for timber production. The clearest separation is the designation of protected areas with severely restricted use or a total ban (Fig. 2.28). Natural processes should take place there undisturbed. According to Germany's National Strategy for Biological Diversity, in order to improve the conditions for the biocoenoses typical of forests, "the proportion of forests with natural forest development should amount to five percent of the forest area" by 2020 (BMUB 2007). The most important instrument is the setting aside of forest areas of various sizes. In the coming years, 2.9% of the German forest area will have to be removed from exploitation. The expected decline in timber harvesting volume is expected to reach 2.3 million m³ per year (Dieter 2011). This means an annual loss of 3–4%. Ott and Egan-Krieger (2012) see a suitable hybrid of both strategies in the integrative silvi-

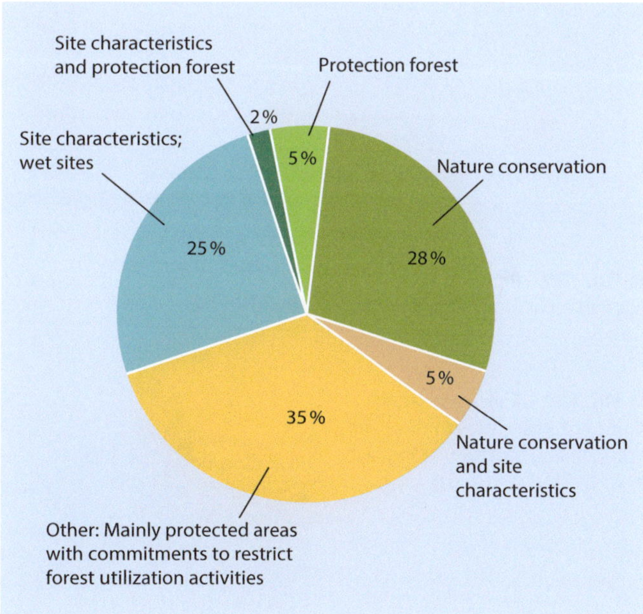

Fig. 2.28 Restrictions on forest use in Germany, 2012. (Source: BMEL 2016a)

culture with partial segregation. The owners of large private forests, who are economically dependent on forest use, suspect that "the conflict between multifunctionality and segregation is only superficially a substantive problem". In fact, it is "a distribution conflict in which political competence, i.e. responsibility for one third of the German territory, is being fought over" (Borchers 2010).

2.2.3.2 Close to Nature or Far from Nature?

In Central Europe, woody biomass is produced in managed forests. They differ from the natural forest essentially in the composition of the tree species and in energy consumption, which is all the higher the further away the forest management is from being nature-compatible (Burschel and Huss 2003). Every forest management measure requires energy. Particularly high energy costs are required for harvesting. However, they are also necessary for soil tillage, fertilisation, breeding and the use of chemical agents. The construction and maintenance of an infrastructure (forest roads, machines, etc.) also require an energy input that every forestry operation must provide. In this sense, plantations are the most artificial forest management systems. In global terms, their cultivation area accounts for 7% of the total forest area (FAO 2010). As the world's demand for wood grows, its importance for the supply of wood increases more and more (Altwegg and Meier 2008).

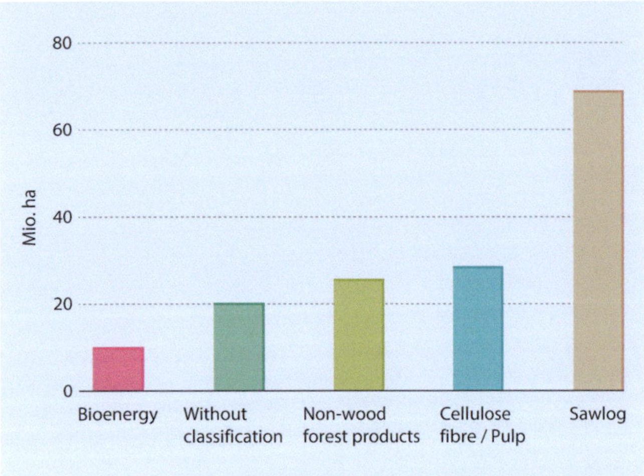

Fig. 2.29 Products from forest plantations measured per hectare of planted area. (Data from Kanninen 2010)

Contrary to popular opinion, pulp for the paper and packaging industry is not the only thing produced on plantations. The most important yield is made up of sawn timber, even though it is not usually used to make high-quality products such as veneers (Fig. 2.29).

The share of industrial roundwood for the global supply, which comes from plantations, was around 5% in 1960 and 30% in 2005 (Bauhus et al. 2010). A share of 75% is estimated for the year 2050. Typical for plantations is the clear-cutting management method. This is the simplest, but ecologically most controversial, forestry system. In it, the final use takes place through large-area clearing. The use and regeneration, mostly by planting, takes place uniformly and temporally concentrated on individual cutting areas. This method is therefore also assigned forests cut by compartments. Other compartment-cutting systems are strip felling systems and shelterwood cutting systems on larger areas. Timber utilization of smaller tree groups, together with initiated regeneration by progressive shelterwood cutting ("Femeln"), represents the transition to compartment-free systems. A mosaic of different development phases is being created with this management system (Fig. 2.30). The natural soil processes are disturbed to a far lesser degree than they are with compartment-cutting processes. The selection forest ("Plenterwald") is a typical single-tree management system. This system allows for the development of an uneven-aged,

Fig. 2.30 Forest management systems: clear-cut (top) and group felling (bottom). (Modified from Weidenbach 2001)

all-aged continuously covered forest with permanent regeneration over a small area.

In the existing silviculture concepts in Germany, clear-cuts are only provided for in exceptional cases. These concepts aim for mixed forests through natural regeneration without the extensive removal of trees. In Germany, there are currently three silvicultural concepts, the practices of which all go beyond the standard of "good forestry practice" prescribed by forestry law:

- The concept of "close-to-nature forestry management (CTNFM)" is represented by most state forest administrations. It is anthropocentrically defined and aims for the use of the forest ecosystem by humans. It includes demands such as increasing the proportion of native deciduous tree species and mixed and multi-layer stands, restricting the cultivation of non-native tree species, making use of natural regeneration and increasing the proportion of old and dead wood.
- The concept of "nature-conform forest management" is based on the so-called permanent forest idea with a consistent use of individual trees. It follows primarily economic goals, too. Stock management is carried out according to the rule "the bad is harvested first."
- The concept of "forest management to protect natural processes" (Fichtner et al. 2013) or "nature conservation forest management" (Röder et al. 1996) is preferred by nature conservation associations. Its primary objective is the protection of dynamic processes in the forest ecosystem. Process protection does not require a focus on human interests, but rather one that is exclusively on nature. The targeted wood reserves and deadwood content should be emphasised. Both should account for at least 80% of the amount of virgin forest in the preferentially cultivated areas ("intensity level III") (Sturm 1995).

◘ Table 2.9 summarises the management characteristics of different silvicultural concepts, compared to the basic variant of good forestry practice.

2.2.4 Possibilities to Increase Wood Production

An increase in wood production is possible, on the one hand, through afforestation of formerly treeless land. This includes initial afforestation, as well as the cultivation of trees on agricultural land as short rotation coppices or in agroforestry systems.

Another option is intensification of use. Additional wood is "mobilized" from existing forest stands. The existing wood reserves are therefore increasingly being skimmed off. This can also include tree components that were previously left in the forest. Furthermore, surpluses of volume increment can be used more consistently. Either the growing stock remains constant or will be permanently reduced to a lower level through early harvesting, as practised in spruce stands of some private forest enterprises in Germany.

A third possibility is to intensify production. The productivity per unit area, i.e., the increase in biomass over a defined period of time, is determined, for example, by mixed stands. In addition to the amount of biomass, the wood value plays an important role. This aspect is particularly important in breeding activities, but also in the shortening of the production period and the use of foreign tree species.

2.2.4.1 Development of New Cultivation Areas

Since 1950, the forest area in Germany has increased by around 1 million ha. This corresponds to almost 18,000 ha of new forest per year. However, this trend has slowed considerably in recent decades. In the period between the last Federal Forest Inventories of 2002 and 2012, around 110,000 ha of new forest were created through first afforestation, but, at the same time, the loss of forest area amounted to around 60,000 ha (BMEL 2014). On average, 4000–5000 ha of new forest areas are available each year. While the loss of forests is mainly due to construction measures, and the affected areas are thereby lost in the long term, new forest areas are created primarily at the expense of agricultural land through compensatory measures and the abandonment of agriculturally unprofitable areas.

With the establishment of trees as short rotation coppice ("SRC") or in agroforestry systems, no agricultural land is legally lost. They are, from a legal point of view, agricultural cultivation systems, but under certain circumstances lead to the at least temporary reduction of land for typical agricultural crops. In order to limit competition for productive land, SRC should preferably be cultivated on marginal yield sites, including grassland. Agroforestry systems offer protection against wind and radiation through their rows of trees. This is intended to lead to increased moisture in the soil in the intermediate spaces used for agricultural purposes, and thus at least partly compensate for the losses of agricultural crops caused by tree cultivation. Both of these cultivation systems not only serve the production of wood, but also have a number of peculiarities with regard to their climatic, hydrological and soil ecological effects that make them comparable to a forest ecosystem (◘ Fig. 2.31).

Short rotation coppices consist of fast-growing tree species that can be harvested after a short production period, usually two to ten years. After each harvest they produce new shoots, which can be cut off again after the next production period. Short rotation coppices are often described as a modern form of coppice forest utilization. However, they show a strong tendency towards plantations with corresponding industrial cultivation methods. After soil and herbicide treatment, selected tree varieties are used, mostly willows, poplars or robinias. The complete biomass is harvested on a larger areas, including branches but without leaves and roots, with harvesting machines. With regard to energy use, the use of an area as a short rotation plantation is extensive compared to agricultural use, and intensive compared to forest management systems. Large cultivation potentials were expected in

Table 2.9 Compilation of management characteristics for forest management types. (Source: Röder et al. 1996)

		Management Type			
		Process-oriented	Nature-conform	Close-to-nature	Good forestry practice
Appropriate tree species for local forest site	Structural characteristics of forests		X	X	X
Geared to the natural forest community			X	X	
Only indiginious tree species		X			
Natural regeneration preferred		X	X	X	
Uneven aged stand structures		X	X	X	
Higher tree ages and mature stands		X	X	X	
Mixed stands		X	X	X	
High timber stocks		X	X		
Large logs and grade logs			X	X	
Multi-storied stand structures		X	X		
Minimum proportion of dead wood		X	X		
Economy at least equally important to other forest functions	Forest goals		X	X	X
Sustainable wood production		X	X	X	X
Significant reduction of timber utilisation		X			
Timber production is not the main target		X			
Large logs and grade logs			X	X	
Ecological stability and vitality			X	X	X
Stability is not generally obligatory		X			
Protection of species and biotops; biological diversity		X	X	X	X
Protection of natural processes		X			
Tree-by-tree utilization is standard	Type of forest operations and treatments		X		
Tree-by-tree and small-scale utilization		X		X	
Long term regeneration		X	X	X	
Natural regeneration is preferred		X	X	X	
Early and repeated interventions				X	
Permanent selection and stock maintenance			X		
Rare interventions		X			
Maintenance measures are largely avoided		X			
Natural processes are used / biological automation		X	X	X	
Restrictive clear cutting				X	
No clearcuts		X	X		

Table 2.9 (continued)

		Management Type			
		Process-oriented	Nature-conform	Close-to-nature	Good forestry practice
Silviculture is preferred over technical aspects	Technology used	X	X	X	
Utilisation of large harvesting machines is assessed negativly		X			
Density of skidding trails is higher restricted		X			
Defined skidding trails inside stands		X	X	X	
Changes in soils are similar to natural processes		X			
Intensive soil treatments are used only in individual situations					X
No soil treatments		X			
Forest stands and soils, landscape and environment are treated with care		X	X	X	X
Chemistry usage severely restricted		X	X	X	X
Complete avoidance of herbecide use		X			

Fig. 2.31 Combined agroforestry/short rotation system: Hedge strips of poplars in Thuringia. (© R. Unseld)

the East German federal states. The areas currently stocked with SRC are listed in the official area statistics, together with the grass type *miscanthus*, under the category "Plants for solid fuels." Since 2010, the area under these crops has more than doubled, from 4000 then to 11,000 hectares today (FNR 2015). Nevertheless, the extent of cultivation has so far fallen well short of expectations.

Agroforestry means the use of land for woody crops and the cultivation of field crops or grassland or agricultural livestock on the same management unit, i.e., on the field or plot (Unseld et al. 2011). Well-known examples are orchard meadows and hedge strips, such as the so-called "Knicks" in Schleswig Holstein. The trees or shrubs are mostly planted in rows and can be used for energy purposes, for trunk wood production and for fruit production, although combinations of these uses are also possible. Accordingly, the period for the production of woody biomass is variable. It can amount to a few years for energy wood production in coppice strips or 40–60 years for valuable wood production from tree species such as cherry, walnut and black walnut.

Both short rotation coppice and agroforestry will meet the essential objectives of a future bioeconomy (Hüttl 2012). In addition to an increase in production per unit area, they offer possibilities for the cultivation of degraded soils. Compared to conventional agricultural use, they improve soil and water protection, carbon storage and biodiversity, increase adaptability, e.g., to changes in water availability, and diversify the landscape.

2.2.4.2 Intensification of Use

The Federal Forest Inventory of Germany records the timber stocks in regard to all types of forest ownership. According to the results of this inventory, stocks have particularly increased in small private forests with forest areas of less than 20 ha (BMEL 2014). Half of the forest in Germany is covered by private forests. Half of these belong to owners with less than 20 hectares of forest. Measured in terms of timber felling, however, the small private forest is used more extensively than the public forest or the large private forest (◘ Fig. 2.32). The most frequently cited reasons are population decrease in rural areas, and thus the number of forest owners who manage their forests themselves, and the increasing fragmentation of ownership through inheritance and sale. Overall, small private forest owners are an inhomogeneous group. Their management is often regarded as "unproductive" in comparison to professional forest enterprises: "The fact that many people do very different things in a confined space creates a colourful mosaic of the most varied interests ... It is therefore time to perceive the supposed construction site „small private forest" as it actually is: colourful, exhausting for the service providers, but extremely important for the regional significance of forest" (Schraml 2014).

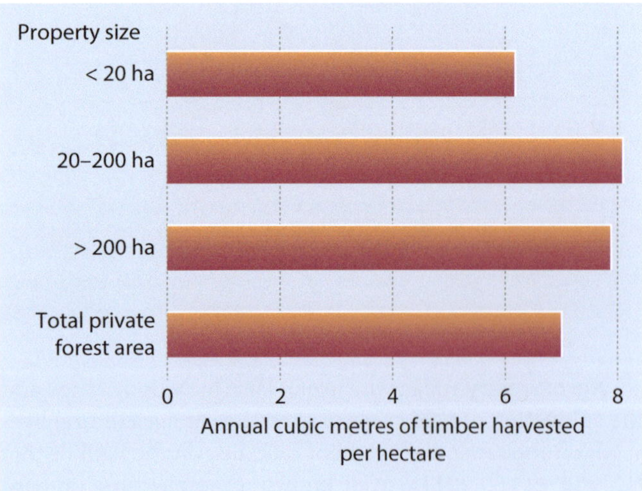

◘ **Fig. 2.32** Timber harvesting by property size of private forests in Germany. (Source: BMEL 2014)

Measures to mobilize wood in small private forests are intended to increase the use of reserves, or at least to achieve a greater absorption of growth. These measures consist, above all, in improving the organisation of forest owners, for example, in forest cooperatives. In addition, there will be an increase in advisory services and an improvement in logging logistics.

As in agriculture, land consolidation procedures are also possible in forests (Gaggermaier and Koch 2011). So far, these have only been used in isolated areas with very small forest parcels, where rational management is almost impossible. Their implementation is more complex than in agriculture, due to the high administrative effort and the more complicated value determination with corresponding compensation.

Another intensification possibility is the use of trunks and wood with a diameter of less than 7 cm. For a long time, wood under this limit was not used, mainly for cost reasons. Due to both the increasing demand for wood chips for energy production and technical innovations in timber harvesting, the complete use of trees ("whole tree harvesting") has also become of economic interest in recent years. A major disadvantage of this use is that, by removing particularly nutrient-rich non-timber components, the natural nutrient recirculation into the ground is diminished. The adaptation of biomass removal to local conditions and soil properties is therefore of great importance (Meiwes et al. 2008). In every forest, it should orient itself according to a simple formula (Kölling and Borchert 2013):

Max. nutrient removal = weathering + input -leaching

Particularly problematic is the use of small woody tree pieces on forest soils with a low nutrient content. These soils are dependent on a natural nutrient element return from remaining tree parts with high bark portions. Extensions to the site maps can provide assistance in assessing possible intensification of use (◘ Fig. 2.33).

In this context, it is important to stress the non-proportional relationship between the yield from additional timber to be harvested and the resulting additional nutrient reduction (◘ Fig. 2.34): "If only 5% of the proceeds are foregone, 70% of the stored phosphorus can be saved and returned to the soil" (Kölling and Borchert, 2013).

2.2.4.3 Intensification of Production

If you define intensification of production as an increase in the growth of wood biomass within a limited area, a higher productivity results from the mixing of tree species. This has already been established many times in individual studies, as well as in global meta-analyses. Even multi-layer forest stands with a wide variation in diameter, which can be created by spatially increasing the stand structure, can increase woody biomass (Dănescu et al. 2016). There are also a few studies that, under certain conditions, see an advantage of pure stands. A meta-analysis puts the productivity gain in mixed stands at 24% compared to similar pure stands of the same tree species (Zhang et al. 2012). In nitrogen-limited soils, which are today mostly found in the subtropics and tropics, nitrogen-binding tree species cause increased growth rates

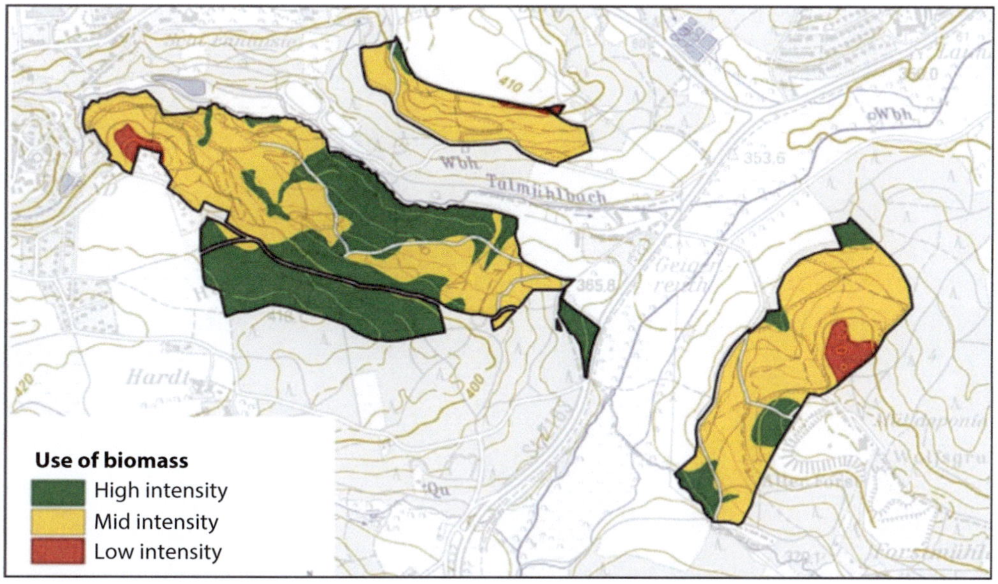

Fig. 2.33 Example of a traffic light map" for biomass use showing three levels of use intensity with respect to crown and bark use. (Source: Meiwes et al. 2008)

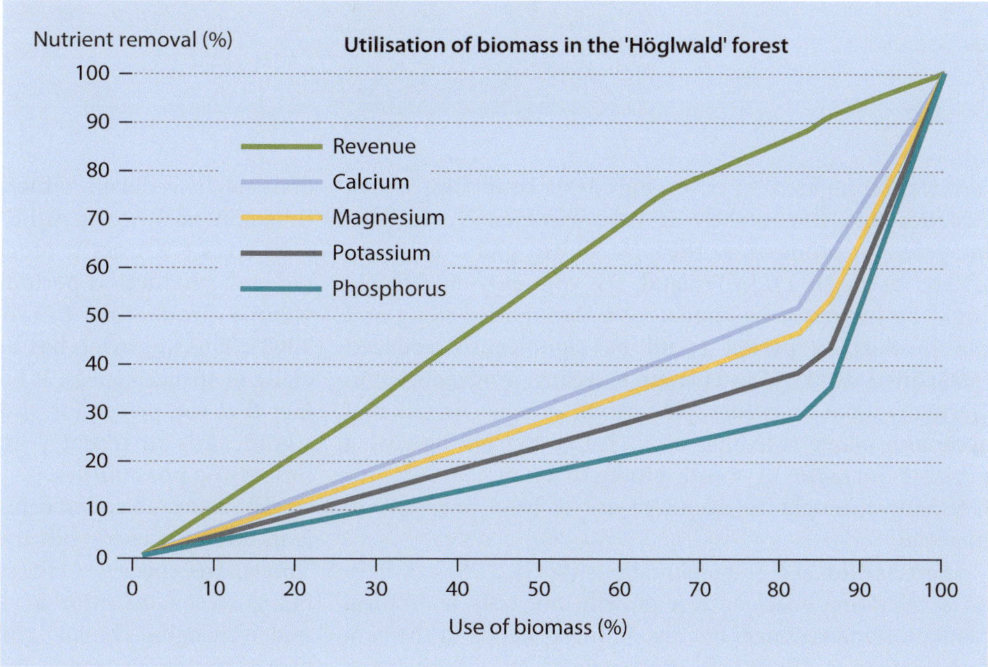

Fig. 2.34 Distribution of nutrient removal and revenue at varying intensities of biomass use on a nutrient-rich productive site in Bavaria. (Source: Kölling and Borchert 2013)

for the entire population. In Central Europe, the ecological advantages of combining a fast-growing pioneer tree species with slow-growing, shade-tolerant species into time-limited tree mixtures have been recognized for a long time (Burschel and Huss 2003). Frost protection for sensitive tree species and support for natural pruning of the target trees for the extraction of valuable wood have been the main objectives so far. Only recently has the targeted use of the biomass of the rapidly growing pioneers been investigated (Unseld and Bauhus 2012). In recent years, research has also been carried out in forest stands where there was no such clear ecological interference between the various tree species. Here, too, mixed stands of beech, spruce and, in some cases, fir or oak showed higher productivity. The additional growth in mixed stands is estimated at 20–30%, with beech making a high contribution throughout. The mixing effects are most obvious in poor site conditions. This is probably due to better utilisation of soil resources. (Del Río et al. 2013; Forrester et al. 2013; Pretzsch 2013; Pretzsch et al. 2013) Research into the interactions between tree species and the causes of increased productivity currently leaves many questions unanswered. In order to reduce cultivation risks, improve soil quality and increase biodiversity compared to pure stands, mixed stands are urgently recommended, from a silvicultural perspective anyway.

The production of biomass can also be increased if new varieties of a tree species can be planted extensively, for example, after a clear-cut or storm damage. In Germany,

Fig. 2.35 Soil milling in strips on a former clear cut in central Germany. (© R. Unseld)

however, compared to other European countries, North America and Brazil, there is little experience with this increase in production through forest plant breeding (Dieter et al. 2011). In Finland, the volume yield of pine could be increased by almost 20% through breeding and the production period could be significantly reduced (Haapanen 2011). The current breeding programmes in Germany focus on spruce, pine, larch, Douglas fir, oak and sycamore maple (Meißner et al. 2015). Experts assume a possible increase in wood productivity of 10–30% for Germany through the exclusive use of bred propagation material.

Fertilisation and soil meliorations (Fig. 2.35) can have an accelerating effect on tree growth, not only in tropical plantation management but also in forest stands in the cool temperate zone. The growth-promoting effects of fertilisation measures have been observed, for example, in forests with former litter utilisation, in poplar stands and in young trees (Klädtke 2003; Fröhlich and Grosscurth 1973; Colye and Coleman 2005). Soil meliorations in the forest are mainly measures for improving the soil structure. This can increase the chances of survival and the growth rates of various tree species after clear-cutting (Rantala et al. 2010; Knoche und Martens 2012, Unseld et al. 2016). For reasons of nature conservation, forest drainage measures only play a role in rare cases.

There are two different approaches to reducing the production period (Reif et al. 2010). In the first variant, the production target is maintained. It is usually expressed by the desired diameter of a tree ("target diameter"). This is to be achieved in the shortest possible time by appropriate thinning methods. In the second variant, the target diameter is reduced, which also shortens the production time and, at the same time, increases the number of final crop trees ("Z trees"). Stocks are lower at the end of the reduced production period; however, the average of total volume production can be increased (Borchers et al. 2008). This approach has been used for some years, especially in spruce stands. It has become attractive due to the fact that the prices for smaller diameter logs have risen significantly in recent years as a result of new technical processing possibilities.

The cultivation of more productive tree species in Germany is mostly associated with the use of so-called "strangers" or "foreign tree species." At the same time, their cultivation is also being accelerated with a view to maintaining production under changing climatic conditions (▶ Sect. 2.2.2.2). In the case of coniferous woods, the Douglas fir (Fig. 2.36) serves as a substitute for spruce, and, for deciduous trees, the Northern red oak serves as a substitute for indigenous oak species (Bolte and Polley 2010).

2.2.5 Strategies for the Rational Use of Wood

The substantial implementation of a bioeconomic transformation will not be possible without woody biomass. This will lead to an increasing timber demand at odds with a supply of harvestable wood that will, at best, remain largely constant over the next 40 years, and will be accompanied by an increase of competing demands for forest areas, e.g., through urbanization, that could actually lead to a decreased volume (Fig. 2.37).

Fig. 2.36 Douglas fir with beech understorey in the submontane zone of the Black Forest. (© R. Unseld)

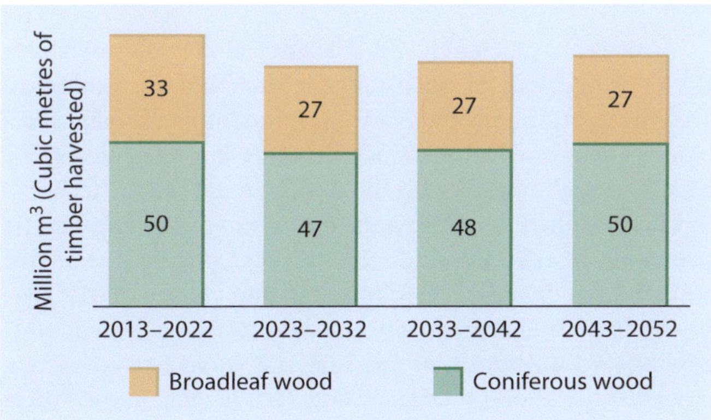

Fig. 2.37 Raw wood potential in Germany up to the year 2052 based on harvest volume of 2012. (Source: BMEL 2016b)

The already existing intensive forestry and the rather limited possibilities for additional silvicultural intensification will leave little room in Germany for a noticeable increase in the volume of timber while maintaining the various sustainability criteria. Under the current political and economic conditions, there is no incentive to provide the necessary arable land for possibilities that can be implemented in the short term, such as timber production on agricultural land. Many intensification possibilities, such as breeding or the cultivation of tree species with strong growth, are only effective in the medium to long term. Some intensification measures, such as the extraction of existing timber reserves, are only temporarily effective until a new level of use has been reached.

These restrictive framework conditions are aggravated by an ever-increasing global demand for timber. Increasing domestic demand in Germany is also being met more and more by imports. An additional domestic demand for wood for bio-economic use could lead to further trade flows, providing a mass of raw material from unsustainable forestry over long transport distances. This can be countered by allowing trade only in certified timber and, at the same time, rehabilitating degraded forest areas worldwide (Bioökonomierat 2016).

Despite the foreseeable supply limits for the most important resource, wood, one of the aims of the bioeconomic approach is to achieve an ecologically sustainable mode of management. Their implementation requires the application of sufficiency, efficiency and consistency strategies (Huber 1995) (▶ Chap. 10). Applied to forestry, they could be formulated as follows:

- Sufficiency strategy: demand for voluntary or prescribed frugality in the extraction and consumption of wood products: "More quality instead of quantity."
- Efficiency strategy: intelligent and rational utilisation of wood as a raw material from processing to combustion.
- Consistency strategy: compatibility of nature and technology to the greatest degree possible; use of nature-compatible technology in forests and in the processing and (re-)use of wood with the lowest possible damage emissions; adaptation of material and energy flows to the regenerative capacity of the forest ecosystem.

To pursue only one of these strategies is hardly sustainable (Höltermann and Oesten 2001), although a willingness to cultivate sufficiency is a basic prerequisite for any ecological modernisation effort (Paech 2016). Sufficiency in forestry under the original concept of sustainability was implemented at an early stage as a harvesting restriction. These restrictions are intended to prevent overuse of wood biomass and conserve soil resources. This sufficiency has been extended in recent years with the implementation of concepts for maintaining old trees and dead wood for the realisation of biodiversity goals. Up to now, sufficiency in forests has predominantly been decreed by the state or can be effectively implemented with compensation payments from public funds.

On the other hand, it is much more difficult to demand sufficiency on the part of the consumer. With continuously increasing consumption of short-lived wood-based products and an oil-price-dependent, flexible conversion of stoves from "pleasure burners to wood heaters" (Mantau 2012b) in private households, an opposite trend is visible. This has already led to a significant increase in prices for the formerly economi-

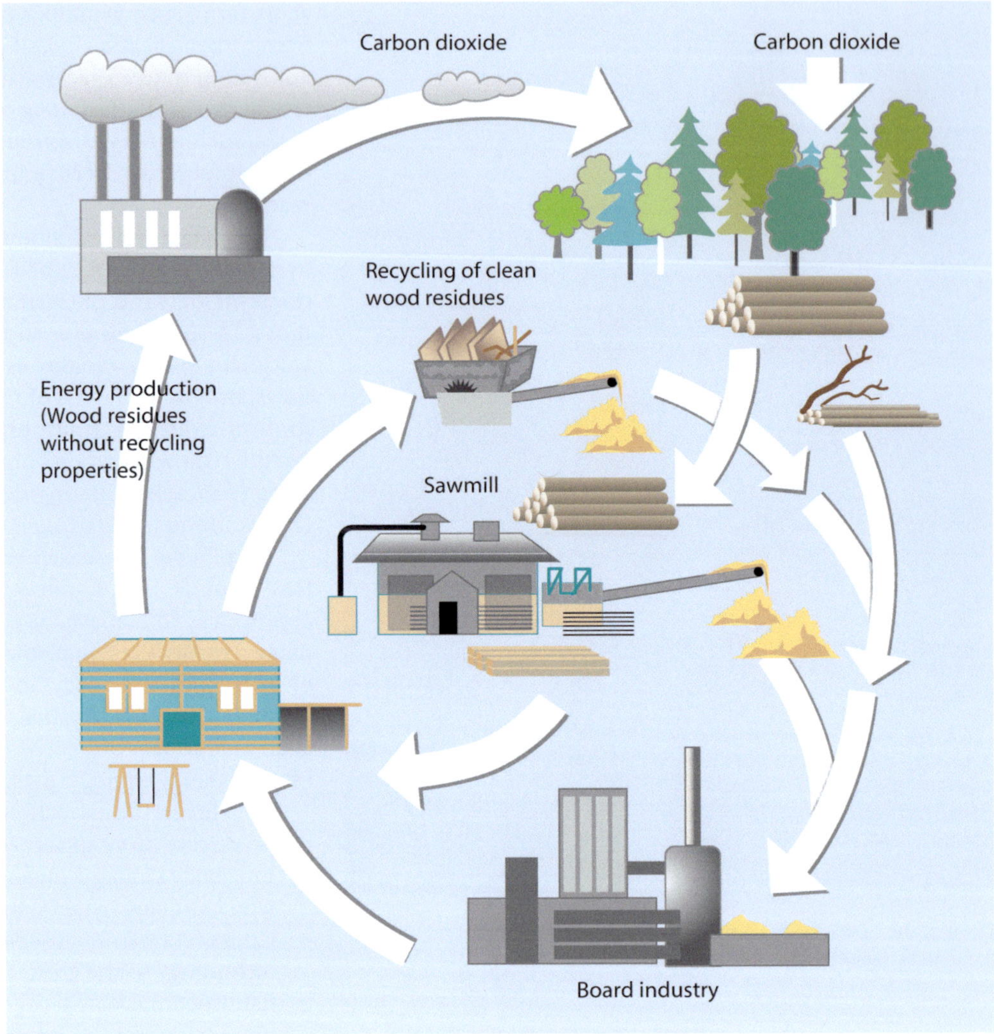

Fig. 2.38 Example of wood use in a cascading system

cally unattractive assortment of industrial wood and firewood. New bio-economic developments of products that could possibly remain in the economic cycle for a longer period of time are facing strong competition here, partly because wood is also intended to form the backbone of biomass production for the transition to renewable energies. However, it can only fulfil this task if the efficiency of energy use is significantly increased and modern heating technology permits optimum energy yield (Mantau and Sörgel 2006). It is questionable, however, whether efficiency increases alone are sufficient for the purpose of using wood to a significant extent for energy system transformation, unless massive energy savings are made at the same time. In the long run, wood could have no relevant share in a constantly increasing primary energy consumption (Luick and Hennenberg 2015). Nevertheless, efficiency aspects should be given greater consideration, even when forest areas are used. Losses in production could be reduced if nature conservation was to be understood as the management of forests that is "target-oriented and resource-efficient when production factors are scarce" and suitable controlling instruments are used (Seintsch 2015).

However, sufficiency and efficiency alone will only solve the core problem of unsustainable demands on ecological systems in rudimentary form: "The fundamental basis for a concept of ecological sustainability is its supplementation by a consistency strategy with the development of new technology and product developments that do not run counter to the principles of action described for an ecologically sustainable way of doing business" (Höltermann and Oesten 2001). One possibility for the more efficient and more consistent processing and utilisation of raw material wood is a consistent cascade utilisation (► Chaps. 7 and 9). It is seen, in the Forest Strategy 2020 (BMELV 2011), as "the first step in dealing with increasingly scarce resources."

In cascade use, multiple uses of a wood product take place over different levels of use (Fig. 2.38). This is intended to ensure that the raw material is used in a particularly efficient and thorough manner, and that it remains in the economic cycle for as long as possible. Under German waste timber regulation (AltholzV 2002), for example, wood and woody materials have to be collected in recycling depots at the end of their useful life. Consequently, further material or energy re-use becomes possible. From an environmental point of view, however, waste wood should continue to be used primarily for material purposes, thereby increasing the size of the cascade.

The consistent supply of raw materials already begins in the forest. Ecological methods of land use management in forests are “cardinal cultural-historical examples of consistent material flows” (Bode 1997). For the primary productive sector, energy productivity is therefore the yardstick for consistent management. The continuous (cover) forest of natural forestry is regarded as the forestry ideal (▶ Sect. 2.2.3.2). Also worth mentioning are the energy balances of the management of short rotation plantations (Burger 2010). They show that even wood from intensive forest systems is the ecologically more rational supply option for raw materials containing fibres compared to raw materials from agriculture. Their synergetic integration into traditional land use systems such as biotope network systems or erosion control strips is therefore desirable (Bioökonomierat 2016).

2.3 Biomass from Fisheries and Aquaculture

Johannes Pucher

2.3.1 The Importance of Aquatic Organisms

Biomass of aquatic organisms is, for humans, an important source of easily digestible and high-quality animal-based and plant-based proteins, essential amino acids, unsaturated fatty acids (especially omega-3 fatty acids), vitamins and minerals (FAO 2014). These resources are used as food, food additives, animal feed and cosmetic additives.

For the supply of aquatic food, humans use an abundance of different organisms from the strains of vertebrates, arthropods, molluscs and echinoderms, as well as micro- and macroalgae. Marine mammals will not be discussed here. The large variety of aquatic species live in numerous habitats of varying salinity (from freshwater to seawater) in different climate zones (polar zone to tropics). This includes all trophic levels, from primary producers (trophic level 1) to top predators (trophic levels 4 to 5), and all feeding types, from planktivorous, herbivorous, omnivorous, detrivorous to carnivorous/piscivorous organisms. Most of these aquatic organisms are poikilothermal. They therefore do not use energy to maintain their body temperature. The temperature of the surrounding water thus becomes an essential factor for their biological activity and productivity.

Among the **aquatic vertebrates**, mainly fish (and, to a much lesser extent, frogs) are consumed. The group of consumed fish includes all trophic stages with different feeding types (planktivor, herbivor, omnivor, piscivor). They live in a wide range of habitats, which differ in environmental factors, like, for example, salinity, temperature, pH value, water hardness and proximity to the ground or surface. In addition, fish are also divided into shoal-fish and loners.

The **aquatic arthropods** consumed worldwide mainly consist of crustaceans such as shrimps and crabs of various sizes. Larger crustaceans are mostly ground living organisms of higher trophic level. Smaller crustaceans, by contrast, are of lower trophic levels, living near the ground or found in open waters. Depending on the type, they are swarm-formers or loners that can, however, also form swarms for spawning.

The **molluscs** include mussels, snails and cephalopods. Bivalve molluscs are mostly planktivorous organisms of lower trophic levels that live sessile on structures or conditionally mobile in sediments. Snails are predominantly ground living and of a rather low trophic level. However, individual types may also be predatory. Among the cephalopods are squids, octopuses and cuttlefish, which usually live predatorily on the ground or in open water.

Echinoderms such as sea urchins and starfish are ground-living organisms of low trophic level, but can also belong to predatory feeding types. They are mostly found in marine habitats.

Algae and aquatic macrophytes are phototrophic organisms of very large species diversity. Algae occur as free-swimming single- or multi-cell organisms or colonized structures. Aquatic macrophytes such as seaweed are mostly sessile organisms that frequently colonise entire coastal regions and strongly influence the habitats in which they are living. Algae and seaweeds are used for human nutrition, especially in Asia, but have recently also been increasingly used in Western cuisine.

Overall, aquatic food is an important source for human nutrition (◘ Fig. 2.39). In 2009, aquatic animal-based foods accounted for 16.6% of the animal protein supply for humans and 6.5% of the total protein supply (FAO 2012b).

Due to rising human population, rising living standards, higher production and improved transport possibilities, the annual per capita consumption of aquatic organisms has risen worldwide from 17 kg (2000) to 19.2 kg (2012), and is rising further. However, there are strong continental and regional differences in both consumption patterns and production (FAO 2014). Especially in developing and emerging countries in Southeast Asia and Central Africa, aquatic foods are of paramount importance, as they represent an affordable source of animal proteins for the poor (FAO 2012b) and are also increasingly consumed by wealthier populations. In 2012, the global production of aquatic organisms as food and feed accounted for 136 million tonnes (FAO 2014). An increase to 152 million tonnes is expected by 2030 (World Bank 2013). The relative share of various aquatic food and feed products in the global supply has changed considerably in recent decades (◘ Fig. 2.40).

Fig. 2.39 The large spectrum of aquatic organisms is of great importance for human nutrition. (© singidavar/Fotolia.com)

2.3.2 The Fishing Industry

Up until the 1990s, marine capture fishery (Fig. 2.40 and 2.41) and, to a lesser extent, inland capture fishery were the main producers of aquatic-based food and feed (not including algae and marine mammals). Since then, however, the quantity of fish and other aquatic organisms landed through capture fishery has stagnated and can no longer meet global demand by itself. It has stabilised fairly steadily at an annual capture fishery production of around 90 million tonnes (FAO 2015). Since the 1990s, the reduction in production due to increasing overfishing of several fish populations and the increase in production due to improved fishing techniques have been in balance (Pauly 2009; FAO 2012b).

Marine capture fishery is the main producer of higher trophic and higher priced aquatic organisms for the global food markets. Among the most frequently landed organisms are fish such as tuna, groupers, sea bass, pollock, cod, hake, meagre, snappers, flatfish and sea bream, molluscs such as squid, octopus and cuttlefish, and crustaceans such as shrimps, lobsters and crabs (Tacon et al. 2010; Neori and Nobre 2012). Inland fisheries can be very important regionally for food supply and globally account for a small but growing share of total fishery production compared to marine fisheries. Between 1990 and 2013, that share rose from 8% to 14% (Fig. 2.40).

In addition to the production of food, capture fishery is essential for the production of fishmeal and fish oil. These resources are used worldwide as a high-quality feed resource and are mainly produced from pelagic small fish or by-catch. Fishmeal and fish oil are traditional feed sources for animal-based proteins and unsaturated fatty acids, in particular, the two major omega-3 fatty acids eicosapentaenoic acid (EPA) and docosahexaenoic acid (DHA). In the 1990s, capture fishery of pelagic fish for the production of fishmeal and fish oil reached its peak at 30 million tonnes per year. Since then, production has fallen to around 15 million tonnes in 2010, with climatic influences such as El Niño leading to strong annual fluctuations in the catches (FAO 2012b). The ever-increasing demand for these two essential feed resources, coupled with further declines in catches, is leading to drastic price increases and an increasing share of fishmeal and fish oil produced from by-products of fish processing. The use of fishmeal and fish oil has also changed considerably. In the 1960s, 98% of both feed components were still used to feed pigs and chickens. In 2010, 73% of the fishmeal produced worldwide was used in aquaculture feed instead (Shepherd and Jackson 2013). This is due to the sharp rise in the price of fishmeal and fish oil, which can be used more efficiently in aquaculture production (Naylor et al. 2009).

Under the current consumption pattern, global demand for aquatic food and feed can no longer be met by the stagnating fishing industry. The demand gap can only be closed through aquaculture production, which is growing accordingly.

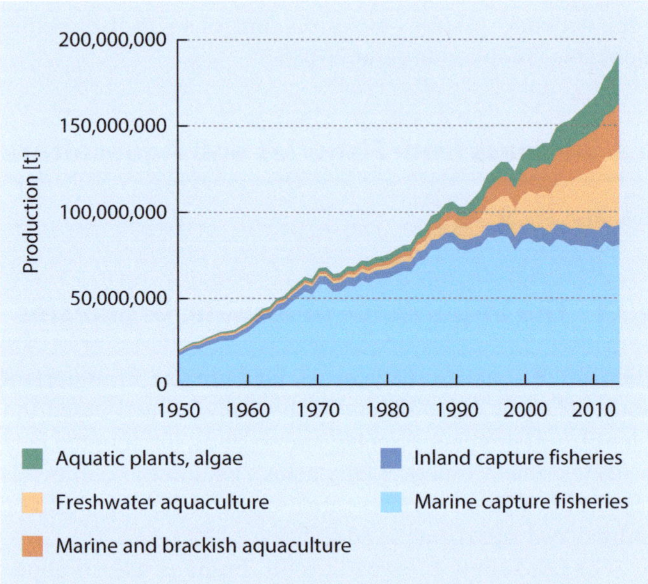

Fig. 2.40 Temporal development (1950–2012) of global production of aquatic biomass resources for the supply of food and feed. (Source: FAO 2015)

Fig. 2.41 Fishing boats in Vietnam. (© Johannes Pucher)

2.3.3 Aquaculture

Aquaculture is defined as the controlled rearing of aquatic organisms (fish, bivalve molluscs, crustaceans and algae), while an increase in production can be influenced by certain measures (e.g., stocking, feeding, protection against predators, reproduction, breeding) (FAO 1997).

In 2013, around 70.2 million tonnes of animal-based biomass and 27.0 million tonnes of plant-based biomass were produced worldwide in aquaculture systems (FAO 2015). Aquaculture production in freshwater has particularly expanded (Fig. 2.40). Over the past three decades, total aquaculture production has grown at an average annual rate of 8.8%, making the aquaculture sector the fastest growing food production sector (FAO 2012b). More than half of the fish consumed as food worldwide comes from aquaculture (Naylor et al. 2009).

One of the most important prerequisites for efficient aquaculture production is the knowledge and control of the life cycle of the species to be cultivated. Only this knowledge can open up the possibility of domesticating the targeted species. In this process, breeding, as well as development of more suitable feeds and adapted production technologies, can lead to better feed utilisation, higher resistance to stress and diseases and the optimisation of other phenotypic traits (e.g., meat content) in aquaculture species.

Nowadays, a large number of aquatic species are cultured globally. In aquaculture, the species produced in highest quantitites are predominantly of lower trophic levels (Tacon et al. 2010; Neori and Nobre 2012). Among the top ten produced species are carp such as the planktivorous silver carp *(Hypophthalmichthys molitrix)*, the herbivorous grass carp *(Ctenopharyngodon idellus)*, the omnivorous common carp *(Cyprinus carpio)* and the planktivorous bighead carp *(Hypophthalmichthys nobilis)*, as well as the omnivorous Nile tilapia *(Oreochromis niloticus)*, which is a representative of the cichlids. All of these fish species have a trophic level below 2.5. They are important species in the extensive and semi-intensive pond polyculture that is often used, especially in Asia (▶ Sect. 2.3.3.2) (Milstein 1992; Kestemont 1995; Neori and Nobre 2012).

The only carnivorous fish species among the ten most globally produced aquatic species are the Atlantic salmon, *Salmo salar,* and the rainbow trout, *Oncorhynchus mykiss.* These two salmonids are typical species of intensive coldwater aquaculture (▶ Sect. 2.3.3.1) and are probably the most domesticated fish species in the world. Generally, carnivorous species have a higher market value than low trophic species. While, in Asia, many carp species are produced and consumed, Western countries show a higher consumption of carnivorous fish, which often have to be imported.

The Food and Agriculture Organization of the United Nations (FAO) classifies all aquaculturally produced species from freshwater, brackish water and marine water into about 520 species or groups of species (excluding marine mammals and algae).

Each of these species has its specific requirements in terms of culture conditions, water qualities (temperature, oxygen levels, salinity, water hardness, pH value), nutrition and stocking densities. Depending on the type of nutrition (planktivor, herbivor, detrivor, omnivor, carnivor/piscivor), the needs differ in the type of feed, feed composition and feed quality. Omnivorous and herbivorous species are physiologically adapted to plant-based protein and energy sources, whereas piscivorous species are physiologically adapted to high-quality animal protein sources. Meeting these different demands of organisms in feeding and culture condition requires the use of a wide range of technologies. In general, lower trophic species require lower feed quality than higher trophic species. This makes aquaculture of low trophic species a very sustainable and efficient way of producing high quality animal food for humans.

A distinction is made between forms of aquaculture production according to their degree of intensity. The production intensity is classified according to the following criteria: production quantity per area or water volume and time, stocking density, need for technical knowhow, management requirements, complexity, costs of setting up the necessary infrastructure, running costs, workload, risk of diseases and technical failure, use of veterinary substances, quantity and quality of feed and fertilisers, as well as the dependence of the organisms on natural feed resources (Edwards et al. 1988; Tacon 1988; Prein 2002).

2.3.3.1 Intensities of Aquaculture Production

Basically, aquaculture is divided into three different levels of intensity:

Extensive aquaculture means production systems in which one or more species of fish or crustacean are cultured exclusively by means of natural feed resources in water. The species concerned are predominantly of low trophic level. They feed on aquatic resources such as algae, zooplankton, benthos, prey fish, bacterial biomass or decomposing organic substances (detritus). They are not supplied with feed or fertilisers from external resources. However, due to the low stocking densities per pond area (or water volume) and the limited availability of natural feed in the water, this type of near-natural aquaculture is only moderately productive. But, this type of aquaculture requires only a minimal use of technology and financial investment. Because almost every natural feed resource in water is based on the primary production of algae, extensive aquaculture requires large areas of water. Near-nature extensive aquaculture often plays an important role in preserving the biodiversity of flora and fauna; further, these systems act as water protection areas and flood buffer zones. Extensive aquaculture has the positive effect of binding nutrients dissolved in water for the production of high-quality food. Thus it acts as a nutrient sink and tends to counteract the increase in nutrient loads (eutrophication) in water bodies. Especially in developing countries, extensive aquaculture plays an important role in food security for the poor population, because it produces essential food with minimal use of resources. However, the increasing pressure on the use of freshwater resources makes it unlikely that this type of inland aquaculture will be significantly expanded in the future. The multiple use of water, e.g., by industry or agriculture, also increases the risk of contamination by pesticides, heavy metals and other environmental contaminants. This limits the prospects of extensive aquaculture for the production of safe food.

However, the potential for expansion is greater in extractive aquaculture, which is a special form of extensive aquaculture. In extractive aquaculture, filtering species are grown in mesotrophic or eutrophic waters. The best example of an expanding extractive aquaculture is the cultivation of mussels in coastal zones. Here, mussels are attached to ropes (◘ Fig. 2.42) or kept in net bags near the coast, filtering algae, zooplankton and other suspended matter from the water. Consequently, they extract nutrients from the water and bind them into biomass. Mussels may also filter potentially harmful bacteria and viruses that can create food safety issues.

Another example of extractive aquaculture is the cultivation of macroalgae in open water. The phototrophic organisms are fixed to the ground or to certain structures and absorb the nutrients dissolved in the water. Consequently, they also counteract the eutrophication of water bodies.

In **semi-intensive aquaculture**, the production of natural feed resources in the pond is promoted by a targeted addition of organic or inorganic fertilizers and an adapted water management. This means that a higher level of fish biomass can be cultured per pond area than in extensive aquaculture and fish reach marketable sizes more quickly. Natural food resources such as algae, zooplankton, benthos and microbes are very rich in protein and have a high-quality protein composition. However, their protein levels exceed the needs of the fish. This means that the fish use the excess proteins from the natural feed as a source of energy. However, this is an inefficient use, because it would make more sense to use the high-quality proteins from the natural feed as building units for the growth of the fish. In order to promote greater efficiency in the use of proteins from natural feed for conversion into fish biomass, semi-intensive aquaculture often uses carbohydrate-rich, high-energy feed resources as an external supplemental feed resource (Viola 1989; De Silva 1994). As in extensive aquaculture, the ponds of semi-intensive aquaculture are usually stocked with different fish species (polyculture) that use different natural feed resources. This minimizes the competition between the species, as they feed on different types of natural feed. Semi-intensive aquaculture is the intermediate transition from extensive to intensive aquaculture, so that, depending on the system, technical aids such as aeration devices and automatic feeders are also used. In addition to the production of fish as food, semi-intensive aquaculture also has the function of increasing and stabilising biodiversity. It serves as a habitat for amphibians, plants and migratory birds, protects surface and groundwater and offers local recreation for the population. In Germany, for example, traditional carp ponds represent a form of semi-intensive aquaculture and are an important part of the cultural landscape. Fertilization and supplementary feeding with grain are used to increase the productivity of natural feed resources and fish production. Worldwide, semi-intensive aquaculture of low trophic fish has immense importance for the food supply in general and for the poor population in particular. Although an expansion of this type of aquaculture may be regionally

◘ **Fig. 2.42** Mussel production on longlines is an example of extractive aquaculture. (© Johannes Pucher)

possible, in view of the increasing land use pressure, a higher global production of aquatic biomass will need to be realized through intensification of aquaculture production, rather than by expansion of less intense, space-demanding aquaculture systems (Tacon et al. 2010).

Intensive aquaculture means aquaculture systems in which aquatic species are mostly kept in monocultures and produced as cost- and space-efficiently as possible. In these systems, the cultured species are fed only external feed resources, normally in pellet form. The feed is adapted to the special nutritional requirements of the cultured species. This makes this type of aquaculture one of the most feed-efficient productions. In intensive aquaculture, the aim is to adjust and stabilise the culture conditions (e.g., oxygen content, temperature, flow rate, salinity, pH value) as precisely as possible to the needs of the cultivated species. Such aquacultures are characterised by the use of technical devices such as aerators, filter units, disinfection equipment, pumps and water quality sensors. This makes them technically very expensive and requires well-trained personnel and high financial investment. Intensive aquacultures usually have high stocking densities and use large amounts of feed. This intensity, in turn, has the advantage that the cultured organisms can be better protected against harmful external influences (e.g., pathogens) and that more efficient management of feed, water and hygiene measures is possible. Intensive aquacultures are more productive and easier to control against external influences than extensive or semi-intensive cultures. They are used on a large scale to supply international markets.

In intensive aquaculture, a distinction is made between net cages, flow-through systems and recirculation systems. Net cages are installed in rivers, lakes or the sea (◘ Fig. 2.43). In net cages, the natural water flow is used to supply the fish with fresh water and to remove the excreta. Thus, net cage aquacultures have a direct influence on the environment. Flow-through systems are land-based aquacultures, in which water flows through the tanks, supplying the cultured fish with oxygen and removing their faeces and metabolites that are dissolved in the water. The outflowing water or effluent can be purified of particles and dissolved nutrients, but the effluent water is no longer used for fish culture. Circulation systems are mostly used for the culture of warm water organisms. Here, too, water flows through the fish tanks, but the water is treated and recycled for repeated use in the fish tanks. The first step is to filter the faeces and feed residues from the outflowing water mechanically. In addition to faeces, aquatic organisms excrete soluble ammonium from the digestion of feed proteins via the gills. However, ammonium is toxic to fish. When the water is recirculated, it must be oxidised through nitrification in so-called nitrification units into nitrate, which is far less toxic to aquatic organisms. Only nitrification makes a circulatory system a water-efficient form of fish farming. Due to its productivity, efficiency and controllability, intensive aquaculture is considered to have the highest growth potential of all forms of aquaculture. Intensive aquacultures are normally land based or near the cost, but there is an increasing number of attempts to implement aquacultures further out in the sea (offshore).

◘ **Fig. 2.43** Fish farming in net cages off the Croatian coast. (© Nightman1965/Fotolia.com)

However, an increase in production and expansion of intensive aquaculture will increase the need for feed, thus increasing the demand for classical animal- and plant-based feed resources, and consequently for land, water and fertilisers for their production (Tacon and Metian 2008). The limited availability and rising price of fishmeal and fish oil as classic feed components, especially for piscivorous aquaculture species, is forcing the sector to minimise the use of these resources and replace them with alternative feed resources without conflicting with human feed resources. Due to a suitable amino acid spectrum, vegetable protein from soybeans has become an important substituent in many feedstuffs, even for piscivorous organisms in aquacultures. Other alternative plant- or animal-based protein sources are also increasingly used worldwide in feed for aquacultures (Hardy 2010; Hernández et al. 2010). The fishmeal content is thus reduced to a physiological minimum as a source of essential amino acids, which are only present in low concentrations in plant-based protein sources. Increasingly, artificially synthesized essential amino acids are added to animal feeds to further reduce the proportion of fishmeal. There are major scientific efforts in progress to zero the fishmeal content in aquaculture feeds and to identify further sources of suitable replacement components.

In the search for alternative feed components, many potential animal-based feed resources (e.g., insects, by-products) and plant-based feed resources are being researched and increasingly used. As a potential resource of unsaturated fatty acids to replace fish oil, algae (especially microalgae) are increasingly being investigated. Similar to animal species, microalgae can also be cultivated in intensive production units. The different algae species differ in their culture conditions, nutrient and light requirements, and in their ingredients. The type of nutrient solution and the environmental conditions in the production cycle allow for limited control of the compound composition of the cultured algae. Efficient production units adaptable to the respective algae species are currently being researched and, in some

2

cases, are already in industrial use. Although microalgae can be easily produced in open systems with sunlight, the risk of contamination by unintended algae species and algae consumers is very high, which affects both product quality and productivity. Closed algae production systems in reactors, hose or pipe systems that are exposed to either solar or artificial light are easier to control.

2.3.3.2 Integrated Aquaculture Systems

In order to increase the efficiency of aquacultures in the use of feed and water, multiple uses of nutrients are being realised in different types of integrated aquaculture system.

One example of the multiple use of nutrients in aquaculture is the **biofloc aquaculture**, which is a mixture of intensive and semi-intensive aquaculture. In this aquaculture, shrimps or fish are kept in tanks with minimal water exchange. In addition to the feed, an organic carbon source (e.g., molasses) is added to the tanks in combination with strong aeration. The result is a type of activated sludge in which heterotrophic microorganisms transform the nutrients excreted by the cultured species (especially nitrogen and phosphorus) into biomass. The microorganisms use the added carbon source as an energy source. Strong aeration provides the habitat with sufficient oxygen and keeps the microbial biomass in the water column moving. Larger flakes develop themselves, the so-called bioflocs, which, on the one hand, are taken up again by shrimps or fish as feed and, on the other hand, assimilate the toxic ammonium into biomass. Such systems promise efficient nutrient and water use combined with high productivity per volume of water.

Recently, **integrated multitrophic aquaculture (IMTA)** systems have gained importance. IMTA systems combine different aquatic species of different trophies to utilize nutrients more efficiently. One example of an IMTA system is the combination of fish production with the cultivation of detrivorous or filtering aquatic species. The detrivorous species, such as crustaceans, or filtering species, such as mussels, use the fish's excrements directly or absorb the emitted nutrients via algae or microbes. This principle makes it possible to partially bind the nutrients emitted from fish production into the associated cultured species, minimize the environmental impact and, at the same time, produce an additional high-priced food product. However, the benefits of such systems in application have inspired much controversy in the scientific community.

Another type of modern integrated aquaculture is the combination of fish and plant production. This is a special form of IMTA and is called **aquaponics**. In such systems, the dissolved nutrients excreted by the fish serve as fertilizer for plants for human consumption. Plants and fish are kept in common or separate circuits. These systems promise to produce fish and vegetables in a more nutrient-efficient manner. However, the scientific community is still discussing whether the integration of the two productions is more efficient than the respective individual productions, since the latter can be better adapted to the different needs of the target organisms. Increasingly, such systems are being operated in urban and periurban regions in order to utilize off-heat and to produce food close to urban consumers. Currently, aquaponics are being run at a smaller scale and supply suitable niche markets.

However, the idea of combining various trophic organisms - which, in Germany, is being implemented in highly engineered closed systems in halls and buildings - is not new. In many countries of Southeast Asia, pond aquaculture has long been an essential part of the traditional integrated agriculture. This type of integration is known as **integrated agricultural aquaculture (IAA)** and combines plant production on fields, horticulture, animal husbandry and pond aquaculture in order to establish resource-efficient nutrient cycles. This traditional form of agriculture is widespread in Asia. Today, it is particularly important for small-scale farmers in developing Asian countries. It forms the livelihood of the poor population and ensures a regional supply of food. However, the intensification of individual agricultural production sectors (e.g., rice and maize cultivation) brings an imbalance to the nutrient-efficient agricultural systems that have emerged over hundreds of years. The adaptation of traditional IAA systems to the new conditions of modern agriculture and a globalised world will, in the future, also have to take climate change and the limited availability of water and feed into account. This is essential for the future supply of the world's population with safe and healthy food. Overall, aquaculture can be seen as one component to supply the future demand for safe, nutritious food. However, aquaculture activities need to be sustainably embedded into local conditions and feed resources must be found without conflicting with the direct use as human food.

2.4 Biomass from Waste Management

Klaus-Rainer Bräutigam

Waste is generated during the manufacture, use and application of products. Likewise, products that have reached the end of their useful life, and therefore no longer fulfil their intended purpose, become waste. This waste must be collected in an appropriate manner, transported and then sent for recovery, treatment or disposal. All of these activities take place within an area that is generally referred to as waste management.

In recent decades, waste management has developed into a large and powerful economic sector in Germany (and also in other European countries). Its sales in 2017 amounted to around € 36 billion. It provided around 155,000 jobs in around 1460 companies (Statista 2019a).

In addition to waste, residual materials are also produced. These are substances that are left over from the manufacture of a product, because they are not required as part of that product, but are incorporated into the production of other goods, and therefore do not have to be disposed of as waste.

Part of the waste and residues is of animal or plant origin (biowaste, organic waste fraction, biodegradable waste). It is generated during the treatment, processing and use of biomass in agriculture and forestry, in the production of food and animal feed, in the conversion of biomass into energy sources and in the manufacture of a wide range of products. However, large quantities of waste and residual materials are also generated in the private sector, for example, in the processing and consumption of food and in garden maintenance.

What waste in general, and biowaste in particular, is and how to deal with it is regulated in Germany by the so-called Waste Management Act (Gesetz zur Förderung der Kreislaufwirtschaft und Sicherung der umweltverträglichen Bewirtschaftung von Abfällen (KrWG 2012)) and by the Municipal Waste Disposal Law (Verordnung über die Umweltverträgliche Ablagerung von Siedlungsabfällen - AbfAblV 2001)) (Waste Disposal Ordinance). A separate legal area therefore applies to the treatment of waste and residual materials.

2.4.1 Legal Basis

Waste, in the sense of the KrWG § 3, are all materials or objects that their owner discards, wants to discard or has to discard. Waste for recovery is waste that is recovered; waste that is not recovered is waste for disposal.

Biowastes, within the meaning of the KrWG § 3, are biodegradable vegetable, animal or fungal materials:

1. Garden and park waste
2. Landscape conservation waste
3. Food and kitchen waste from households, restaurants, caterers, retail trade and similar waste from food processing industries
4. Waste from other sources comparable in nature, quality or material properties to the wastes referred to in points 1 to 3

The core element of the KrWG in § 6 is the five-level waste hierarchy. The following order of precedence applies to waste management measures:

1. Prevention
2. Preparation for recycling
3. Recycling
4. Other types of recovery, particularly energy recovery and backfilling
5. Disposal

According to § 11 KrWG, biowaste that is subject to a handover obligation pursuant to § 17 (1) must be collected separately starting January 1, 2015 at the latest. In the case of waste for recovery, the handover obligation only applies to waste from private households, and only to the extent that private households do not intend to recycle it themselves. Waste producers and owners who are not private households must recycle waste themselves for recovery in accordance with § 7 (2) KrWG.

In 2010, the connection rate to separate collection of organic waste bins in Germany was around 52%. Around 40 million inhabitants did not use an organic waste bin. By 2015, the year in which the separate collection obligation began, an increase in the connection rate to around 65% had been expected (Umweltbundesamt 2014). Presently, the connection rate is estimated to be about 70%. With a further increase in the connection rate and with a sorting of waste according to type, the amount of bio-waste available for high-quality recycling will also increase.

Biowaste from areas of origin other than private ones is not subject to the obligation to hand it over to the public waste management authorities, and is therefore not subject to the obligation of separate collection. Rather, this waste must be recycled by the waste owner or producer. In this case, the industrial waste ordinance (GewAbfV 2002) has to be applied. It stipulates that biowaste must be "kept separate, stored, collected, transported and recycled." If biowaste from other areas of origin is left to the public waste management authorities, the separate collection requirements of the public waste management authority apply to this biowaste (Hennsen 2012).

At the level of legislation of the European Union, the Council directive on the landfill of waste (Council Directive 1999/31/EC) obliged Member States to reduce the amount of biodegradable waste going to landfills to 75% by July 2006, 50% by July 2009 and 35% by July 2016, compared to the year 1995 (some countries have exemptions until 2020). One of the aims of this regulation is to prevent the release of methane, produced during the decomposition of organic substances, which contributes to global warming. The Waste Disposal Ordinance transposes this Landfill Directive into German law. Accordingly, municipal solid waste in Germany can only be deposited in landfills if it meets certain criteria. In concrete terms, this means that, since June 1, 2005, the dumping of untreated waste that does not meet these criteria has been banned. This generally applies to biowaste. However, special rules apply to mechanically and biologically pre-treated waste. These may be deposited in Class II landfills, provided that they meet certain criteria regarding the organic content of their dry matter residue (TOC content) and their calorific value.

Due to the fact that the Waste Management Act gives top priority to the avoidance of waste, recycling of waste of bio-genic origin may only be considered if its avoidance is not possible. This fact limits the potential of available waste of biogenic origin in regard to its material or energy use in the bioeconomy.

2.4.2 Generation and Composition of Municipal Waste

The Federal Statistical Office publishes data for the waste balance in Germany at annual intervals. This includes the amount of waste generated, broken down by type of waste, and the whereabouts of each type of waste (deposition, incineration, treatment for disposal), as well as the particular recycling procedure (energy recovery, material recycling). ◘ Figure 2.44 shows the volume of waste generated in Germany in 2016, with a total volume of around 364 million tonnes. Municipal waste accounted for 58 million tonnes out of this total. Around 48% of this (25 million tonnes) was

2

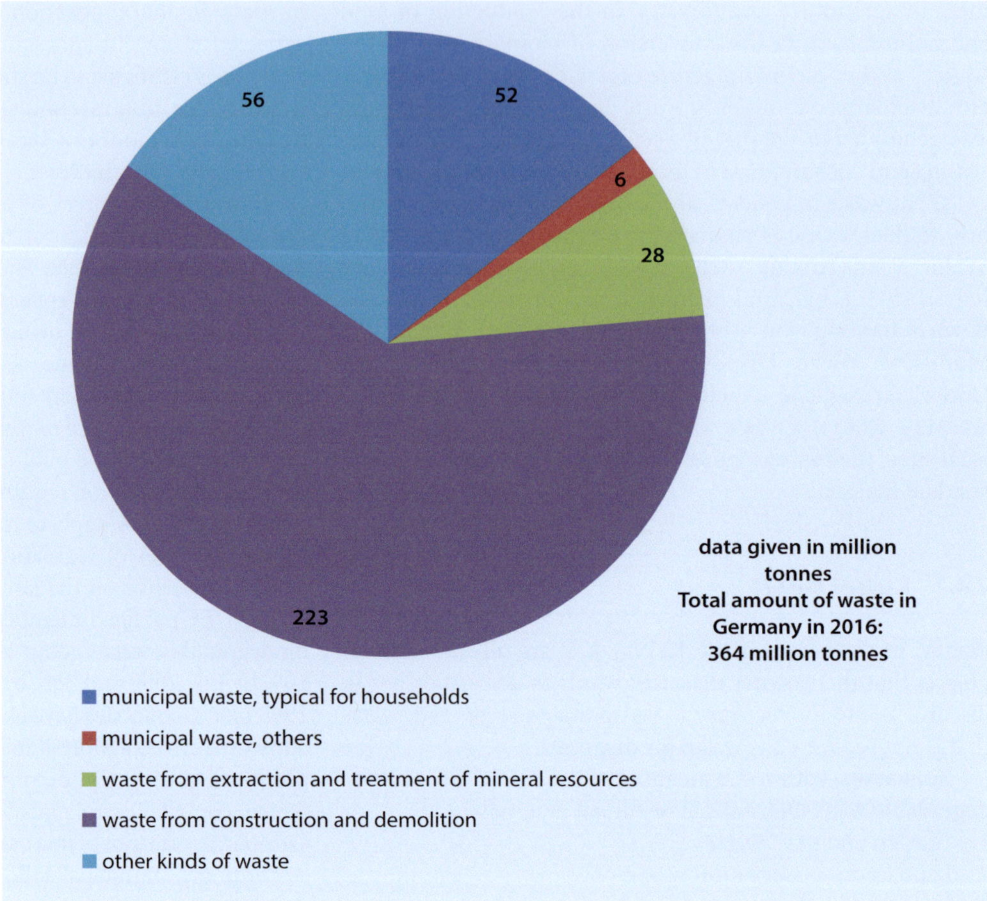

Fig. 2.44 Waste generation in 2016. (Author's own representation based on data from Federal Statistical Office of Germany 2018)

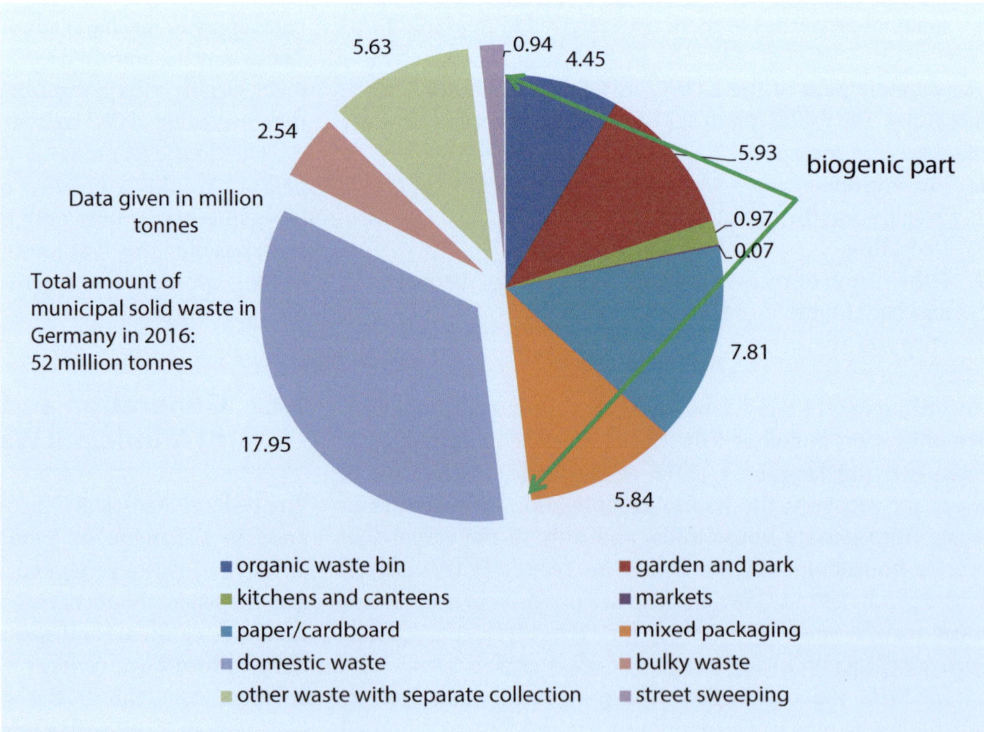

Fig. 2.45 Composition of municipal solid waste in 2016. (Author's own representation based on data from Federal Statistics Office of Germany 2018)

waste that can be counted as biomass - paper and cardboard and mixed packaging material included (Federal Statistical Office 2018a) (▫ Fig. 2.45). However, these figures refer only to separately collected waste, as the statistics do not cover biogenic residues collected by residual waste collection (in particular, household waste).

In 2010, the share of these biogenic substances in residual waste in areas without organic waste bins was around 55%. In areas with organic waste bins, it was significantly lower, at 25–45% (Federal Environment Agency 2014). Assuming a total volume of 14 million tonnes of residual waste, and further assuming that 30% of this is of biogenic origin, this corresponds to a biomass of around 4 million tonnes, which would also be available for high-quality recycling if the bio-waste bin were introduced nationwide with almost complete coverage of the separation of the biogenic fraction (Deutscher Bundestag 2014).

The evolution of the generation of waste of biogenic origin is shown in ▫ Fig. 2.46. Between 2004 and 2017, this volume increased by around 23%. In 2017, the volume of biomass collected separately in organic waste bins was around 4.9 million tonnes (Federal Statistical Office 2018b).

2.4.3 Generation of Residual Material

Residual materials include agricultural by-products, industrial residual materials, e.g., from food production and animal feed manufacture, wood and forestry residual materials, as well as residual materials from other land-areas such as landscape conservation (DBFZ 2015).

Liquid manure is a naturally occurring fertilizer that consists mainly of urine and faeces from farm animals. Liquid manure is produced, in particular, through pig and cattle husbandry. Poultry farming, on the other hand, usually produces manure in solid form. 26.9 million pigs and 12.7 million cattle produce urine and faeces in Germany. Every year, German farmers distribute more than 200 million tonnes of liquid manure on arable land and meadows. Before being used as fertilizer, the organic components contained in the manure that have not been digested by the animals can also be used to produce energy. For this purpose, the liquid manure is degraded by microorganisms in a fermenter of a biogas plant. The resulting methane-rich biogas can then be used to produce bioenergy. After fermentation, all important plant nutrients contained in the liquid manure can be found in the slurry, which can be spread on arable land or meadows with the same technique as liquid manure (Wikipedia 2016a). (Less than one fifth of the excrement (slurry and solid manure) produced in German stables was used to generate energy in 2015 (biogas.org 2016). The main reasons for this are the logistical costs and the high investment costs.

Straw is the crop residue that occurs after the cultivation of cereals and rapeseed (AEE 2013). Of the 30 million tonnes of straw produced annually in Germany, between 8 and 13 million tonnes could be used sustainably for electricity or fuel production. However, so far, no significant quantities of straw have been used to generate energy. Grain straw also plays an important role in the humus balance of soils. This means that part of the straw must remain on the field so that nutrients are not permanently removed from the soil (DBFZ 2013).

Competition also arises from the use of straw as a material: straw is used as bedding in animal husbandry. Demand is expected to continue in the future, depending also on the further development of livestock numbers and husbandry

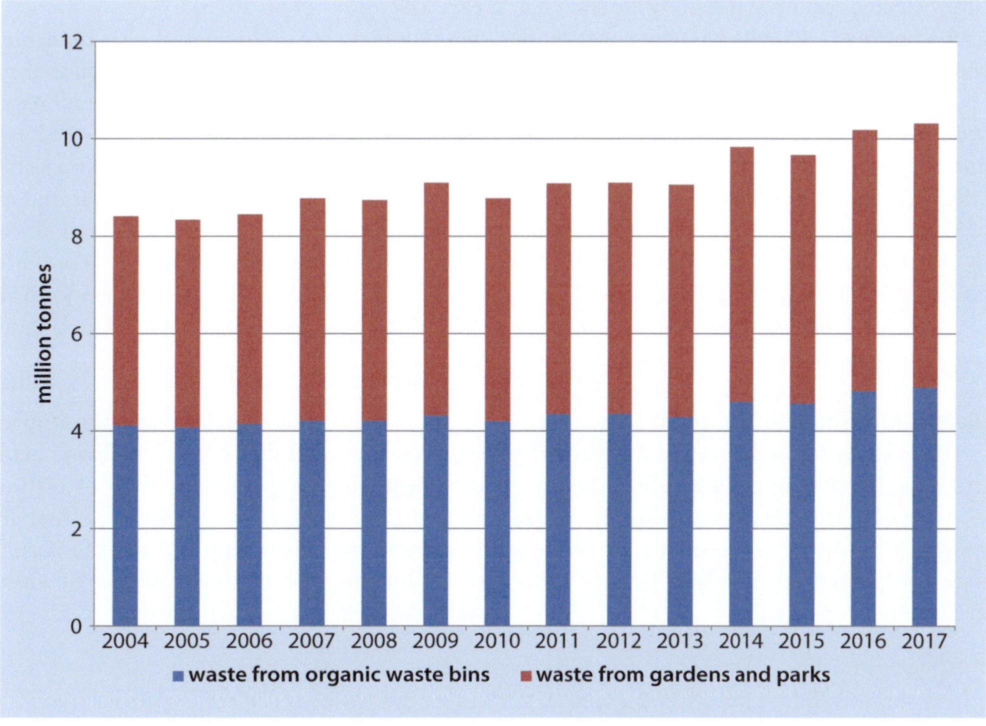

▫ **Fig. 2.46** Development of the volume of waste of biogenic origin. (Author's own representation based on data from Federal Statistical Office of German 2018)

methods. Further material uses lie in the building sector (building and insulating materials). In addition, cereal straw represents, along with wood, a potential supplier of lignocellulose as a raw material for use in biorefineries for the production of various platform chemicals (▶ Chap. 4).

Biogenic residues from industry are, for the most part, already used for feed production and other material purposes. Those that are important in terms of quantity are (weight data given as dry matter content) oil cake from cooking oil and fat production, with about 6 million tonnes (for animal feed), beet pulp and molasses from the sugar and food industries, with about 3.3 million tonnes (mainly for animal feed and application in biotechnology and pharmaceuticals), and bran and flour dust from grain mills, with 1.7 million tonnes (for animal feed). The quantities of biogenic residues from milk processing (whey), beverage production and the production of bakery products are each less than 1 million tonnes (AEE 2013).

The quantities of food waste that are generated in the production of food (agriculture), its processing, trade, preparation and consumption can be investigated with the aid of model calculations. In a corresponding study (Bräutigam et al. 2014), food waste was defined as that which would, in principle, have been suitable for human consumption, taking cultural aspects into account. This does not include, for example, the peels of oranges or potatoes or parts of animals, such as bones, that cannot be eaten. However, it does include, for example, agricultural products that are not sold for economic reasons or because of their appearance, as well as food that can no longer be consumed because it has spoiled or because it is past its expiration date - with careful planning, it would have been possible to consume these foods in good time (▶ Excursus 2.5).

Excursus 2.5 Too good for the bin!

Under the title "Zu gut für die Tonne" ("Too good for the bin"), the German Federal Ministry of Food and Agriculture (BMEL) has launched an information campaign against the discarding of food. The aim is to raise awareness of the value of food throughout the whole food chain, from agriculture, industry and commerce to the consumer. This initiative includes the awarding of a federal prize for special commitment to reducing food waste. The prize was awarded for the first time in 2016. More than 200 applications were received from all over Germany. The "Zu gut für die Tonne" federal prize is divided into four categories: trade, gastronomy, production, and society and education. In addition, there is a sponsorship award for ideas that are at the very beginning of their implementation.

For the model calculations within the study, statistical data on the production, import and export of the different food groups, as well as the nature of their use, were taken from the *Food Balance Sheets* of the *Food and Agriculture Organization* (FAO). The waste produced at each stage of the food chain was determined on the basis of the percentages given in the literature. Selected results of the study are shown in ◘ Figs. 2.47 and 2.48. They give – in spite of some uncertainties - good indications about the occurrence and composition of food waste.

◘ Figure 2.47 shows the share of individual stages of the food chain in total food waste for the year 2011 for Germany and for the countries of the EU-28 (averaged over all countries). In total, around 18.9 million tonnes of food waste (231 kg per person) are produced in Germany. Around 28% of this comes from agriculture and 45% from preparation and consumption in the private sector.

◘ Figure 2.48 shows the share of different food groups in the total amount of food waste in private households. In Germany, fruit, vegetables and cereal products are in first place, with a share of 26% each.

The following reasons exist for throwing away food that can actually or originally be consumed in private German households: The product is mouldy, it has no taste (anymore), it has been in the fridge for too long, its expiration date has passed or there are leftovers that you no longer wish to use. In addition, the vast majority of households surveyed in a non-representative survey considered the amount of discarded food that was originally still edible to be significantly lower compared to the amount actually disposed of with household waste. The amount of food that ends up as waste could therefore be greatly reduced (Jörissen et al. 2015).

This could also be achieved by taking measures at the political level to reduce the volume of food waste (Priefer et al. 2016). Such measures could concern, for example, the level and structure of charges for waste disposal. Special measures could remove nonsensical rules on the appearance and quality of food on the market. They could question the term '*use-by-date*' and instead orient practice around the English term '*best before*.' Measures could improve the possibility of passing on food that is no longer needed, but in good condition, to food banks (food sharing). Most important, however, is to raise people's awareness that throwing away food is not only detrimental for economic reasons, but also has a negative impact on the environment. As many studies show, it is always better to reduce the amount of food thrown away than to dispose of it together with biowaste and then produce a fuel from it in a biogas plant. This also complies with the Waste Management Act, according to which the avoidance of waste takes priority over recycling.

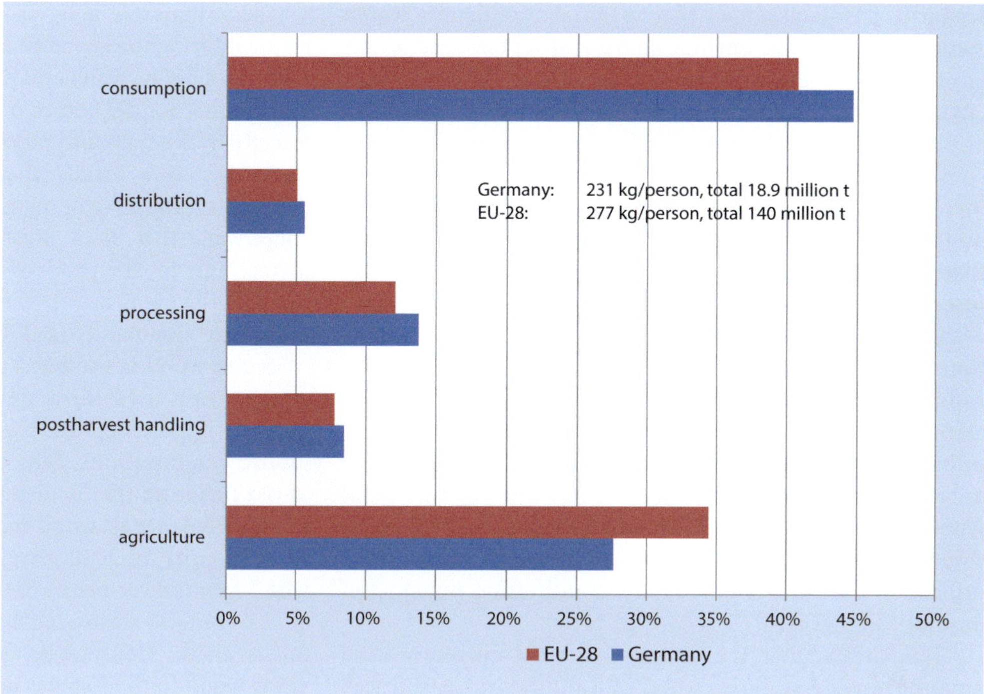

Fig. 2.47 Share of the individual stages in the food processing chain in the total amount of food waste. (Author's own representation based on data from Bräutigam et al. 2014)

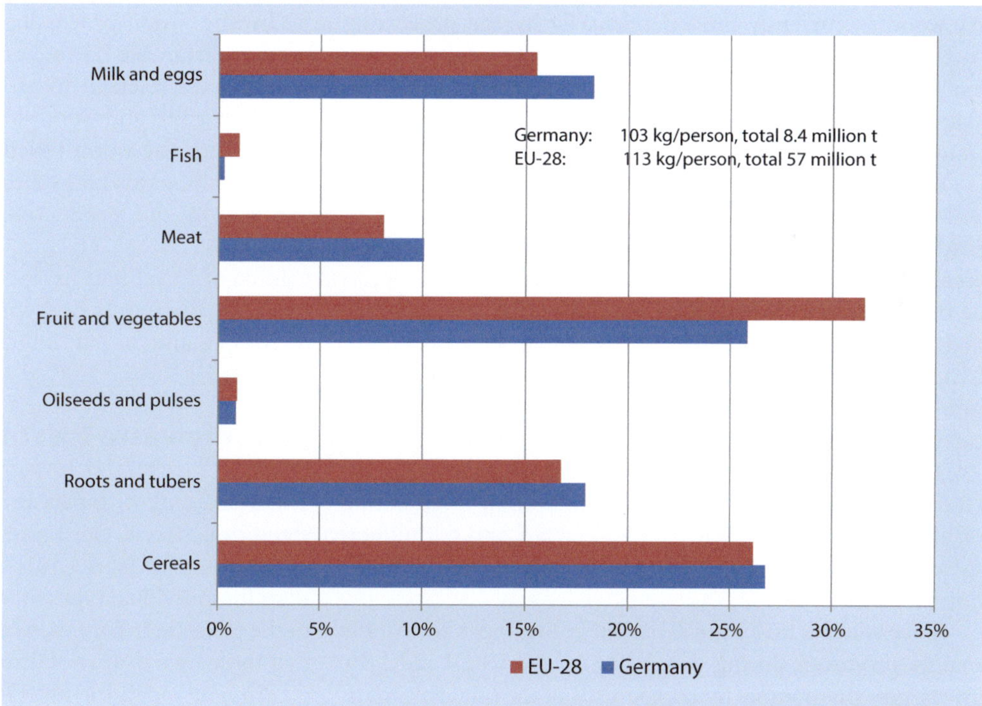

Fig. 2.48 Food waste generation in private households - Share of different food groups in total food waste generation. (Author's own representation based on data from Bräutigam et al. 2014)

2.4.4 Forestry (Residual) Biomass

During harvesting and thinning in the forest, residual materials, such as wood from treetops and branches, are left over (▶ Sect. 2.3). This wood cannot be used as timber. Its volume in Germany for 2010 was estimated at 84.6 million m³. The current use of residual forest wood is around 8.0 million m³ (AEE 2013).

For the use of residual forest wood, different scenarios have been created (Mantau 2012a). For the upper revenue scenario, it was assumed that material production would come to the fore. The environmental restrictions place fewer limitations on the use of wood, in part because new environmentally-friendly technologies are being developed. The possible negative impacts of residual wood use are not considered to be as serious as the negative impacts of the increased use of fossil fuels or other building materials. Based on these assumptions, a utilisation potential of slightly more than 40 million m^3 has resulted. This upper potential therefore does not correspond to the entire residual forest wood.

In the lower volume scenario, mobilisation for the utilization of wood for energy purposes is subject to strong environmental requirements. The use of residual wood is associated with negative environmental impacts. Large area units are removed from wood production. The forest owners are developing a rather cautious attitude towards the use of wood. In this scenario, too, mechanization processes continue to progress for cost reasons, but have little effect on the utilization potential. This results in a utilization potential of around 13 million m^3.

This means that the above-mentioned currently used amount of residual forest wood of 8 million m^3 is still well below the usable amount from the lower revenue scenario. It can be assumed that the expansion of the use of residual forest wood is currently limited primarily by the processing costs. The extent to which the potential will be exploited ultimately depends on the prices that can be achieved, the available technology, the will of the forest owners and the political framework conditions.

Landscape conservation material refers to grass-, herb- and wood-like organic residues from the maintenance of roadside areas, waterwayside areas, and nature conservation areas, as well as public recreation areas and cemeteries. It can be divided into green waste (grass and herbaceous portion) and landscape maintenance wood (woody portion). Landscape conservation material is mainly produced in the municipalities. Garden wood is not usually included in the category of landscape maintenance material.

The volume of landscape conservation material in Germany is estimated at around 7.25 million m^3. Depending on the degree of mobilization, between 5.5 and 6.5 million m^3 are usable. The volume currently in use is around 4.5 million m^3 (AEE 2013).

Since wood is no longer debarked in the forest, the bark is mainly produced during processing in the wood industry. The largest share (more than two thirds of the total volume of softwood and hardwood bark) is accounted for by sawmill operations. A further 20% is generated in the wood and pulp industry (Mantau 2012a).

The quantity of bark in 2010, in the volume of 4.7 million m^3, was used for material purposes, primarily as bark mulch, or in wood-fired heating plants or wood-fired cogeneration plants to generate heat and/or electricity (AEE 2013).

Black liquor is a by-product of pulp production. It is formed during the separation of lignin and cellulose and is a mixture of lignin, water and the chemicals used for extraction. Black liquor hardly comes on the market, but rather is used directly in the pulp and paper industry to generate heat and electricity. This means that the quantity produced also depends on the development of the pulp industry. It is quite conceivable that black liquor will, in future, also be used increasingly for the production of raw chemical materials in biorefineries (▶ Chap. 4). The annual volume amounts to 3.7 million m^3 (Mantau 2012a).

Scrap wood is the term used to describe wood that has already been used for a specific purpose and is no longer needed. It accumulates, for example, in the construction industry (renovations, demolition), as packaging material or as old furniture that is disposed of as bulky waste. The raw material balance of wood recorded a scrap wood volume of 14 million m^3 in 2010. Scrap wood is mainly used in large wood-fired power plants for electricity and heat production or co-fired in waste incineration plants or conventional power plants. Households also occasionally burn their own waste wood in their stoves and fireplaces. About one fifth of waste wood is used in the woodworking and processing industry, e.g., for chipboard production.

In the wood-processing industry, all wood residues resulting from the cutting and processing of roundwood in sawmills are referred to as sawing by-products. These are mainly sawdust and wood shavings. They are an essential raw material for the wood-based panel industry in the production of chipboards and other materials, and are used as raw material in the production of wood pellets and paper (Wikipedia 2016b). The raw wood material balance records a volume of 15 million m^3 for the year 2010. As these are by-products, the supply of sawing by-products depends mainly on the demand for cut timber.

2.4.5 Biowaste Treatment Plants

The availability of biowaste is very variable, but, due to its significant size, it can represent a substantial input for the bioeconomy (◘ Table 2.10). It should be noted that the quantities for individual types of waste and residual material given in the table are in tonnes, while others are in cubic metres, so that these figures are not directly comparable.

Already today, the utilization paths of waste and residues of organic origin are manifold (◘ Table 2.11). In addition to the direct utilization of materials, biowaste treatment plants currently represent an important utilization path for waste of biogenic origin. Biowaste treatment plants directly produce either compost or - via an upstream fermentation process in a biogas plant - methane, which is used to generate energy. In addition, a fermentation residue is produced. Both the com-

Table 2.10 Waste and Residue Generation - Summary

	million tonnes		million m³
organic waste bin	4	residual forest wood	8
garden and park waste	5	landscape wood	4.5
paper, cardboard,	8	bark	4.7
mixed packagings	6	black liquor	3.7
waste from kitchen and canteen	1	old wood	14
biowaste from the residual waste bin (estimated value)	4	sawdust	15
slurry and solid manure	200		
straw	30		
oil shot	6		
waste materials from the sugar industry	3		
residues from grain mills	2		
residues from milk processing, manufacture of beverages and bakery products	2		
total	270	total	49.9

Table 2.11 Overview of wastes and residues of biogenic origin and their utilization paths - Summary

Type of waste/remaining material	exploitation paths
Slurry and manure	Agriculture, biogas plant
Straw	Remaining in the field
	Bedding for animal husbandry
	Insulating material
	Biorefineries (future)
	Electricity and fuel production
Residues from the food and feed industry	Feed production
	Biotechnology
	Pharmaceuticals
Wood	Power generation
	Chipboard production
Bark	Bark mulch
	Power generation
Sawdust	Chipboard production
	Pellets (energy produc-tion)
	Papermaking
Organics in household waste	Compost
	Biogas (methane)

post and the fermentation residues from the biogas plant are used in agriculture and horticulture.

The biowaste delivered to biowaste treatment plants (Fig. 2.49; Federal Statistical Office 2018a) amounted to around 15.6 million tonnes in 2016. Garden and park waste (4.8 million tonnes) and waste from organic waste bins (4.4 million tonnes) accounted for the largest share of this. Of the total waste delivered to bio-waste treatment plants, around 7.5 million tonnes are sent to composting plants and around 6.6 million tonnes to fermentation plants (Statista 2019b). In total, around 3 million tonnes of compost were produced, of which around 61% went to agriculture and around 17% to earthworks. 100% of the fermentation resi-dues from the biogas plants are put into agriculture.

For biogas plants, the input substrate of municipal bio-waste plays only a minor role (around 3% of the total mass-related substrate input). Renewable raw materials account for the largest share (52%), followed by slurry and manure from animal husbandry with around 43% (Scheftelowitz et al. 2015) (▶ Chap. 8).

Based on the volume of waste generated in the individual countries of the European Union (plus Norway and Switzerland) and information on the quantities that go into biological treatment (EUROSTAT 2019), it is possible to estimate the proportion of organic matter in household waste recycled via composting and fermentation (European Environmental Agency 2013) in the individual countries (Fig. 2.50). Switzerland and the Netherlands lead, with a share of more than 80%. Germany has a value of about 50%, with the lowest values being found, for the most part, in the countries of Eastern Europe. Overall, it can be seen that, in most countries, there is still great potential for high-quality recycling, such as composting and fermentation or other bio-economically relevant paths of use.

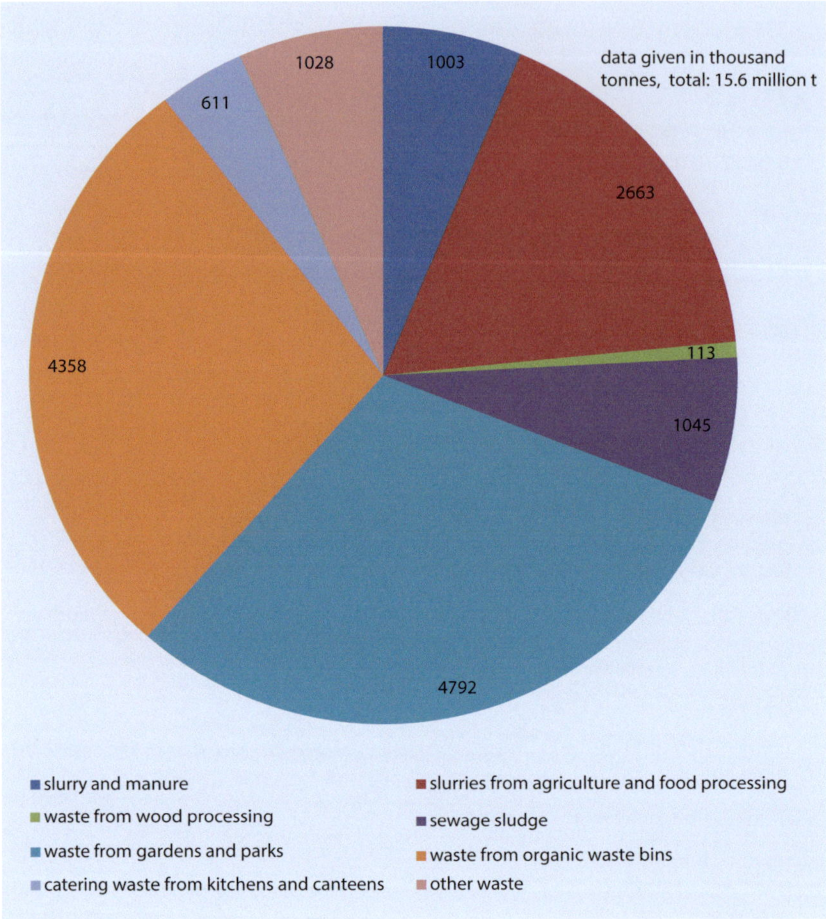

Fig. 2.49 Composition of biowaste delivered to biowaste treatment plants. (Author's own representation based on data from Federal Statistical Office of Germany 2019b)

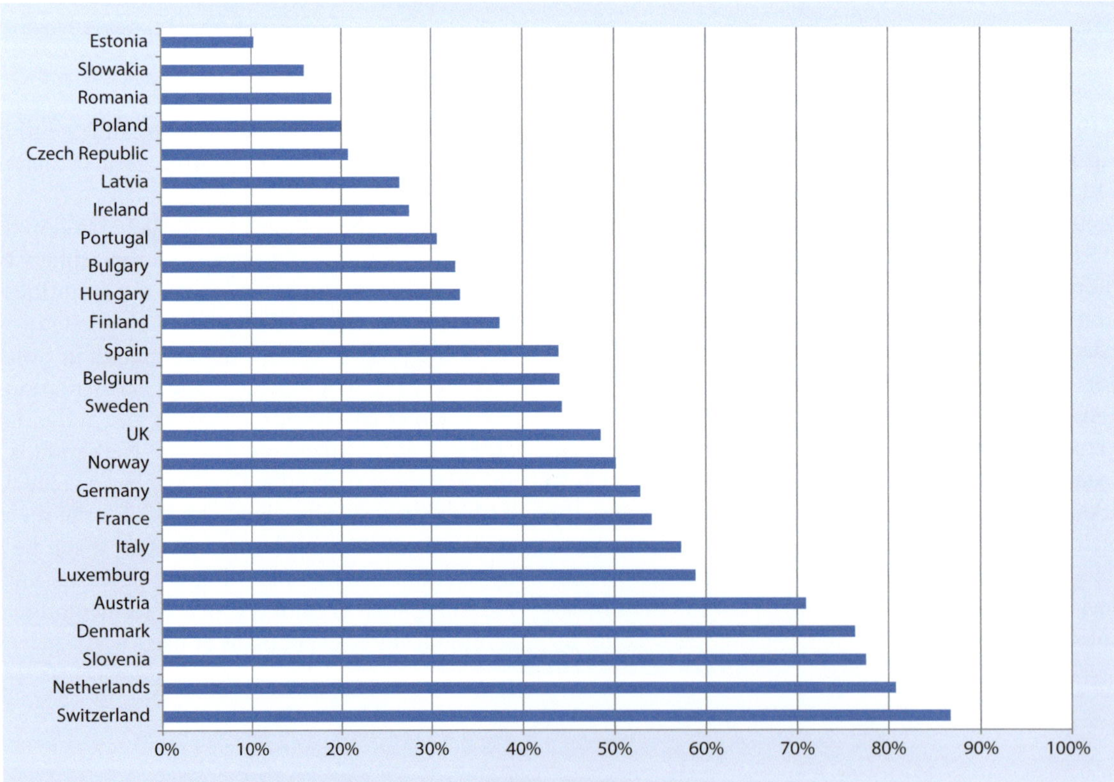

Fig. 2.50 Share of recycling of the organic fraction in household waste in EU countries, Norway and Switzerland via composting and fermentation. (Author's own calculations based on data from EUROSTAT 2019 and EEA 2013)

References

Abfallablagerungsverordnung (AbfAblV): Verordnung über die umweltverträgliche Ablagerung von Siedlungsabfällen; Date of issue 20.2.2001; BGBl. I p 305

AEE (2013) Agentur für Erneuerbare Energien: Renews Spezial: Reststoffe für Bioenergie nutzen - Potenziale, Mobilisierung und Umweltbilanz; Ausgabe 64 vom April 2013

AKS Arbeitskreis Standortskartierung (2003) Forstliche Standortsaufnahme. Begriffe, Definitionen, Einteilungen, Kennzeichnungen, Erläuterungen. 6. Auflage. IHW Verlag, Eching; 352 p

AltholzV (2002) Altholzverordnung vom 15. August 2002 (BGBl. I S. 3302), die zuletzt durch Artikel 96 der Verordnung vom 31.August 2015 (BGBl. I S. 1474) geändert worden ist. 16 p

Altwegg M, Meier P (2008) Timberland. Holz - die grüne Anlageklasse. Betriebswirtsch. Verlag, GWV Fachverlage GmbH, Wiesbaden. 106 p

ANL Bayerische Akademie für Naturschutz und Landschaftspflege (1994) Begriffe aus Ökologie, Landnutzung und Umweltschutz. Informationen 4, Frankfurt/Main, Laufen, 139 p

Bauhus J, Van der Meer PJ, Kanninen M (2010) Ecosystem Goods and Services from Plantation Forests. Earthscan Publishing for a Sustainable Future, London/Washington, DC, 240 p

biogas.org (2016). http://www.biogas.org/edcom/webfvb.nsf/id/de_pm-09-16; Accessed: 27.1.2019

Bioökonomierat (2016) Holz in der Bioökonomie – Chancen und Grenzen. Prelimiary Version BÖRMEMO 05 19.04.2016. 8 p

BMEL Bundesministerium für Ernährung und Landwirtschaft (2013) Presseinformation zum Illegalen Holzeinschlag. BMELF und Thünen-Institut 03/2013; 2 p

BMEL Bundesministerium für Ernährung und Landwirtschaft (2014) Der Wald in Deutschland. Ausgewählte Ergebnisse der dritten Bundeswaldinventur. 56 p

BMEL Bundesministerium für Ernährung und Landwirtschaft (2016a) Holznutzung zunehmend eingeschränkt oder ausgesetzt. https://www.bundeswaldinventur.de Accessed: 20.05.2016

BMEL Bundesministerium für Ernährung und Landwirtschaft (2016b) Waldentwicklungs- und Holzaufkommensmodellierung (WEHAM). https://www.bundeswaldinventur.de Accessed: 17.06.2016

BMELV Bundesministerium für Ernährung, Landwirtschaft und Verbraucherschutz (2011) Waldstrategie 2020. Bonn; 36 p

BMUB Bundesministerium für Umwelt, Naturschutz, Bau und Reaktorsicherheit (2007.) Nationale Strategie zur biologischen Vielfalt, 180 p

Bode W (1997) Der Mensch hat ein Problem. Hat der Wald eine Lösung? p 213–257 in: Bode, W. (1997): Naturnahe Waldwirtschaft. Deukalion Verlag; 396 p

Bolte A, Polley H (2010) Nutzungs- und Ertragspotenziale des Waldes. Vortrag 3. Symposium Waldstrategie 2020 Mehr Holz im Einklang mit der Gesellschaft, Berlin 12.04.2010. Online Document www.biooekonomierat.de

Borchers J (2010) Segregation versus Multifunktionalität in der Forstwirtschaft. Forst und Holz 65, Heft 7/8 ; p 44–49

Borchers J, Elbs A, Schindele G (2008) Fichte im Erwerbsforstbetrieb – Baumart mit Zukunft, Holzzentralblatt, Nr.44; p 1229

Born GJ van den, Minnen JG, Olivier JGJ, Ros JPM (2014) Integrated analysis of global biomass flows in search of the sustainable potential for bioenergy production, PBLY report 1509, PBL Netherlands Environmental Assessment Agency

BPP Bundeszentrale für politische Bildung (2010) Jährliche Änderung der Waldbestände. Online-Dokument http://www.bpb.de. Accessed: 12.5.2016

Braeutigam K-R, Jörissen J, Priefer C (2014) The extent of food waste generation across EU-27: Different calculation methods and the reliability of their results. Waste Management & Research 32(8):683–694

Burbiel J (2014) Nanocellulose. Europäische Sicherheit & Technik; Mittler Report Verlag GmbH, Bonn April 2014; p 70

Burger F (2010) Bewirtschaftung und Ökobilanzierung von Kurzumtriebsplantagen. München, Techn. Univ., Diss.,180 p

Burschel P, Huss J (2003) Grundriss des Waldbaus : Ein Leitfaden für Studium und Praxis. Ulmer, E; Auflage: 3., Aufl. 487 p

Chilonda P, Otte J (2006) Indicators to monitor trends in livestock production at national, regional and international levels. Livestock Research for Rural Development 18(8)

Cossalter, Pye-Smith (2003) Fast-wood forestry: myths and realities. Center for International Forestry Research

COUNCIL DIRECTIVE 1999/31/EC: Council Directive 1999/31/EC of 26 April 1999 on the landfill of waste

Coyle DR, Coleman MD (2005) Forest production responses to irrigation and fertilization are not explained by shifts in allocation. Forest Ecology and Management 208:137–152

Dănescu A, Albrecht A, Bauhus J (2016) Structural diversity promotes productivity of mixed, uneven-aged forests in southwestern Germany. Oecologia 182(2):319–333

Dauber J, Brown C, Fernando AL, Finnan J, Krasuska E, Ponitka J, Styles D, Thrän D, Jan Van Groenigen K, Weih M, Zah R (2012) Bioenergy from 'surplus' land, environmental and socio-economic implications. BioRisk 7:5–50

Davis SC, Hay W, Pierce J (2014) Biomass in the energy industry: an introduction. Published by BP p.l.c.; 122 p. www.bp.com/energysustainabilitychallenge

DBFZ (2013) Deutsches Biomasseforschungszentrum GmbH; Potenzial von Stroh im Energiemix bisher unterschätzt; Pressemitteilung vom 21.10.2013; https://www.dbfz.de/presse/pressemitteilungen-2013/potenzial-von-stroh-im-energiemix-bisher-unterschaetzt.html; Accessed: 27.1.2019

DBFZ (2015) Deutsches Biomasseforschungszentrum gemeinnützige GmbH – Biomassepotenziale von Rest- und Abfallstoffen – Status quo in Deutschland; Hrsg.: FNR – Fachagentur Nachwachsende Rohstoffe e.V. Schriftenreihe Nachwachsende Rohstoffe 36; 2015

De Silva SS (1994) Supplemental feeding in semi-intensive aquaculture systems. Farm-made aquafeeds. Proceedings of the FAO/AADCP Regional Expert Consultation on Farm-Made Aquafeeds, Bangkok, Thailand, FAO-RAPA/AADCP, Bangkok, Thailand, 434 p

Del Río M, Schütze G, Pretzsch H (2013) Temporal variation of competition and facilitation in mixed species forests in Central Europe. Plant Biology. https://doi.org/10.1111/plb.12029

Deutscher Bundestag (2014) Drucksache 18/2214 vom 28.7.2014; Antwort der Bundesregierung auf die Kleine Anfrage der Abgeordneten Peter Meiwald, Dr. Julia Verlinden, Annalena Baerbock, weiterer Abgeordneter und der Fraktion BÜNDNIS 90/DIE GRÜNEN - Getrennte Bioabfallsammlung und die Verwertung von Bioabfällen

Diepenbrock W (2014) Nachwachsende Rohstoffe. Verlag Eugen Ulmer, Stuttgart

Diepenbrock W, Eller F, Léon J (2012) Ackerbau, Pflanzenbau und Pflanzenzüchtung: Grundwissen Bachelor, Eugen Ulmer Verlag, Stuttgart, 3. Auflage

Dieter M (2011) Flächenstillegung im Wald – Studie des vTI. Tagung "Wertschöpfung im Wald" am 13. Juli 2011 in Fuld. Online-Dokument: www.netzwerk-laendlicher-raum.de

Dieter M, Englert H, Degen B, Liesebach M (2011) Beitrag der Forstpflanzenzüchtung zum Ausgleich konkurrierender Ansprüche an den Wald. Workshop Forstpflanzenzüchtung Berlin 2011. Gülzower Fachgespräche Band 36:62–68

Dixon J, Gulliver A, Gibbon D (2001) Farming systems and poverty: Improving farmer's livelihoods in a changing world. FAO & World Bank, Rome/Washington DC

Doppler W (1991) Landwirtschaftliche Betriebssysteme in den Tropen und Subtropen. Eugen Ulmer Verlag, Stuttgart

Döring P, Glasenapp S, Mantau U (2016) Energieholzverwendung in privaten Haushalten 2014. Marktvolumen und verwendete Holzsortimente. Abschlussbericht Zentrum Holzwirtschaft Arbeitsbereich: Ökonomie der Holz- und Forstwirtschaft. 37 p

DVFF Deutscher Verband Forstlicher Forschungsanstalten (2014) Offener Brief an das Bundesamt für Naturschutz am 4.6.2014; 7 p
DWD Deutscher Wetterdienst (2015) Zahlen und Fakten zum Klimawandel in Deutschland. Klima-Pressekonferenz des Deutschen Wetterdienstes am 10. März 2015 in Berlin; 36 p
Edwards P, Pullin RSV, Gartner JA (1988) Research and education for the development of integrated crop-livestock-fish farming systems in the tropics. International Center for Living Aquatic Resources Management, Manila
EEA (European Environmental Agency) (2013) Managing municipal solid waste - a review of achievements in 32 European countries. EE Report No 2/2013 - figure 2.10 and table 2.1, p.17 et seqq
Englisch M (2009) Die Möglichkeit den Standort zu erkennen - Forstliche Standortskartierung. BFW-Praxisinformation 19:17–19
Eurostat 2019: Municipal waste by waste management operations. http://appsso.eurostat.ec.europa.eu/nui/show.do?dataset=env_wasmun&lang=eng Accessed: 27.1.2019
EWG (1992) Richtlinie des des Rates der Europäischen Gemeinenschat vom 24. September 1990 über die Nährwertkennzeichnung von Lebensmitteln (90/496/EWG)
FAO - Food and agriculture organization of the United Nations (2010) Global forest resources assessment 2010 main report. FAO Forestry Paper 163; 378 p
FAO- Food and agriculture organization of the United Nations (2012a) The state of the World's Forest 2012. FAO, Rome; 60 p
FAO - Food and agriculture organization of the United Nations (2016) Global Forest Resources Assessments. http://www.fao.org/forest-resources-assessment/explore-data. Accessed: 14.6.2016
FAO (1997) Aquaculture development. FAO, Fisheries Department, Rome
FAO (2012b) The state of world fisheries and aquaculture. FAO Fisheries and Aquaculture Department, Rome
FAO (2014) The state of world fisheries and aquaculture. FAO Fisheries and Aquaculture Department, Rome
FAO (2015) Fisheries and Aquaculture Department, Statistics and Information Service FishStatJ: Universal software for fishery statistical time series
FAO (2003) World agriculture: towards 2015/2030. An FAO perspective. Food Agricultural Organisation. Earthscan Publications Ltd, London
FAOSTAT (2014) Food and Agriculture Organization of the United States - Statistics Division. http://faostat.fao.org/. Accessed: 05.04.2016
Federal Environment Agency (2014) Verpflichtende Umsetzung der Getrenntsammlung von Bioabfällen UBA Texte 84/2014: 19ff
Federal Statistical Office (2018a) Abfallbilanz (Abfallaufkommen/-verbleib, Abfallintensität, Abfallaufkommen nach Wirtschaftszweigen) 2016; Statistisches Bundesamt Wiesbaden 2018; https://www.destatis.de/DE/Publikationen/Thematisch/UmweltstatistischeErhebungen/Abfallwirtschaft/AbfallbilanzPDF_5321001.pdf?__blob=publicationFile; Accessed: 16.1.2019
Federal Statistical Office (2018b) – Genesis online Datenbank –https://www-genesis.destatis.de/genesis/online/data;jsessionid=CB42ACB832FA46C0A6E7EA508CE329A7.tomcat_GO_2_1?operation=abruftabelleAbrufen&selectionname=32121-0001&levelindex=0&levelid=1466164842295&index=4; Accessed: 27.1.2019
Fichtner A, von Stehen S, Schmid M, Sturm K (2013) Der Ökowald als Baustein einer Klimaschutzstrategie. Herausgeber Greenpeace e.V., 39 p
Fischer G, van Velthuizen HT, Shah MM, Nachtergaele FO (2002) Global Agro-ecological
FNR 2008 Studie zur Markt- und Konkurrenzsituation bei Naturfasern und Naturfaserwerkstoffen (Deutschland und EU), Gülzower Fachgespräche 26
FNR Fachagentur für Nachwachsende Rohstoffe (2015) Anbaufläche für nachwachsende Rohstoffe Stand 2015. Online-Dokument: https://mediathek.fnr.de
Foley JA, DeFries RS, Asner GP, Barford C, Bonan G, Carpenter SR, Chapin FS, Coe MT, Daily GC, Gibbs HK, Helkowski JH, Holloway T, Howard EA, Kucharik CJ, Monfreda C, Patz JA, Prentice IC, Ramankutty N, Snyder PK (2005) Global consequences of land use. Science 309:570–574
Forrester DI, Kohnle U, Albrecht AT, Bauhus J (2013) Complementarity in mixed-species stands of Abies alba and Picea abies varies with climate, site quality and stand density. Forest Ecology and Management 304:233.242. https://doi.org/10.1016/j.foreco.2013.04.038
Fröhlich HJ, Grosscurth W (1973) Züchtung, Anbau und Leistung der Pappeln, Mitteilungen der Hessischen Landesforstverwaltung (Band 10), Sauerländer, Frankfurt am Main; 268 p
FSC Forest Stewardship Council (2013) FSC im Wald - ein Leitfaden für Praktiker. 40 p
FTP Forest- based Sector Technology Platform (2013) Strategic Research and Innovation Agenda for 2020. Online.Dokument: http://www.forestplatform.org. 28 p
Gaggermaier A, Koch M (2011) Waldflurbereinigung – Für wen lohnt sich das ? Vortrag anlässlich des Treffens der Forstpolitikwissenschaftler am 13. - 15. April 2011 in Duderstadt, Deutschland. Online https://www.wup.wi.tum.de/fileadmin
Geoportal Baden-Württemberg: Forstliche Standortskartierung im Geodatenviewer. http://www.geoportal-bw.de. Accessed: 23.8.2016
Gewerbeabfallverordnung (GewAbfV): Verordnung über die Entsorgung von gewerblichen Siedlungsabfällen und von bestimmten Bau- und Abbruchabfällen; Date of issue 19.6.2002; BGBl. I S. 1938
Goetzl A (2015) Development in the global trade of wood pellet U.S. International Trade Commission No. ID-039; 28 p
Grewe BS (2011) Wald. EGO - Europäische Geschichte Online. http://ieg-ego.eu/de/threads/hintergruende/natur-und-umwelt/bernd-stefan-grewe-wald/?searchterm=Grewe&set_language=de
Gustavsson J, Cederberg C, Sonesson U, van Otterdijk R, Meybeck A (2011) Global food losses and food waste: extend causes and prevention. Food and Agriculture Organization (FAO) of the United Nations, Rome
Haapanen M (2011) Bedeutung der Forstpflanzenzüchtung beim Nadelholz in Finnland. Tagungsband zum Workshop Forstpflanzenzüchtung in Berlin. Gülzower Fachgespräche Band 36:100–117
Hanewinkel M, Hummel S, Cullmann D (2009) Modelling and economic evaluation of forest biome shifts under climate change in Southwest Germany. Forest Ecology and Management 259:710–719
Hanewinkel M, Cullmann DA, Schelhaas MJ, Nabuurs GJ, Zimmermann NE (2013) Climate change may cause severe loss in the economic value of European forest land. Nature Climate Change 3:203–207
Hardy RW (2010) Utilization of plant proteins in fish diets: Effects of global demand and supplies of fishmeal. Aquaculture Research 41(5):770–776
Hennsen D (2012) Studie zur Umsetzung der Pflicht der Getrenntsammlung von Bioabfällen nach § 11 KrWG - Inhalt, Reichweite und Verbindlichkeit der neuen Vorschrift; Hrsg: Bundesgütegemeinschaft Kompost e. V sowie Verband der Humus- und Erdenwirtschaft e. V. (VHE); September 2012
Hernández C, Olvera-Novoa MA, Hardy RW, Hermosillo A, Reyes C, González B (2010) Complete replacement of fish meal by porcine and poultry by-product meals in practical diets for fingerling Nile tilapia *Oreochromis niloticus*: Digestibility and growth performance. Aquaculture Nutrition 16(1):44–53
Höltermann A, Oesten G (2001) Ein forstwirtschaftliches Konzept als Vorbild für die Strategie der nachhaltigen Entwicklung? In: Der Bürger im Staat. 51. Jhg. Heft 1 „Der deutsche Wald" Landeszentrale für politische Bildung Baden-Württemberg. S.39–45
Huber J (1995) Nachhaltige Entwicklung durch Suffizienz, Effizienz und Konsistenz. In Fritz P, Huber J, Levi HW. (Hrsg.): Nachhaltigkeit in naturwissenschaftlicher und sozialwissenschaftlicher Perspektive, Stuttgart: Hirzel/Wissenschaftliche Verlagsgesellschaft, p 31–46
Hüttl R (2012) Bioökonomie und Agroforstwirtschaft. Vortrag in Cottbus am 6.12.2012. Onlinedokument: http://www.agroforstenergie.de

Imhoff ML, Bounoua L, Rickett T, Loucks C, Harriss R, Lawrence WT (2004) Global patterns in human consumption of net primary production. Nature 429:870–873

IPCC (2013): Summary for Policymakers. In: Climate Change 2013: The Physical Science Basis. Contribution of Working Group I to the Fifth Assessment Report of the Intergovernmental Panel on Climate Change. 28 p

Jeroch H, Drochner W, Simon O (Hrsg.) (2008) Ernährung landwirtschaftlicher Nutztiere - Ernährungsphysiologie, Futtermittelkunde, Fütterung, 2. Aufl., Eugen Ulmer Verlag, Stuttgart

Jörissen J, Priefer C, Bräutigam K-R (2015) Food waste generation at household level: Results of a survey among employees of two European research centers in Italy and Germany. Sustainability 7(3):S. 2695–S. 2715. https://doi.org/10.3390/su7032695

Kanninen M (2010) Plantation forests – Global perspectives. S.1-15. In: Bauhus J, van der Meer P, Kanninen M (eds) Ecosystem Goods and Services from Plantation Forests. Earthscan, London, UK, 227 p

Kestemont P (1995) Different systems of carp production and their impacts on the environment. Aquaculture 129:347–372

Kim S, Dale BE (2004) Global potential bioethanol production from wasted crops and crop residues. Biomass and Bioenergy 26: 361–375

Klädtke J (2003) Düngungsversuch Fichte 337 Pfalzgrafenweiler Waldwachstumskundliche Ergebnisse und Schlussfolgerungen. AFJZ 174 2/3, 31–40

Knoche D, Martens S (2012) Bodenzustand und Wurzelwachstum von Stieleichen-Jungbeständen – Meliorationsversuch „Collm 1996". Bericht www.sachsenforst.de, 23 p

Kölling, Borchert (2013) Nachhaltige Nutzung des Produktionsfaktors Boden - Herausforderung Kronenbiomassenutzung. LWF Wissen 72:47–53

Kölling C, Binder F, Falk W (2013) Risiko und Ertrag in ungewisser Zukunft: Der Klimawandel fordert die Generationengerechtigkeit heraus. LWF Wissen 72:54–58

Kralisch D (2014) Nanozellulose ein bakteriell erzeugtes Hochleistungspolymer. Spektrum der Wissenschaften, Januar 2015, p 78–84

Kreislaufwirtschaftsgesetz (KrWG): Gesetz zur Förderung der Kreislaufwirtschaft und Sicherung der umweltverträglichen Bewirtschaftung von Abfällen; Date of issue 24.2.2012; BGBl. I S. 212

KTBL (2015) KTBL-Taschenbuch Landwirtschaft, Kuratorium für Technik und Bauwesen in der Landwirtschaft e.V. (KTBL) Darmstadt.

Lal R (2005) World crop residues production and implications of its use as a biofuel. Environment International 31:575–584

Lewandowski I (2015) Securing a sustainable biomass supply in a growing bioeconomy. Global Food Security 6:34–42

Luick R, Hennenberg K (2015) Potenziale und ihre Grenzen: Biomassenutzung im Kontext von Energiewende und Bioökonomie. In: Nachhaltiges Waldmanagement und regionale Wertschöpfung (Hrsg.: Fakultät für Umwelt und Natürliche Ressourcen der Universität Freiburg & Forstliche Versuchsund Forschungsanstalt Baden-Württemberg).-Berichte Freiburger Forstliche Forschung (99). Freiburg: p 1–18

Mantau U (2012a) Holzrohstoffbilanz Deutschland, Entwicklungen und Szenarien des Holzaufkommens und der Holzverwendung 1987 bis 2015, Zentrum Holzwirtschaft - Arbeitsbereich: Ökonomie der Holz- und Forstwirtschaft; Universität Hamburg, Oktober 2012

Mantau U (2012b) Standorte der Holzwirtschaft – Holzrohstoffmonitoring. Energie-holzverwendung in privaten Haushalten. Marktvolumen und verwendete Holzsortimente. Universität , Zentrum Holzwirtschaft

Mantau U, Sörgel C (2006) Energieholzverwendung in privaten Haushalten, Marktvolumen und verwendete Holzsortimente. Abschlussbericht., Universität Hamburg - Zentrum Holzwirtschaft, Arbeitsbereich Ökonomie der Holz- und Forstwirtschaft. 37 p

MCPFE Ministerial Conference on the Protection of Forests in Europe (2003). Improved Pan-European Indicators for Sustainable Forest Management. Expert Level Meeting 7-8 October 2002, Vienna, Austria; 6 p

Meißner M, Janßen A, Konnert M, Liesebach M, Wolf H (2015) Vermehrungsgut für klima- und standortgerechten Wald. AFZ Wald 70(11):24–26

Meiwes KJ, Asche N, Block J, Kallweit R, Kölling C, Raben G, v. Wilpert K (2008) Potenziale und Restriktionen der Biomassenutzung im Wald. AFZ - Der Wald 63 (10-11)

Michels J (2013) Die Lignocellulose-Bioraffinerie – Von der Idee zur Realisierung. Vortrag anlässlich des Freiburger Winterkolloquiums

Milstein A (1992) Ecological aspects of fish species interactions in polyculture ponds. Hydrobiologia 231:177–186

Naylor RL, Hardy RW, Bureau DP, Chiu A, Elliott M, Farrell AP, Forster I, Gatlin DM, Goldburg RJ, Hua K, Nichols PD (2009) Feeding aquaculture in an era of finite resources. PNAS 106(36):15103–15110

Nehring S, Kowarik I, Rabitsch W, Essl F (2013) Naturschutzfachliche Invasivitätsbewertungen für in Deutschland wildlebende gebietsfremde Gefäßpflanzen. BfN-Skripten 352, Bundesamt für Naturschutz, Bonn; 204 p

Neori A, Nobre AM (2012) Relationship between trophic level and economics in aquaculture. Aquaculture Economics & Management 16:40–67

Nothdurft A, Wolf T, Ringeler A, Böhner J, Saborowski J (2012) Spatio-temporal prediction of site index based on forest inventories and climate change scenarios. Forest Ecology and Management 279:S. 97–S.111

Nova-Institut (2015). Nachhaltig nutzbare Potenziale für Biokraftstoffe in Nutzungs-konkurrenz zur Lebens- und Futtermittelproduktion, Bioenergie sowie zur stofflichen Nutzung in Deutschland, Europa und der Welt; nova-Institut GmbH, Hürth, Deutschland, 268 S. (http://bio-based.eu/markets)

Ott K, Egan-Krieger von T (2012) Normative Grundlagen nachhaltiger Waldbewirtschaftung; Eberswalder Forstliche Schriftenreihe Band 50; S.7–38

Paech N (2016) Befreiung vom Überfluss – Grundlagen einer Wirtschaft ohne Wachstum. Fromm-Forum 2016(20):70–76

Pauly D (2009) Beyond duplicity and ignorance in global fisheries. Scientia Marina 73(2):215–224

Prein M (2002) Integration of aquaculture into crop-animal systems in Asia. Agricultural Systems 71:127–146

Pretty J, Toulmin C, Williams S (2011) Sustainable intensification in African agriculture. International Journal of Agricultural Sustainability 9(1):5–24

Pretzsch H (2013) Räumliche und zeitliche Variation von Mischungseffekten in Abhängigkeit von Umweltbedingungen. Tagungsband der Sektion Ertragskunde im DVFFA, Rychnov nad Kneznou: 44–55

Pretzsch H, Bielak K, Bruchwald A, Dieler J, Dudzinska ME, Jensen HP, Johannsen AM, Kohnle VK, Nagel U, Spellmann J, Zasada H, Zingg M, A. (2013) Mischung und Produktivität von Waldbeständen. Ergebnisse langfristiger ertragskundlicher Versuche. Allgemeine Forst- und Jagdzeitung 184(7/8):177–196

Priefer C, Jörissen J, Bräutigam K-R (2016, 109) Food waste prevention in Europe - A cause-driven approach to identify the most relevant leverage points for action. Resources, Conservation and Recycling:155–165

Radkau J (2007) Holz - Wie ein Naturstoff Geschichte schreibt. Oekom Verlag, München, 341 p

Rantala J, Veli-Matti S, Hallongren H (2010) Quality, productivity and costs of spot mounding after slash and stump removal. Scandinavian Journal of Forest Research 25(6):507–514

Reif A, Brucker U, Kratzer R, Schmiedinger A, Bauhus J (2010) Waldbau und Baumartenwahl in Zeiten des Klimawandels aus Sicht des Naturschutzes. Abschlussbericht eines F+E-Vorhabens im Auftrag des Bundesamtes für Naturschutz FKZ 3508 84 0200; BfNSkripten 272; 128 p

Röder H, Pausch R, Quednau HD (1996) Begriffliche Analyse von Waldbewirtschaftungkriterien. Tagung des Dt. Verbandes Forstl. Forschungsanstalten, Sektion Biometrie und Informatik; 9. Tagung Tharandt / Oybin, 9.-12. September 1996; 10 p

2

Ruthenberg H (1980) Farming Systems in the Tropics, 3rd edn. Clarendon Press, Oxford, UK

Sabaté S, Gracia CA, Sánchez A (2002) Likely effects of climate change on growth of Quercus ilex, Pinus halepensis, Pinus pinaster, Pinus sylvestris and Fagus sylvatica forests in the Mediterranean region. Forest Ecology and Management 162:32–37

Scheftelowitz M et al (2015) Stromerzeugung aus Biomasse (Vorhaben IIa Biomasse) DBFZ Deutsches Biomasseforschungszentrum GmbH,; https://www.dbfz.de/fileadmin/eeg_monitoring/berichte/01_Monitoring_ZB_Mai_2015.pdf. Accessed: 27.1.2019

Schraml U (2014) Von Bären und Menschen. Reparaturbetrieb Kleinprivatwald. ProWald Heft 11/2014; p 10–12

Seintsch B (2015) Waldbewirtschaftung im Spannungsfeld gesellschaftlicher Ansprüche vor dem Hintergrund der Gesetzesentwürfe zum LNatSchG und LWaldG in Schleswig-Holstein. Vortrag im Rahmen des Herbstwaldgesprächs der Schutzgemeinschaft Deutscher Wald Nov. 2015 in Flintbek. PP-Präsentation; Online-Dokument

Seintsch B, Weimar H (2013) Holzbilanzen 2010 bis 2012 für die Bundesrepublik Deutschland. Johann Heinrich von Thünen-Institut, Thünen Working Paper 9. Hamburg

Seré C, Steinfeld H (1996) World livestock production systems. Current status, issues and trends, Food and Agriculture Organization of the United Nations, Rome

Shepherd CJ, Jackson AJ (2013) Global fishmeal and fish-oil supply: inputs, outputs and markets. J Fish Biol 83:1046–1066

Smeets E, Faaij A, Lewandowski J, Turkenburg W (2007) A Bottom-Up Assessment and Review of Global Bioenergy Potentials to 2050. Energy and Combustion Science 33:56–106

Sommerauer M (2006) Auch die mitteleuropäische Forst- und Holzwirtschaft muss sich dem globalen Markt stellen. Forstzeitung; österreichischer Agrarverlag Wien

Statista (2019a) Statistiken zur Entsorgungswirtschaft, https://de.statista.com/themen/741/abfallwirtschaft-recycling/ Accessed: 16.1.2019

Statista (2019b): Menge des in Behandlungsanlagen eingesetzten Biomülls in Deutschland in den Jahren 2007 bis 2016 (in 1.000 Tonnen) https://de.statista.com/statistik/daten/studie/257365/umfrage/biomuelleinsatz-in-behandlungsanlagen-in-deutschland/ Accessed: 27.1.2019

Sturm K (1995) Naturnahe Waldnutzung in Mitteleuropa, In: Der Dauerwald - Zeitschrift für naturgemäße Waldwirtschaft, August Heft, p 6–21

Tacon AGJ (1988) The nutrition and feeding of farmed fish and shrimp - a training manual. 3. Feeding methods. FAO Field Document, Project GCP/RLA/075/ITA. Field Document 7/E, FAO, Brasilia, Brazil.

Tacon AGJ, Metian M (2008) Global overview on the use of fish meal and fish oil in industrially compounded aquafeeds: Trends and future prospects. Aquaculture 285(1-4):146–158

Tacon AGJ, Metian M, Turchini GM, de Silva SS (2010) Responsible aquaculture and trophic level implications to global fish supply. Reviews in Fisheries Science 18(1):94–105

Unseld R, Hesse A, Fei J, Wessels W, Bauhus J (2016) Entwicklung von Vorwäldern bei verschiedenen Begründungsverfahren. AFZ - Der Wald 5/2016; p 43–47

Unseld R, Bauhus J (2012; ed.) Energie-Vorwälder - Alternative Bewirtschaftungsformen zur Steigerung der energetisch nutzbaren Biomasse im Wald durch Integration von schnell wachsenden Baumarten. Schriftenreihe Freiburger Forstliche Forschung; Band 91; 217 p

Unseld R, Reppin N, Eckstein K, Zehlius-Eckert W, Hoffmann H, Huber T (2011) Leitfaden Agroforstsysteme. Möglichkeiten zur naturschutzgerechten Etablierung von Agroforstsystemen. Bundesamt für Naturschutz (Hrsg.) 43 p

Viola S (1989) Production of commercial feeds for warm water fish. In: Shilo M, Sarig S (eds) Fish Culture in Warm Water Systems: Problems and Trends. CRC Press Inc., Boca Raton, pp 143–162

Vor T, Spellmann H, Bolte A, Ammer C (2015; Hrsg.) Potenziale und Risiken eingeführter Baumarten. Baumartenportraits mit naturschutzfachlicher Bewertung. Göttinger Forstwissenschaften, Band 7; Universitätsverlag, Göttingen

Wegener G, Zimmer B (2001) Holz als Rohstoff. In: Landeszentrale für politische Bildung Baden-Württemberg (Hrsg.); Der Bürger im Staat 51, 1, p 67–72

Weidenbach P (2001) Waldbauliche Ziele im Wandel. Wirtschaftliche, soziale und kulturelle Rahmenbedingungen der Waldentwicklung seit 1800. Der Bürger im Staat.51. Jhg. Heft1; p 30–E38

Weimar H (2014) Holzbilanzen 2012 bis 2013 für die Bundesrepublik Deutschland. Johann Heinrich von Thünen-Institut, Thünen Working Paper 31. Hamburg; 37 p

Weiß JW, Pabst W, Granz S (eds) (2011) Tierproduktion. Lehmanns Verlag, Berlin

Wikipedia (2016a) https://de.wikipedia.org/wiki/G%C3%BClle; Accessed: 27.1.2019

Wikipedia (2016b) https://de.wikipedia.org/wiki/S%C3%A4genebenprodukt; Accessed: 27.1.2019

World Bank (2013) FISH TO 2030: Prospects for Fisheries and Aquaculture. Agriculture and Environmental Service Discussion Paper 03, World Bank Report. 83177-GLB

Zhang Y, Chen HYH, Reich PB (2012) Forest productivity increases with evenness, species richness and trait variation: a global meta-analysis. J Ecol 100:742–749

Food Security and Healthy Nutrition in the Context of the Bioeconomy

Ulrich Schurr

At the weekly market in Santiago de Chile. (Photo André Künzelmann, Helmholtz Centre for Environmental Research – UFZ)

J. Pietzsch (ed.), *Bioeconomy for Beginners*, https://doi.org/10.1007/978-3-662-60390-1_3

"Food first" has become a globally accepted rule in bioeconomic research and practice. This has not always been the case, and the way, how this shall be implemented, is subject to debates. One milestone in the global discussion has certainly been the Communiqué of the *Global Bioeconomy Summit* in 2015, which recognized the important role of the bioeconomy in sustainable development (GBS 2015). The Communiqué derives the relevance of the bioeconomy from the fact that it is an instrument for achieving the sustainable development goals of the United Nations (United Nations 2015). The key driver behind all of these challenges is the massive increase in the world's population in the past decades, and the forecast that 10 to 12 billion people will populate our planet in 2050 (► Chap. 1). First and foremost, this requires the provision of sufficient and safe food for the future.

Food security, however, cannot be achieved through the availability of sufficient biomass to feed the population alone. Rather, the *World Food Summit* 2009 defined that "food security exists when all people, at all times, have physical, social and economic access to sufficient, safe and nutritious food to meet their dietary needs and food preferences for an active and healthy life" (FAO 2009). In this definition, it is very vividly stated that food security does not depend solely on pure primary production, but encompasses many health, economic and social aspects. Therefore, food security calls for systemic solutions. At the same time, this definition already indicates lines of rupture and conflict. For example, ethical questions need to be asked as to how resource-intensive food habits can be accepted and, if necessary, reconciled when food resources are scarce. With this complexity in mind, this chapter does not claim to discuss the topic of nutrition comprehensively. However, it aims at highlighting some essential elements of the topic of nutrition within the framework of a sustainable bioeconomy. Beyond this, we would like to encourage further reflection and reading on this fascinating topic.

3.1 Forms and Consequences of Malnutrition

The diversity of the issues related to food security and malnutrition is highlighted in the annual *Global Nutrition Report*, written by an independent group of international experts with the support of the *International Food Policy Research Institute* (IFPRI) (2016) (Capacci et al. 2013). In general, three forms of malnutrition can be distinguished:

- hunger due to insufficient intake of calories,
- hidden hunger for essential nutrients,
- overeating and obesity.

The Global Nutrition Report 2016 shows that, currently, around 800 million people worldwide suffer from hunger due to low calorie intake, 2 billion people suffer from deficiency symptoms due to a lack of vitamins and micronutrients, and 1.9 billion people are overweight. "Hidden hunger," especially in children in the growth phase, leads to damage with lifelong consequences. In 2016, there were 161 million children worldwide who were chronically malnourished and, at the same time, with a rising trend, 42 million children who suffer from obesity (◘ Fig. 3.1).

The global distribution of malnutrition is alarming: it is still closely linked to poverty. Countries such as India and many African countries, where most of the world's population growth will take place in the coming decades, are particularly affected. At the same time, obesity is undergoing a massive increase in all countries of the world. Therefore, in some cases, all three categories of malnutrition occur simultaneously in the same countries (interactive graphs on all of the countries surveyed can be found online in the Global Nutrition Report (IFPRI)).

3.1.1 Malnutrition

The global community has already set the goal of ending hunger in the past. However, this subgoal of the so-called *Millennium Goals*, according to which the number of hungry people should be halved by 2015 (target year) relative to 1991 (reference year), has not been achieved (◘ Fig. 3.2). Overall, the number of malnourished people has decreased significantly, though 300 million more people remained malnourished than in the target. This is mainly due to the 2 billion increase in the overall number of people on the planet between 1991 and 2015. The measures to reduce hunger could not keep pace with this increase in the world population (FAO 2015). There are also considerable regional differences in the fight against hunger. While a significant decrease in the absolute amount of malnutrition was observed in Asia and Latin America, the total amount increased by 50 million against the trend, especially in Africa (FAO 2015).

This clearly shows the connection between economic development and malnutrition. It is particularly recommended that the more interested reader consults Gap Minder's interactive statistics. This tool also reveals the dynamics with which the current situation has developed historically (Gap Minder) (◘ Fig. 3.3).

3.1.2 Insufficient Supply of Micronutrients and Vitamins

Insufficient supply of micronutrients and vitamins currently affects almost 2 billion people, more than a quarter of the human population on the planet. The causes are manifold, and often local or regional. While hidden hunger is always accompanied by an undersupply of micronutrients and vitamins, additional deficiencies of zinc, iron, iodine, selenium

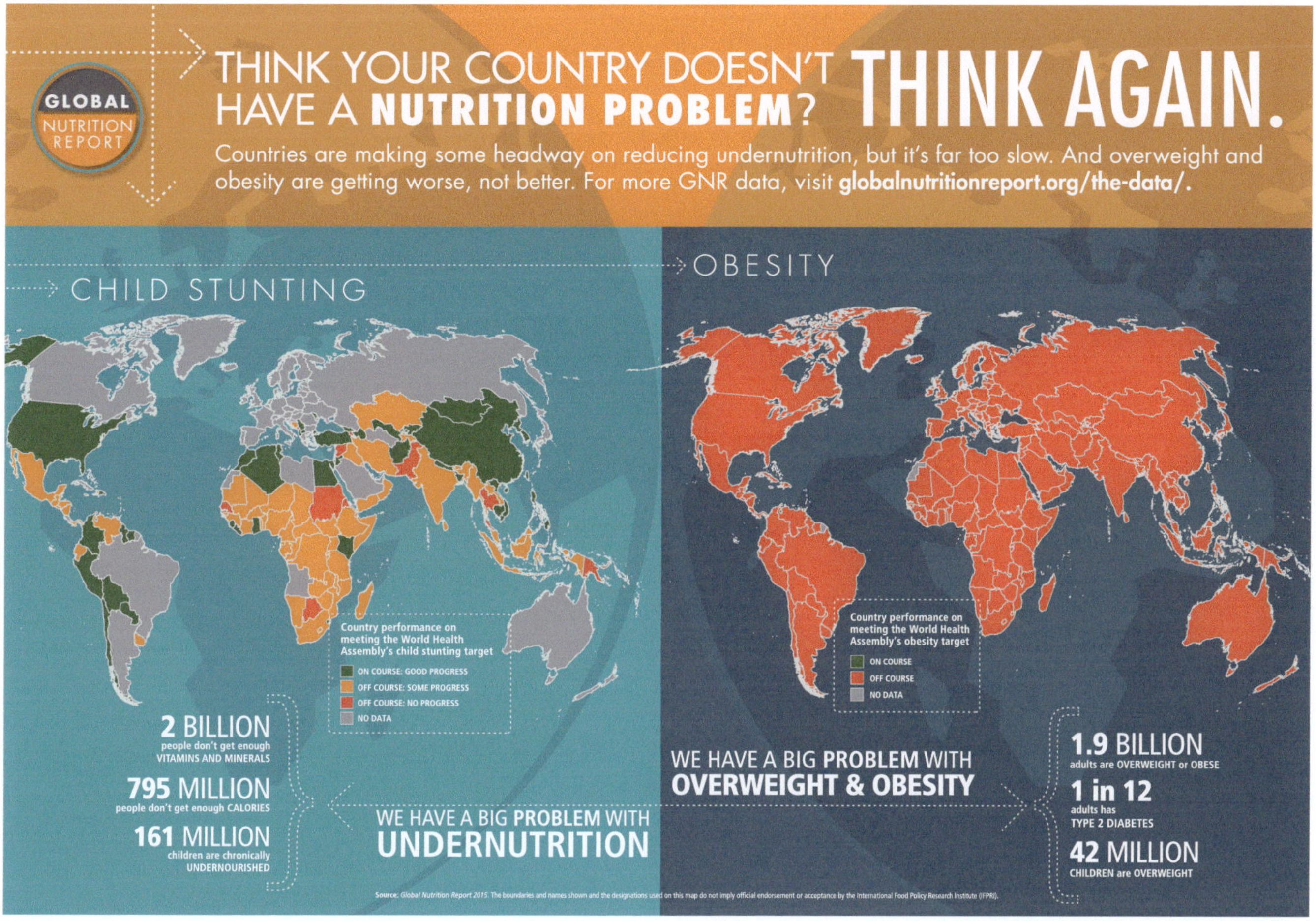

Fig. 3.1 Core statements of the Global Nutrition Report. (International Food Policy Research Institute (2015), ► http://globalnutritionreport.org/)

and, most especially, vitamin A originate from the typical composition of much human nutrition. Common diseases, diarrhoea in particular, also reduce the effective availability of essential nutrients - with drastic consequences for growing children. Here, too, the hotspots are in Africa and Southeast Asia (Muthayya et al. 2013). In micronutrient malnutrition, anaemia and reduced growth are the main symptoms, while vitamin A deficiency weakens the immune system and, in severe cases, leads to blindness and malformations of the skull, skeleton and other parts of the body, with lifelong consequences. In 2010, the deaths of ten million children were attributed to vitamin A deficiency.

Excursus 3.1 The *Golden Rice* Project

Vitamin A deficiency is particularly prevalent in those regions of the world where rice is a major part of the diet. The concentration of vitamin A in the rice grain is very low, because the rice grain lacks the metabolic ability to form the precursor of vitamin A. As a manner of addressing this, projects have been pursued since 1990 that seek to use transgenic approaches to giving rice the ability to produce vitamin A precursors in the grain. This was achieved in 2000 with the introduction of two additional genes from plants and bacteria (Ye et al. 2000). This resulted in rice grains that produce provitamin A. This property has also been transferred to varieties that are mainly used for nutrition in countries with high vitamin A deficiency. Further genetic engineering measures have increased the carotenoid content of rice to such an extent that it could make a significant contribution to the supply of vitamin A, as has been demonstrated in field tests. Parallel to this, patent rights, including precursor patents, were regulated in such a way that breeders from developing countries had free access to this technology. In spite of this success story, there has not, as of yet, been a *Golden Rice* released onto the market, because objections and legal proceedings have so far prevented its approval. As a result, in July 2016, 110 Nobel Laureates and several thousand scientists called upon non-governmental organisations (in particular, Greenpeace) to revisit their position towards the *Golden Rice*.

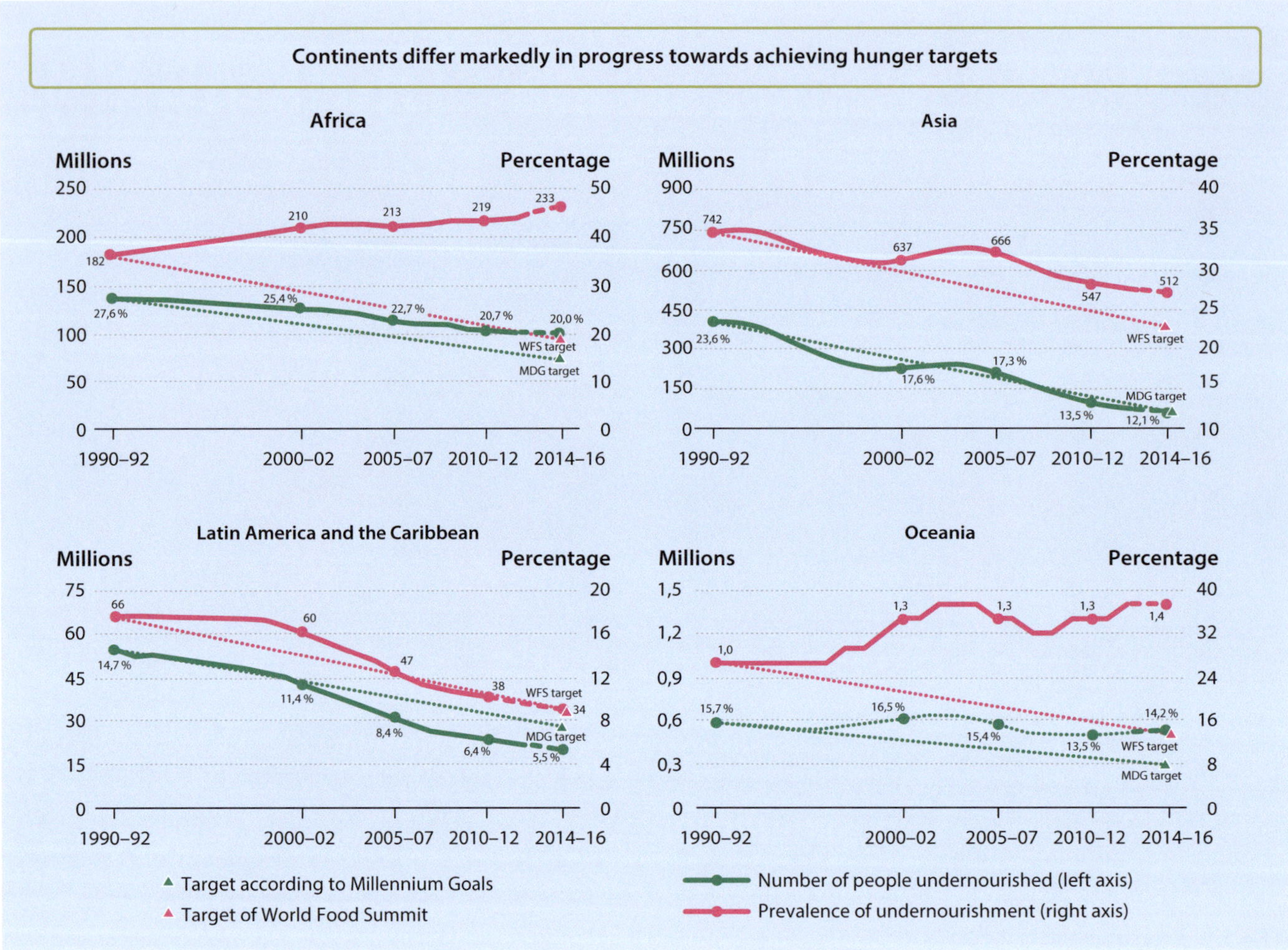

Fig. 3.2 The status of food insecurity in the world. (FAO 2015)

3.1.3 Obesity

According to research by the World Health Organization (WHO), the occurrence of obesity has more than doubled since 1980. In 2014, more than 1.9 billion adults were overweight (Body Mass Index, BMI > 25), of which more than 600 million were obese (BMI < 30) (Fig. 3.4). According to this study, 41 million children under the age of five were already obese. Whereas obesity used to be a phenomenon almost solely confined to industrialised countries, it is now increasingly being observed in emerging and developing countries. This applies not only to adults, but to children as well, in greater numbers, in fact. The number of overweight and obese children in Africa almost doubled between 1990 and 2014. Overeating is mainly attributed to changes in dietary and lifestyle habits. In these cases, excessive calorie intake as a result of strongly increased availability of fats and carbohydrates on the one hand and a way of life characterized by low physical activity on the other hand both boost this syndrome.

The main consequences of obesity are cardiovascular diseases, diabetes and osteoarthritis. Some forms of cancer are also correlated with overeating. According to WHO, overeating and obesity now account for more deaths than malnutrition worldwide, with the exception of sub-Saharan Africa. To combat nutritional deficiencies and their consequences, the WHO initiated the "Global Strategy on Nutrition, Physical Activity and Health" in 2003, issuing a number of recommendations. In these publications, WHO emphasises the need for each individual to adopt personal responsibility, a message sent in conjunction with the food industry and political organizations. Food deficiencies are also strongly rooted in individual behaviour, cultural conditions, eating and exercise habits, food supply (and its composition and provision, inter alia, by the food industry) and political guidelines (regulation). Having published the *Global Action Plan for Prevention and Control of Non-Communicable Disease* in 2013 (WHO 2013), both national and global targets have been designated to reduce diseases associated with poor nutrition. This has led to numerous policy documents and decisions addressing nutrition and physical activity, including the WHO European Region Strategy for Physical Activity (2016–2025), adopted in 2015, also by the European Health Ministers.

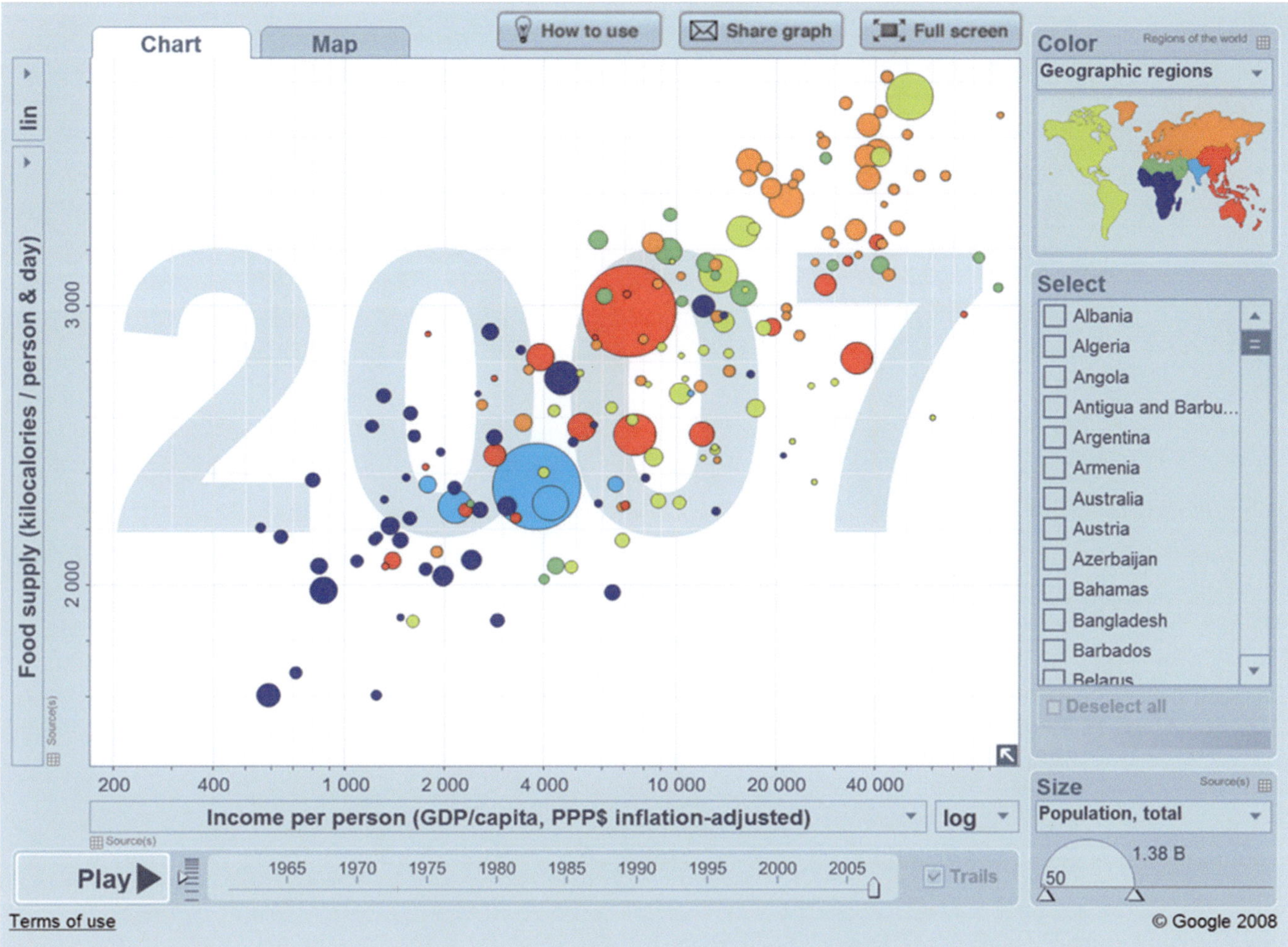

Fig. 3.3 Relationship between the development of malnutrition and personal gross national product. (Free material from ▶ gapminder.org)

3.2 Supply of Food and Food Losses

One of the central goals of the German government's bioeconomy strategy is "... a sustainable bio-based economy oriented towards the natural cycle of materials, to supply the food that feeds the world adequately and healthily …" (BMBF 2010). In line with the particularly important topic of nutrition, three fields of action for research work have been outlined, with the objectives "Secure global nutrition," "Design sustainable agricultural production" and "Produce healthy and safe food" (German Research Strategy Bioeconomy 2030). The role of agricultural production in the bioeconomy has already been generally discussed in ▶ Chap. 2. Three aspects that influence and describe the nutrition system in a special way will be taken up again here.

3.2.1 Food Production

The expected significant increase in the world population in the coming decades calls for a substantial expansion of food production. Due to said sharp increase, it will be necessary to produce as much biomass for food in the next 50 years as has been produced in the entire history of mankind. This would make it necessary to double the yield per year for the largest crop species (Tilman et al. 2011). All forecasts assume a sharp increase in the number of calories required (Bodirsky et al. 2015). This massive challenge is countered by the current trend of a slowdown in the annual yield of large crop species (Ray et al. 2013) (◘ Fig. 3.5). Without substantial changes in the rate by which yield is increased through breeding and sustainable crop management, significant consequences for the environment with additional forest clearance, a sharp rise in the amount of greenhouse gas production as a result of agricultural use and even greater use of fresh water have to be expected (Tilman et al. 2011).

In parallel with the increase in the population, the availability of food per capita (Bodirsky et al. 2015) needs to be improved. However, this does not only relate to the amount of energy (calories), but also to the protein requirement, since both are closely linked to an increase in prosperity (Tilman et al. 2011). The greater consumption of meat and fish that comes along with increasing income is reflected in a stronger increase of the protein requirement compared to the

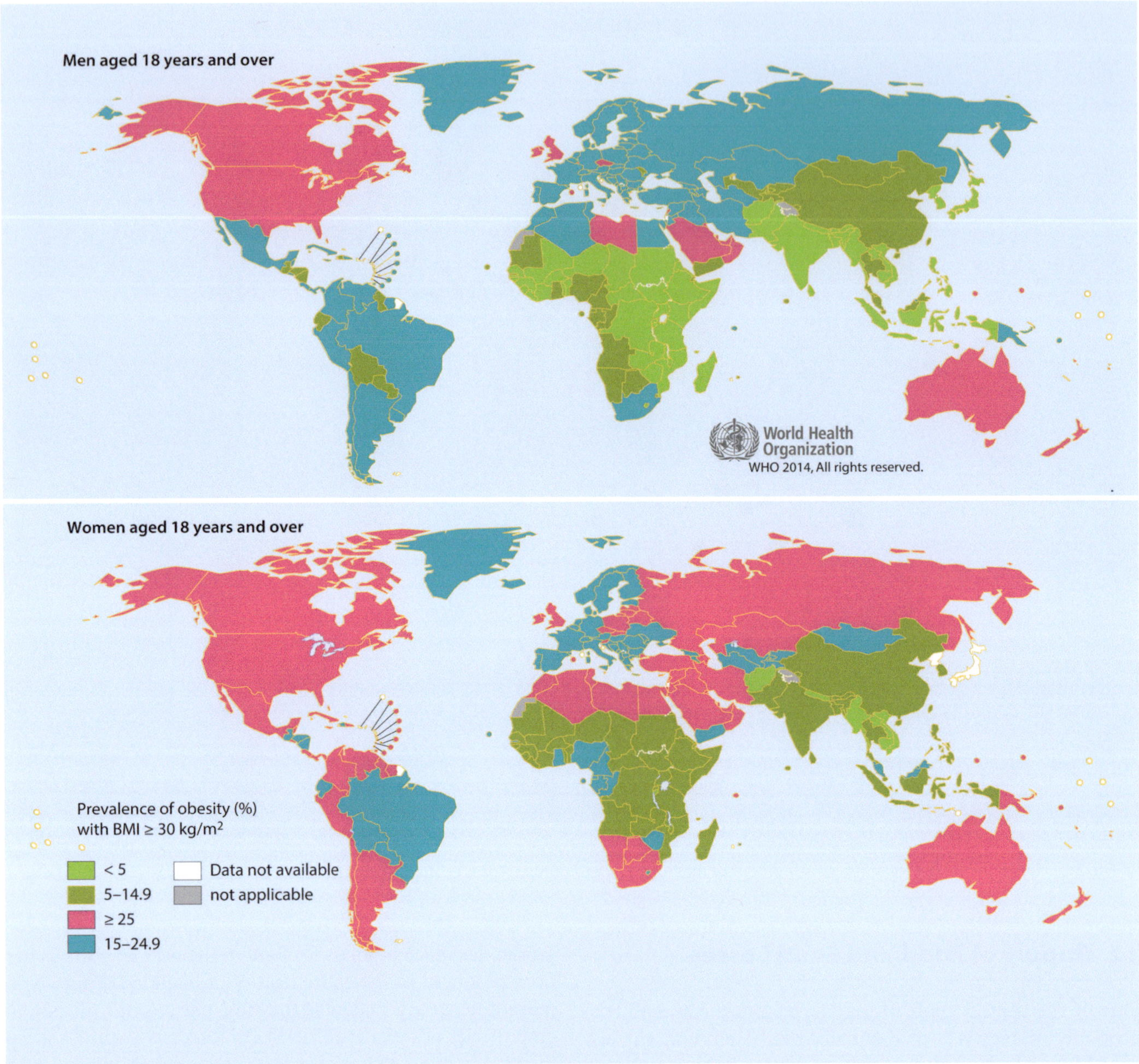

Fig. 3.4 Frequency of obesity in men and women by world region. (WHO 2013)

calorie requirement. Healthy nutrition is, however, closely linked to the increased consumption of vegetables and fruit. Yet, the production and trade of horticultural products is fundamentally different from that of large *cash crops*. The variety of vegetables and fruits is much greater than that of cereals, which are the primary source of carbohydrates. In Germany alone, about 40 types of vegetable are considered to be so relevant that they are statistically recorded by the agricultural offices (BMELV 2013). They were cultivated in 2013 on a total area of more than 110,000 ha; 3.2 million tonnes of vegetables were harvested. The number of fruit species is in a similar range. Global diversity is even greater, as fruits and vegetables are also subject to a high degree of cultural diversity. Vegetables and fruits are mostly sold as fresh goods, while corn, rice, wheat and soybeans can be harvested and traded in dry form. Thus, the supply of healthy horticultural products is either associated with much more complex storage and transport processes – or it is represented by perishable goods that cannot be transported far. In the global perspective, this means that research, agribusiness, horticulture and trade would have to deal with a much larger variety of products to provide healthy food beyond the mere supply of carbohydrates. This is one of the major challenges for a healthy diet for the world's population in the future.

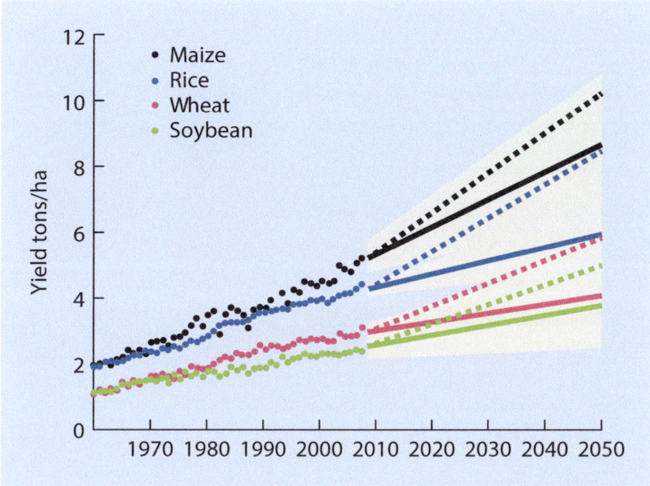

Fig. 3.5 Extrapolation of the global yield. Harvest of the four major crops - maize, rice, wheat and soya beans - from 1961 to 2008 (points). The development expected up to 2050 on this basis is depicted by solid lines. The dotted lines correspond to the growth rate of approximately 2.4% that would be necessary to double the yield of these crops without having to cultivate new land or develop new areas. (Ray et al. 2013)

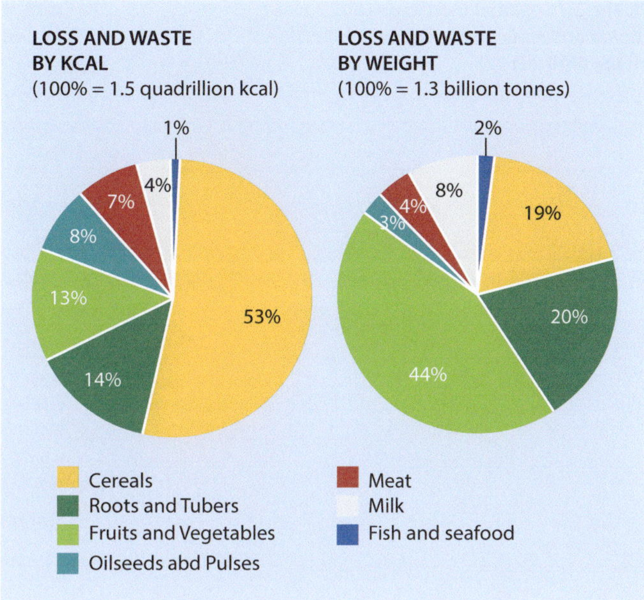

Fig. 3.6 Global food losses by type of food. (WRI 2013)

3.2.2 Animal Feed, Efficiency of the Use of Resources and Meat Consumption

Today, more than 50% of the plant biomass produced in agriculture is used for animal feed (▶ Chap. 2; FAO 2012). In particular, the production of protein is of central importance. The world's largest source of protein for animal feed is soybean, of which about 270 million tons were harvested in 2012. About 75% of this was used for the production of animal feed. Livestock species differ considerably in regard to the efficiency with which they convert feed into biomass. The feed utilisation coefficient (▶ Chap. 2) varies from about 6.8 kg of feed per kg of meat in cattle to 1.1 kg of feed per kg in fish. The far greater degree of efficiency in fish is - in addition to their health-promoting fatty acid profile - one of the most important arguments for the significant expansion of aquaculture worldwide. However, it must also be borne in mind that unsustainably sourced fish meal is predominantly used as animal feed today.

Meat and milk consumption will be a very significant factor in regard to the increase in biomass demand in the future. In recent years, milk consumption in developing and emerging countries has risen by an average of 3.4–3.8% per year, and meat consumption by 5–6%. Nevertheless, the majority of meat and milk production (37% and 40%, respectively) is still concentrated in the industrialised countries (FAO 2012). In those countries, annual meat consumption in 2005 was around 80 kg per capita, while, in the developing and emerging countries, it was around 28 to 30 kg per year. The extent to which developing and newly industrializing countries with high population growth adapt to Western food habits, and whether these will change, will be a major influencing factor in the future demand for animal feed.

3.2.3 Biomass and Food Losses

Beyond the (additional) production, it will be important to avoid a loss of biomass (▶ Sect. 2.4). The reasons for food loss and the quantity lost depend on many factors: Following an analysis of the *World Resource Institute* (WRI 2013), the loss of cereals is highest in terms of calories, at more than 50%, while, e.g., only 7% of meat-based foods end up not being used. In terms of quantity, fruits and vegetables show the highest loss rate, with more than 40% (Fig. 3.6).

Different regions of the world differ drastically in the quantity that is lost: While, in North America, on average, 1520 kcal per capita and day are not used, in Europe and the industrialized regions of Asia, that number is about 750 kcal per capita and day and, in Latin America and Southeast Asia, it is less than 450. In the indivual parts of the value chain from production to the customer, food losses vary in magnitude. In terms of the value chain, industrialised countries and developing countries also differ greatly. While, in developing countries, the largest losses occur in production and storage, the largest loss in industrialized countries takes place in the hands of the consumers (Fig. 3.7).

3.3 Food Insecurity – A Multi-Faceted Syndrome

To ensure food safety, much more must be done than "just" increasing production. This alone poses considerable challenges for agriculture, breeders, the agricultural industry and researchers. At the same time, however, the entire world food system must be adapted - in order to improve yields, conserve natural resources, minimise losses and ultimately pro-

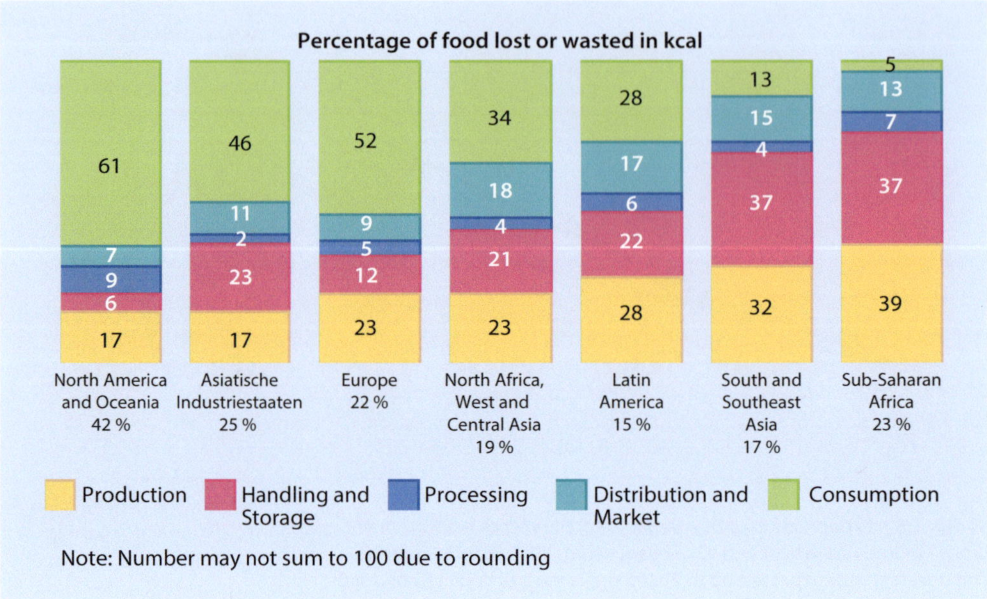

Fig. 3.7 Global food losses by region and stage in the value chain. (WRI 2013)

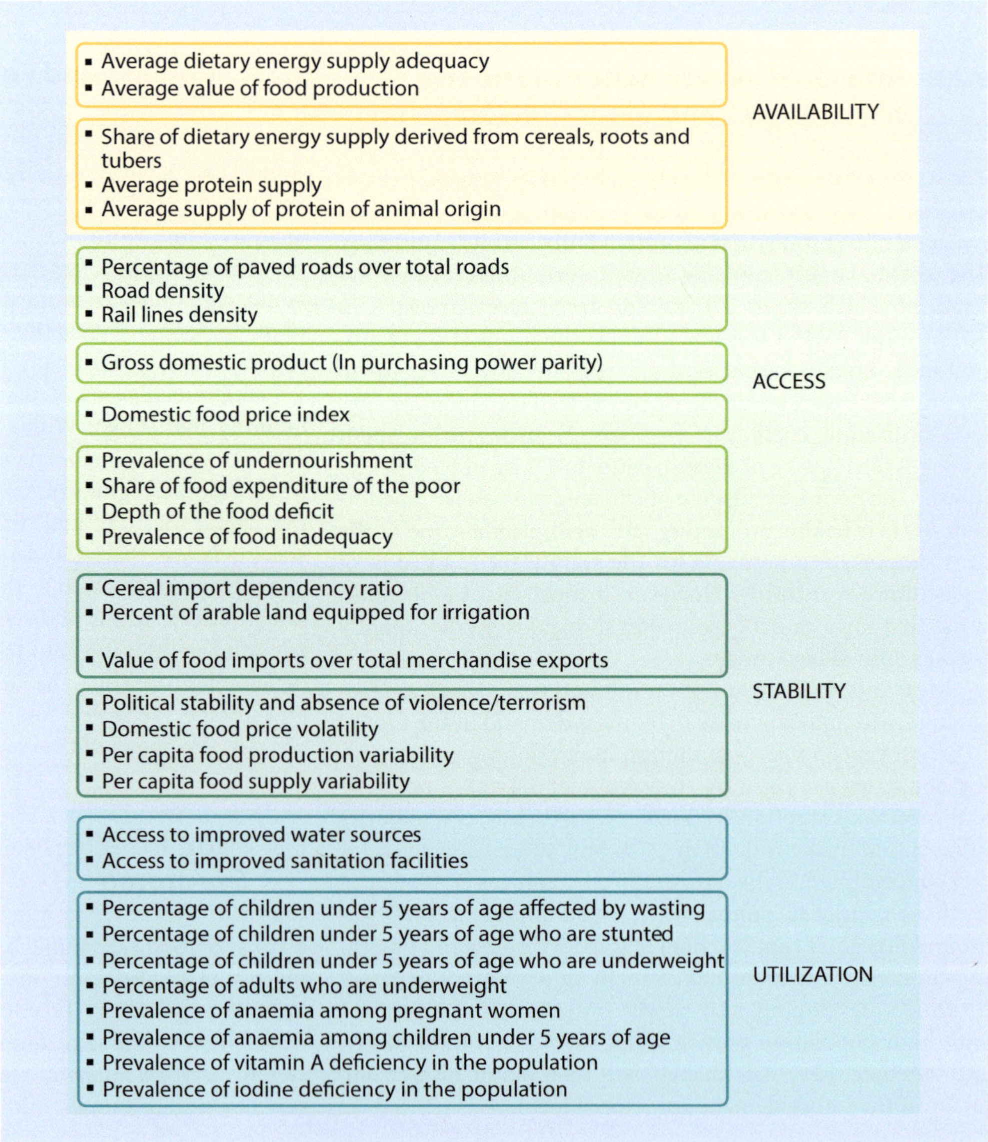

Fig. 3.8 Indicators and dimensions of food security. (WRI 2013)

vide a sufficient amount of healthy food. To achieve the full definition of "food security," the "four dimensions of food security" must be addressed: availability, access, stability and *utilization* (■ Fig. 3.8).

Even if enough food biomass is produced globally, there must also be proper trade conditions, for example, to transport the required quantities to the markets. This is where food production meets logistics, which make a major contribution to availability both regionally and globally. For example, fresh fruits and vegetables can only be delivered to consumers if there are options for refrigerated transport. On the other hand, large quantities of cereals can only be handled cost-effectively by ship.

References

BMBF (2010) Nationale Forschungsstrategie BioÖkonomie 2030. http://www.bmbf.de/pub/nationale_Forschungsstrategie_Biooekonomie_2030.pdf. Accessed 25.09.2016

BMELV (2013). http://www.bmel.de/SharedDocs/Downloads/Broschueren/Der-Gartenbau-in-Deutschland.pdf?__blob=publicationFile

Bodirsky BL, Rolinski S, Biewald A, Weindl I, Popp A, Lotze-Campen H (2015) Global food demand scenarios for the 21st century. PlosONe 10(11). https://doi.org/10.1371/journal.pone.0139201

Capacci S, Mazzocchi M, Shankar B, Traill B (2013) The triple burden of malnutrition in Europe and Central Asia: a multivariante analysis

Communiqué of the Global Bioeconomy Summit (2015). Making Bioeconomy Work for Sustainable Development (2015); http://gbs2015.com/fileadmin/gbs2015/Downloads/Communique_final.pdf. Accessed 25.09.2016

FAO (2009) World Summit on Food Security – Declaration

FAO (2012) World agriculture: towards 2015/2030 – The 2012 Revision. http://www.fao.org/docrep/016/ap106e/ap106e.pdf. Accessed 25.09.2016

FAO (2015) The State of Food Insecurity in the World. FAO, Rome

International Food Policy Research Institute (2015) Global Nutrition Report 2015

International Food Policy Research Institute (2016) Global Nutrition Report 2016

Muthayya S, Rah JH, Sugimoto JD, Roos FF, Krämer K et al (2013) The global hidden hunger indices and maps: an advocacy for action. PLoS One 8(6). https://doi.org/10.1371/journal.pone.0067860

Ray DK, Mueller ND, West PC, Foley JA (2013) Yield trends are insufficient to double global crop production by 2050. PLoS One 8. https://doi.org/10.1371/journal.pone.0066428

Tilman D, Balzer C, Hill J, Befort B (2011) Global food demand and the sustainable intensification of agriculture. Proc Natl Acad Sci USA 108:20260–20264

United Nations, Sustainable Development Goals (2015). http://www.un.org/sustainabledevelopment/sustainable-development-goals/. Accessed: 25.09.2016

WHO (2013) Global Status Report on noncommunicable diseases 2014. http://www.who.int/nmh/events/ncd_action_plan/en

WRI (2013) Reducing Food loss and waste http://www.unep.org/pdf/WRI-reducing_food_loss_and_waste.pdf. Accessed 25.09.2016

Ye X, Al-Babili S, Klöti A, Zhang J, Lcca P, Beyer P, Potrykus I (2000) Engineering the Provitamin A (beta-Carotine) biosynthetic pathway into (caroteinoid-free) rice endosperm. Science 287:303–305

The Use of Biomass for the Production of Fuel and Chemicals

Jochen Michels

Demonstration plant for the production of 2G-ethanol based on straw in Straubing, Bavaria. The Clariant Produkte Deutschland GmbH plant has a production capacity of 1000 t ethanol per year. (© Clariant)

J. Pietzsch (ed.), *Bioeconomy for Beginners*, https://doi.org/10.1007/978-3-662-60390-1_4

4.1 The Current Raw Material Base of the Chemical Industry

Today, basic chemical building blocks are obtained almost exclusively from crude oil. In petrochemical refineries, crude oil is split into different fractions, most of which are further processed into fuels (kerosene, petrol, diesel) or fuel oil (heavy and light). Approximately 10% of the quantity of refined crude oil is naphtha, which is also known as "light petrol." Naphtha is used in subsequent processes to produce synthesis building blocks and platform chemicals for industry.

Naphtha is a fixed co-product of petroleum processing. Its price and that of its derivatives thus depend directly on the price of oil. This price is very volatile and is determined by political and economic events that are difficult to predict (◘ Fig. 4.1). This is also linked to the uncertainty about whether the demand for oil is already exceeding oil production due to continuously increasing global energy consumption *(peak oil)* or whether the demand for oil is rather falling due to economic crises, and therefore too much oil is being produced.

In fact, according to the Federal Institute for Geosciences and Natural Resources in Germany (BGR), the static range (the predicted range of reserves at constant consumption) of oil reserves since 1945 has always been within a corridor of between 20 and 50 years. The reason for this is that new deposits have been discovered time and again, and their exploitation has been more efficient technically. In the past, the volume of oil reserves could thus be increased again and again despite massive increases in production (BGR 2012). It can be assumed that this range estimate will last for many years to come.

For some years now, oil production from dense rock has been economically viable through the combination of horizontal drilling and hydraulic fracking. The fracking boom began in 2014 in the USA, where, due to fracking, more crude oil is currently produced than is consumed. In 2015, the USA even replaced Saudi Arabia as the world's largest oil producer, which led, both directly and indirectly, to an oversupply of crude oil on the world market, with a corresponding drop in prices. On the supply side, this was further strengthened by Iran's return to the world market as a major oil producer, and, on the demand side, by the economic slowdown in China and the other BRIC states' slide into recession. The price of crude oil was quoted at below USD 40 per barrel at the beginning of 2016. It is likely to remain low for some time to come, unless crises, natural disasters or other events revive speculation.

The chemical industry can therefore expect its most important raw material, naphtha, to remain available for decades to come, but without knowing at what price. In view of this uncertainty, industry is looking for ways to broaden its carbon base by using other raw materials to replace naphtha,

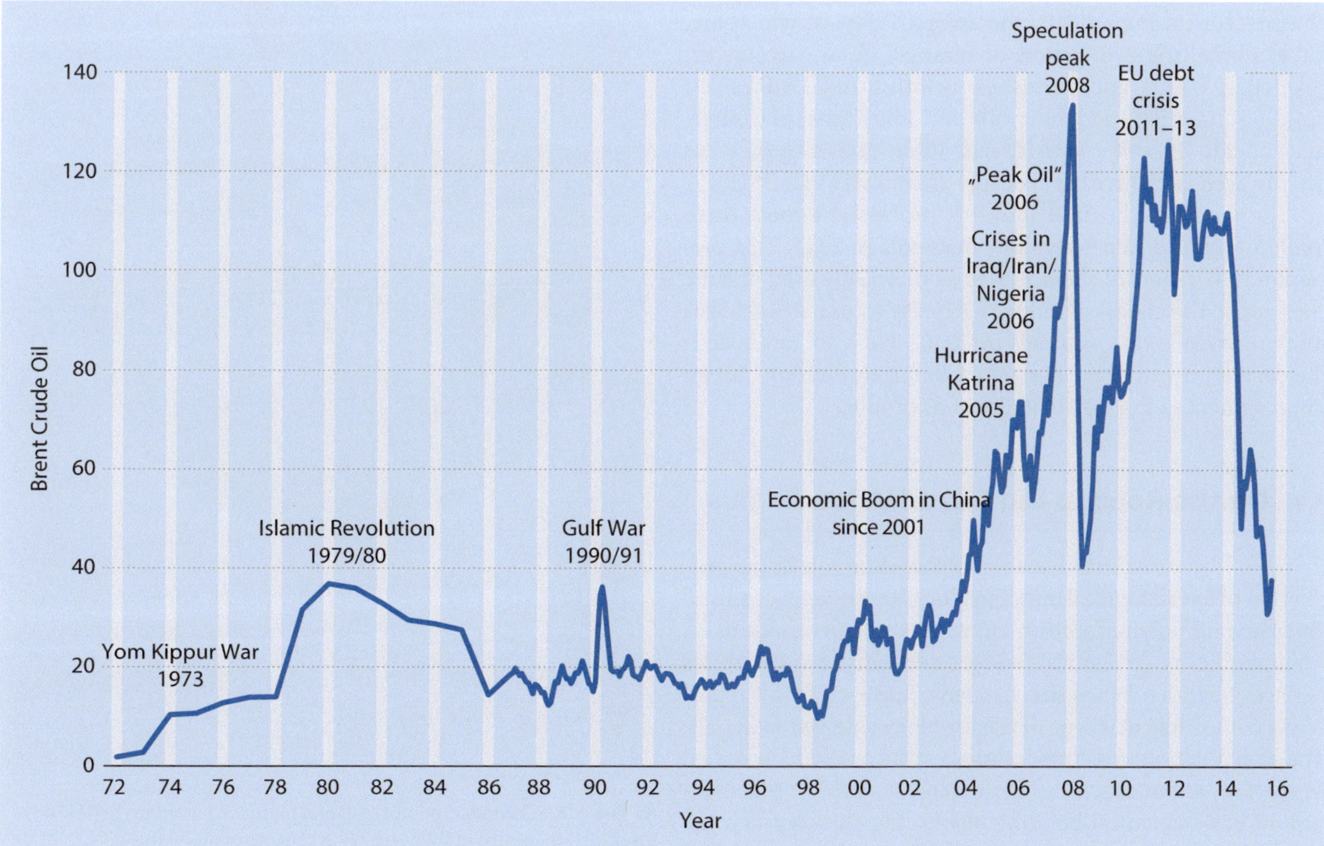

◘ **Fig. 4.1** Historical oil price development (not adjusted for purchasing power) of the North Sea variety of Brent crude in the context of political and economic events. (1972–1986, annual values from BP 2016; from 1987, average monthly values from EIA 2016)

4

at least in part. These alternatives may also include the raw fossil materials coal and natural gas. However, it is not sustainable in the long term to continue using raw fossil material sources. Against the background of the climate agreement of the 2015 UN Conference on the Environment in Paris, which provides for limiting global warming (in relation to the pre-industrial age) to well below 2 °C, or if possible, even 1.5 °C, net greenhouse gas emissions must be reduced to zero by the middle of the century (UNFCCC 2016a). At present, however, these emissions still amount to more than 37 billion tonnes of CO_2-equivalents annually (UNFCCC 2016b), whereby the emission peak has not yet been reached. It is therefore predictable that the price of emission certificates will have to rise significantly through political intervention, which will increase the pressure to develop alternative sources of raw materials that cause no or less greenhouse gas emissions. Renewable raw materials are an attractive alternative in particular, because they are the only renewable carbon source for material use in the chemical industry (VCI 2015).

Renewable raw materials obtained from biomass have always been the basis for many of the chemical industry's products. The reasons for this are not primarily ecological; rather, renewable raw materials have technical and economic advantages over raw fossil materials in certain areas. One of their important advantages, for example, is that nature has already generated an intermediate synthesis and produced complex chemical components in them. The main products concerned are starch, cellulose, sugar, oils and fats, and active pharmaceutical ingredients, the assorted uses of which are, for example, the production of plastics, fibres, detergents, cosmetics, paints and varnishes, printing inks, adhesives, building materials, hydraulic oils and lubricants, and pharmaceuticals. In other words, renewable raw materials are mainly used in the area of specialty chemicals (VCI 2013).

The German chemical industry processed almost three million tonnes of renewable raw materials in 2011. This corresponds to a share of about 13% of all materially used organic chemicals. The chemical industry expects to use at least 50% more renewable raw materials by 2030 (ibid.). Its rationale is the overall growth of the Specialty Chemicals division, not an improvement in the sustainability of its products.

4.2 On the Road to Bio-Based Value Chains

Renewable raw materials have so far played no role in the production of basic chemical building blocks in the petrochemical and standard polymer sectors. However, this will change if, for example, bio-based plastics are in greater demand and biomass replaces naphtha. When this happens, however, it will represent a completely new type of use for renewable raw materials. The aforementioned natural "advance synthesis,"which is used in specialty chemicals, is not in demand at all in the manufacture of basic chemical building blocks. On the contrary, the highly functionalised building blocks provided by nature first have to be defunctionalised as far as possible in order to be used as bio-based platform chemicals in the established family trees of chemical production. This is necessary due to the fact that the overwhelming majority of terrestrial biomass consists of polymers.

4.2.1 The Chemical Classification of Renewable Raw Materials

The main components of terrestrial biomass are cellulose, hemicelluloses, lignin, lipids, proteins, sugars and starch – with more than 90% of this biomass being formed by cellulose, hemicelluloses and lignin, and thus from living or dead plant material (Fig. 4.2). Plant cell walls are formed out of a scaffold of cellulose and hemicelluloses. During the lignification process, lignin is additionally incorporated into the framework. This composite of cellulose, hemicelluloses and lignin is also known as lignocellulose.

However, only the remaining 10% of the total terrestrial biomass can potentially be used as food. From this ratio, it can be deduced that the biggest potential in terms of volume for the use of biomass in the production of fuels and chemicals

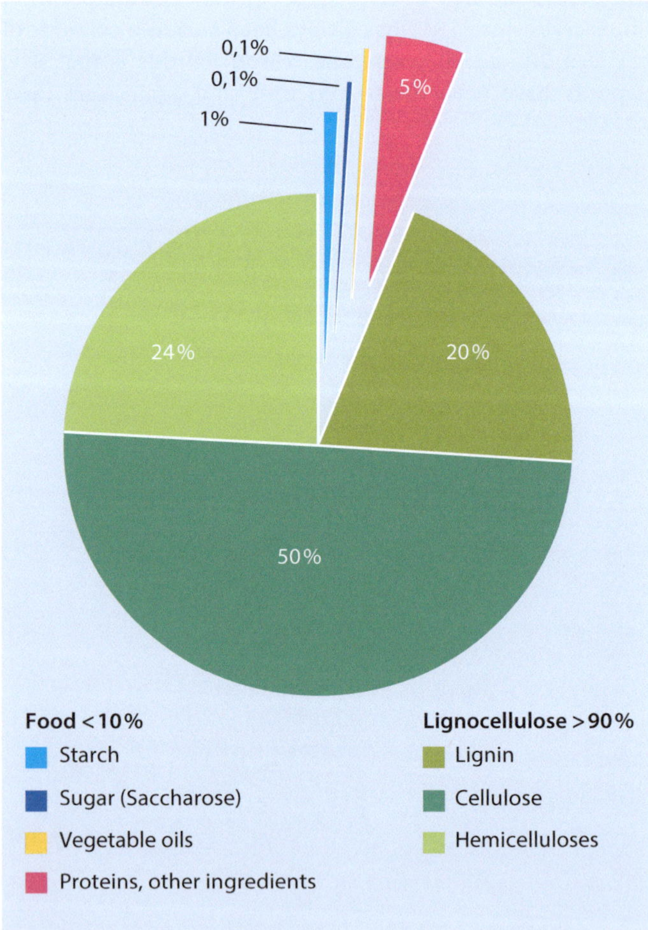

Fig. 4.2 Composition of terrestrial biomass according to GDCh Division Environmental Chemistry and Ecotoxicology (modified). It includes both vegetable and animal biomass, which grows in the countryside. However, animal biomass accounts for only about 1% of terrestrial biomass. (acatech 2012)

lies in the utilization of lignocellulose. Total terrestrial biomass production is 120 billion tonnes per year (acatech 2012).

Taking into account that more than 99% of the terrestrial biomass consists of chemical polymers, a separate step of depolymerization is usually required in the process for producing fuels or chemicals.

The main chemicals that make up biomass can be divided into four groups or classes of substances:

Carbohydrates

These include various sugars, such as glucose, fructose and sucrose, but also polymers such as starch and cellulose, both of which consist of multiple glucose monomers.

Lignins

This scaffold substance in woody plants consists of various aromatic rings that are polymerised into a complex network.

Lipids

Vegetable oils and fats belong to this group. Waxes, membrane-forming lipids, steroids, carotenoids and fatty acids are also included.

Proteins

Proteins are made up of amino acids and perform a variety of tasks; this group includes enzymes and both structural and storage proteins.

Carbohydrates

Carbohydrates are – nominally – composed of carbon (C) and water (H_2O), and are commonly referred to as sugars. In a general chemical formula, the monomers of carbohydrates (monosaccharides) can be expressed as $C_n[H_2O]_n$. Hexoses or C6 sugars have the molecular formula $C_6H_{12}O_6$, while pentoses or C5 sugars have the molecular formula $C_5H_{10}O_5$, accordingly.

D-glucose or dextrose is a monosaccharide (simple sugar) and is the best-known hexose. It is also the basic building block for storage substances such as starch and structural substances such as cellulose. Fructose also belongs to the hexoses and is a chemical isomer of glucose. For instance, in the chemical industry, monosaccharides are reduced to sugar alcohols. Glucose and fructose are reduced to sorbitol, a sugar substitute for the food industry.

Disaccharides consist of two molecules of monomeric sugars, which are linked via a glycosidic bond (oxygen bridge). The best-known disaccharide is **sucrose** (granulated sugar), composed of one molecule each of glucose and fructose. Sugar cane and sugar beet contain large amounts of sucrose, which are drawn out through hot extraction or pressing. The processing of sugar beet is described in ▶ Sect. 4.3.1 (Sugar Biorefinery) in more detail.

Glucose and sucrose are of great importance in the chemical industry as substrates for most biotechnological processes. In addition to the simple alcohols (such as ethanol and butanol) and organic acids (such as acetic acid, propionic acid, and butyric acid) produced through classical fermentations, fine chemicals (like propanediol, citric acid, succinic acid, lactic acid, amino acids and vitamins) and pharmaceutically important peptides and proteins (such as insulins and antibodies) are also produced through modern biotechnological processes, partly using genetically modified organisms.

However, there are many other oligomeric carbohydrates of different chain lengths, branched and unbranched and built up from different hexoses, whose importance as renewable raw materials is rather low. The most important carbohydrates for the bioeconomy are the polymeric carbohydrates starch and cellulose.

Starch is the reserve material of starch plants (such as wheat, rye, corn, potatoes and rice) and consists of polymers made up entirely of glucose molecules. There exist two different types of starch (◘ Fig. 4.3):

- Amylose is composed of unbranched α-D-glucose chains that are linked only by α-1,4-glycosidic bonds. The polymer thus forms a helical structure.
- Amylopectin is composed of branched α-D-glucose chains, which are also linked by α-1,4-glycosidic bonds within the linear ranges. At the branching points, additional glucose chains are attached by a α-1,6-glycosidic bond.

Starch is used in the chemical industry for impregnating paper or gluing corrugated board. As so-called modified starch, it serves as a thickening agent in the food industry. Furthermore, sweeteners are produced from (partially) hydrolysed starch.

For biotechnological applications, starch has to be enzymatically hydrolyzed to glucose first, since starch is almost insoluble in cold water and many of the biotechnologically important microorganisms (such as baker's yeast and *E. coli*) can only hydrolyze starch slowly or not at all. For industrial purposes, special enzymes (amylases) are used that are able to break the α-1,4- and α-1,6-glycosidic bonds at high temperatures.

Cellulose accounts for 50% of the total terrestrial biomass, and is therefore the most common overall vegetable polymer. Like starch, cellulose is composed entirely of glucose molecules. The glycosidic bond in the linear polymer is also linked between the first and fourth C atoms. However, the linking OH group on the first C atom is in the beta position (β-D glucose), and the bond is therefore β-1,4-glycosidic (◘ Fig. 4.4). But this small difference has dramatic consequences: The cellulose polymer is not helical, but rather arranged planar, hence flat in one plane. As a result, adjacent polymer chains can closely attach to each other and stabilise themselves with hydrogen bonds; plus, some crystalline structures can be formed, making cellulose a very resistant, water-insoluble fibre.

In addition to its use as pulp in the paper industry, cellulose is further processed in the chemical industry into regenerated cellulose and cellulose derivatives.

- Regenerated cellulose is formed when cellulose is first dissolved in a solvent and then regenerated by precipitation. Man-made cellulosic fibres are produced, for example, by way of the viscose process, films (cellophane) and (household) sponges by way of the Lycocell process.
- Like starch, cellulose derivatives are chemically modified through etherification or esterification. Some well-known esters are cellulose acetates (membranes, filter

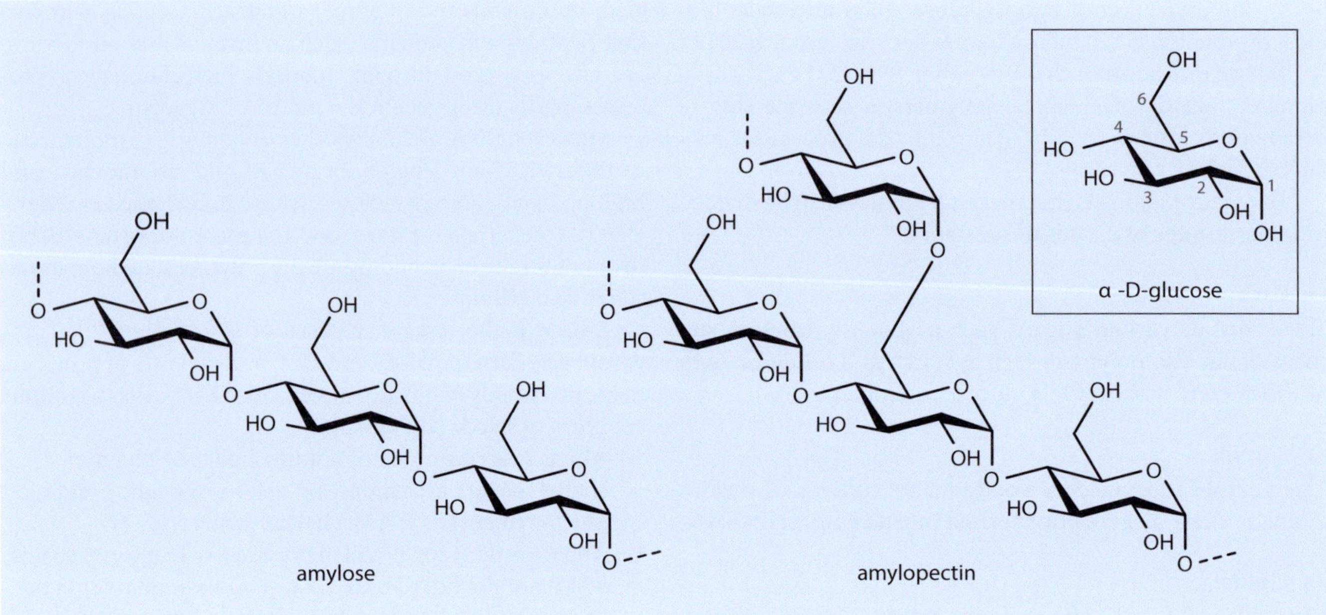

Fig. 4.3 Structure of starch. Clippings of an amylose polymer and an amylopectin polymer in chair conformation are respectively depicted. The box shows the α-D-glucose monomer, also in chair conformation. At the first C-atom (1), the free hydroxy group is in alpha position; in this representation, the bond therefore points downwards. The carbon atoms are numbered consecutively

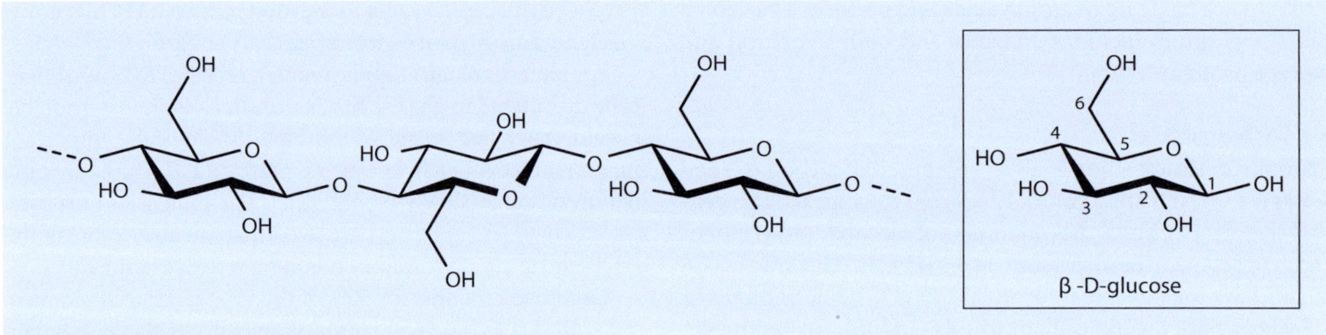

Fig. 4.4 Structure of cellulose. A clipping of a cellulose polymer in chair conformation is depicted. Cellulose consists of unbranched chains of β-D-glucose, which are linked via β-1,4-glycosidic bonds. The box shows the monomer β-D-glucose in chair conformation. At the first C-atom (1), the free hydroxy group is in beta position; in this representation, the bond therefore points to the right. The carbon atoms are numbered consecutively

materials) and cellulose nitrates (membranes, celluloid, gun cotton). Cellulose ethers are used in wallpaper paste (carboxymethylcellulose) and binding agents (hydroxyethylcellulose, methylcellulose).

The use of cellulose as a renewable raw material source is very attractive due to the available quantities and the fact that the raw material does not compete with food production. However, cellulose fibers are much more difficult to break down into glucose molecules than starch. It is possible, though, to hydrolyse cellulose by short treatment with concentrated sulphuric acid. However, many other hydrolysis by-products are formed during these processes, which not only reduce the glucose yield but also often interfere with subsequent fermentation processes.

Cellulose can also be hydrolyzed using enzyme cocktails containing cellulases. This process has also been economically feasible for some years now, as the prices for these enzymes have fallen very sharply. However, such hydrolyses take up to 48 hours, while acid hydrolysis takes place within seconds. There are practically no hydrolysis by-products. Enzymatic hydrolysis processes are used, for example, in a lignocellulose biorefinery (► Sect. 4.3.4) to generate fermentable sugars.

Plant cells also contain other polymeric carbohydrates that serve as scaffold substances. These **hemicelluloses** account for almost a quarter of the total terrestrial biomass. Hemicelluloses consist of unbranched chains of pentoses, i.e., C5 sugars. Hemicelluloses are often spoken of in the plural, because they contain mixtures of pentoses with altered composition, and therefore do not represent a uniform molecule. The most common monomers are D-xylose and L-arabinose.

The hydrolysis of hemicelluloses succeeds under much milder conditions than the hydrolysis of cellulose. Appropriate enzyme cocktails containing hemicelluloses are also available. However, only a few industrially relevant microorganisms are

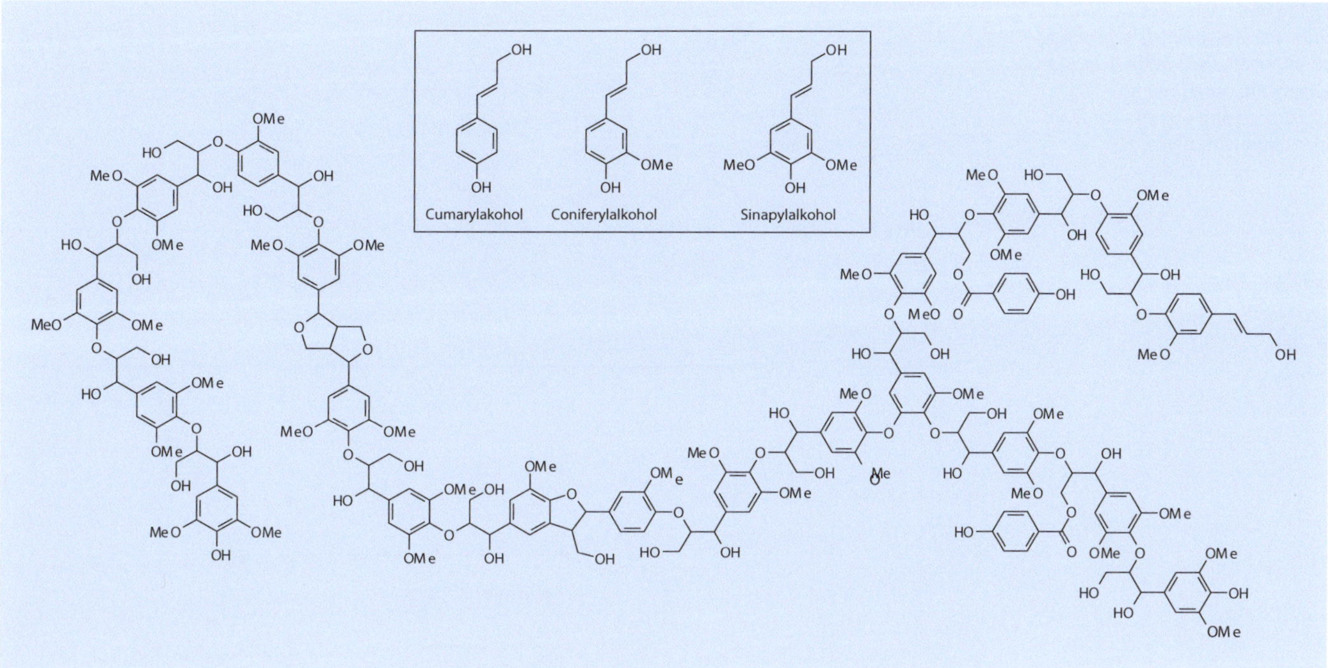

Fig. 4.5 Model of native poplar lignin. The phenylpropanoid building blocks of which lignin is composed are shown in the box. "Me" represents a methyl group (–CH_3). (According to Vanholme et al. 2010)

able to utilize pentoses in any way. Co-fermentation of C6 and C5 sugars would be especially advantageous in the production of bioethanol from lignocellulose (► Sect. 4.3.4) in order to increase overall yield. Either C5-utilizing yeasts can be used here or the necessary enzymes for the pentose utilization pathway can be incorporated via the genetic engineering methods of baker's yeast (Karhumaa et al. 2006).

Lignin

The third largest fraction of terrestrial biomass is represented by lignins. These polymers are found in wood and woody plants and provide the necessary stiffness for the scaffolding substance, perhaps comparable to the role of cement in reinforced concrete.

From a chemical point of view, lignins are composed of phenylpropanoid building blocks. These consist of a benzene ring carrying a propane side chain with hydroxy, as well as methoxy groups and other residual chains as substituents. The most common phenylpropanoids are cumaryl, coniferyl and sinapyl alcohols. The three-dimensional network of lignin (Fig. 4.5) is composed of 30 to 60 of these units (relative molecular mass about 5000 to 10,000 Da) (Faix 2008).

Today, lignins are the most important by-product of the pulp industry. With an annual production of more than 150 million tonnes of pulp from wood (FAO 2015), roughly 75 million tonnes of lignin are produced as a by-product (2–2.7 tonnes of wood are required per tonne of pulp; 2 tonnes of wood contain approx. 0.5 tonnes of lignin). However, 98% of the lignins are burned to generate energy or recover process chemicals (Gosselink 2011).

Moreover, the increasing production of cellulose ethanol from lignified plant residues instead of sugars (second generation ethanol) also leads to an increasing accrual of lignin. The expected amount of lignin produced annually as a by-product of ethanol production from lignocellulose will be extrapolated to more than five million tonnes in 2020 for Europe alone if the EU directive on the "promotion of the use of energy from renewable sources" is implemented (Gosselink 2011). Some application examples of lignin are reported in ► Sect. 4.3.4.

Lipids (Here: Fats and Fatty Oils)

Fats and fatty oils (neutral fats) belong to the lipids and are the storage materials of oil plants such as rape, soya, sunflower, oil palm and coconut, but they can also be found in cereal germs, for example, in corn germ oil or wheat germ oil. Chemically, these oils are designated as triglycerides or triacylglycerols, and are esters of glycerol, a trivalent alcohol, and three, often different fatty acids (Fig. 4.6). Since fatty acids are built up in biosynthesis from C2 building blocks (acetyl residues), natural fatty acids are always even numbered and often unbranched. Fatty acids that contain double bonds in the carbon chain are called (poly)unsaturated fatty acids, while those that do not contain double bonds are called saturated fatty acids. In the food industry, the position of the double bond(s) is counted from the end opposite the carboxy group. The terminal C atom at this end is called "Omega," the last letter of the Greek alphabet, regardless of the chain length. The particularly praised omega-3 fatty acids thus carry the first double bond at the third C atom counted from that end.

Usually, omega-3 fatty acids are obtained from fish oil (e.g., salmon oil). They not only play a special role in human nutrition, but also in fish farming (aquaculture), since fish cannot produce omega-3 fatty acids themselves. Algae are the true producers of omega-3 fatty acids, which only accumulate in

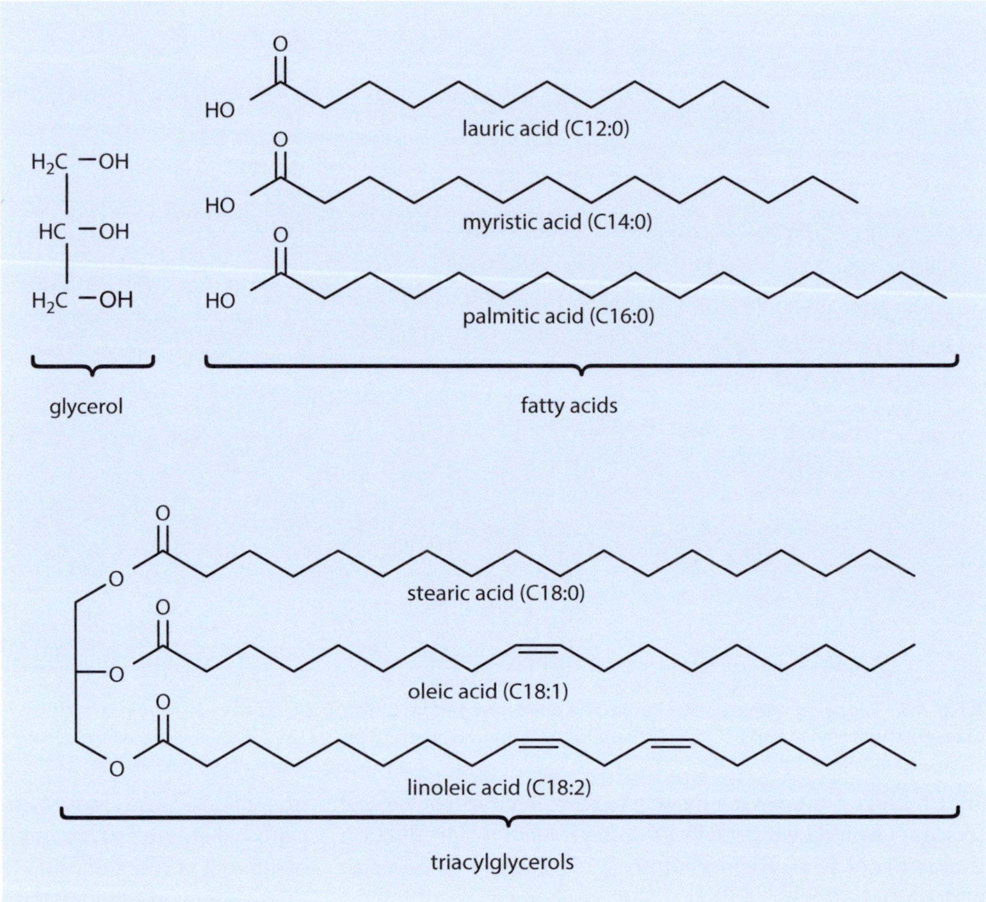

Fig. 4.6 Glycerol, fatty acids and triacylglycerol. Glycerol is a trivalent alcohol. In the triacylglycerol shown, three fatty acids (saturated and unsaturated) are esterified with glycerol

(predatory) fish via the food chain. Therefore, aquacultures are often fed with so-called by-catch. There have already been initial attempts to produce the valuable fatty acids using algae biotechnology and feed them directly to the fish.

The composition of the esterified fatty acids on the glycerol molecule in terms of chain length, number of double bonds and other functional groups determines the properties of the oil. Thus, triacylglycerols with a high proportion of unsaturated, short-chain fatty acids, such as palm oil, palm kernel oil and coconut oil, are solid at room temperature, while oils with unsaturated longer-chain branched fatty acids, such as rapeseed or sunflower oil, are liquid at room temperature. For margarine production, these double bonds are eliminated by hydrogenation, the so-called hardening process.

In the chemical industry, fatty acids from triacylglycerols can be used for the synthesis of surface-active substances (surfactants). These are used, for example, in detergents and cleaning agents, cosmetics and pharmaceuticals, as well as in textile and leather auxiliaries. As a rule, triacylglycerols with short-chain saturated fatty acids, such as lauric acid, are preferred here. These are obtained from either palm kernel oil or coconut oil. Another major field of application is biolubricants, which are also produced from the fatty acids derived from triacylglycerols. Engine and gear oils, hydraulic oils, lubricating oils and metalworking oils can now be produced as bio-based. Biolubricants and oils are often more durable than fully synthetic lubricants, and are also more environmentally friendly, thanks to their biodegradability. Therefore, biolubricants can be found to be in application where these properties are particularly required, e.g., in wind turbines. Further fields of application are as additives for paints and varnishes or lubricants and plasticizers for plastics.

In Germany, a large portion of the triacylglycerols are nowadays processed into biodiesel by transesterifying them with methanol. In the process, 10% of glycerine by mass is always generated as a by-product. Possible applications of glycerol as a platform chemical are discussed in ▶ Sect. 4.2.2 under the heading "C3 molecular building blocks." However, fatty acids containing double bonds, branches or functional groups are of particular importance for the production of fine chemicals. At these molecular locations, chemists can specifically introduce chemical groups or split the molecule. Some of these applications are described in ▶ Sect. 4.3.3.

Proteins

Animal proteins, in particular, are, to a small extent, used today as raw materials for technical chemistry. Gelatine and collagens are used in adhesives, while the former is also used as a coating for photographic and printing papers. Casein is mainly used as a coating for glossy papers and in adhesives, as well as in paints and varnishes. Vegetable proteins are mainly by-products of grain processing, but are only of minor importance as raw materials for the chemical industry. Wheat gluten and soy protein isolates, for example, are used as adhesives and binders (FNR 2014).

Newer applications that are currently under development are for biodegradable plastics. Proteins have thermoplastic properties. This opens up perspectives for new technical applications with native and fiber-reinforced materials made from these plastics.

Enzymes are proteins of particular importance for the bioeconomy. They catalyze, for example, the hydrolytic cleavage of the polymers and oils presented above in order to make raw materials such as sugar and fatty acids available. The biotechnological production of such enzymes has increased considerably in recent years, mainly because some of these enzymes, such as lipases, proteases and cellulases, are being used more and more in the detergent industry. Other enzymes can catalyse very specific chemical reactions on molecules. This allows for the replacement of complex chemical synthesis steps with resource-saving biotechnological process steps (▶ Chap. 5). This "biologization of industry" not only promises to produce lower energy and water consumption, but (fossil) solvents and auxiliary chemicals will also be able to be saved.

4.2.2 Platform Chemicals from Fossil and Renewable Raw Materials

Organic platform chemicals are simple, basic synthesis building blocks of the chemical industry from which the family trees of industrial chemicals are derived. Examples of important platform chemicals used in the chemical industry can be found in ◘ Fig. 4.7 (FCI 2009). In principle, many of these synthesis components can be produced from renewable raw materials as well. At least, the chemical family trees can partially be covered by available bio-based alternatives.

▪ C1 Building Blocks

Methanol is the most important C1 platform chemical in the chemical industry. More than 21 million tonnes are produced annually (Khirsariya and Mewada 2013). Methanol is synthesized via the syngas pathway from coal, naphtha, and, most of all, natural gas. Methanol is the most versatile platform chemical of all: It can be used to produce a large number of other platform chemicals, such as dimethlyether (DME), acetic acid, propylene, olefins and aromatics, as well as fuels such as diesel and gasoline and their additives (MTBE) (◘ Fig. 4.8). Its versatility is such that methanol could be used to access a large part of the spectrum of industrial organic chemistry, thus replacing crude oil and natural gas in the future (Bertau et al. 2015).

Besides hydrogen, carbon monoxide is the main component of synthesis gas (syngas), a gas mixture that is mainly produced through the thermochemical gasification of crude oil or coal. It is an important platform chemical for Fischer-Tropsch synthesis and methanol synthesis.

Methane is the main component of natural gas. In the chemical industry, it is the source of methanol and the other C1 building blocks. Methane is converted into synthesis gas by steam reforming. Methane is also used for the synthesis of chloralkanes or as a reducing agent in ammonia synthesis (de Jong et al. 2011).

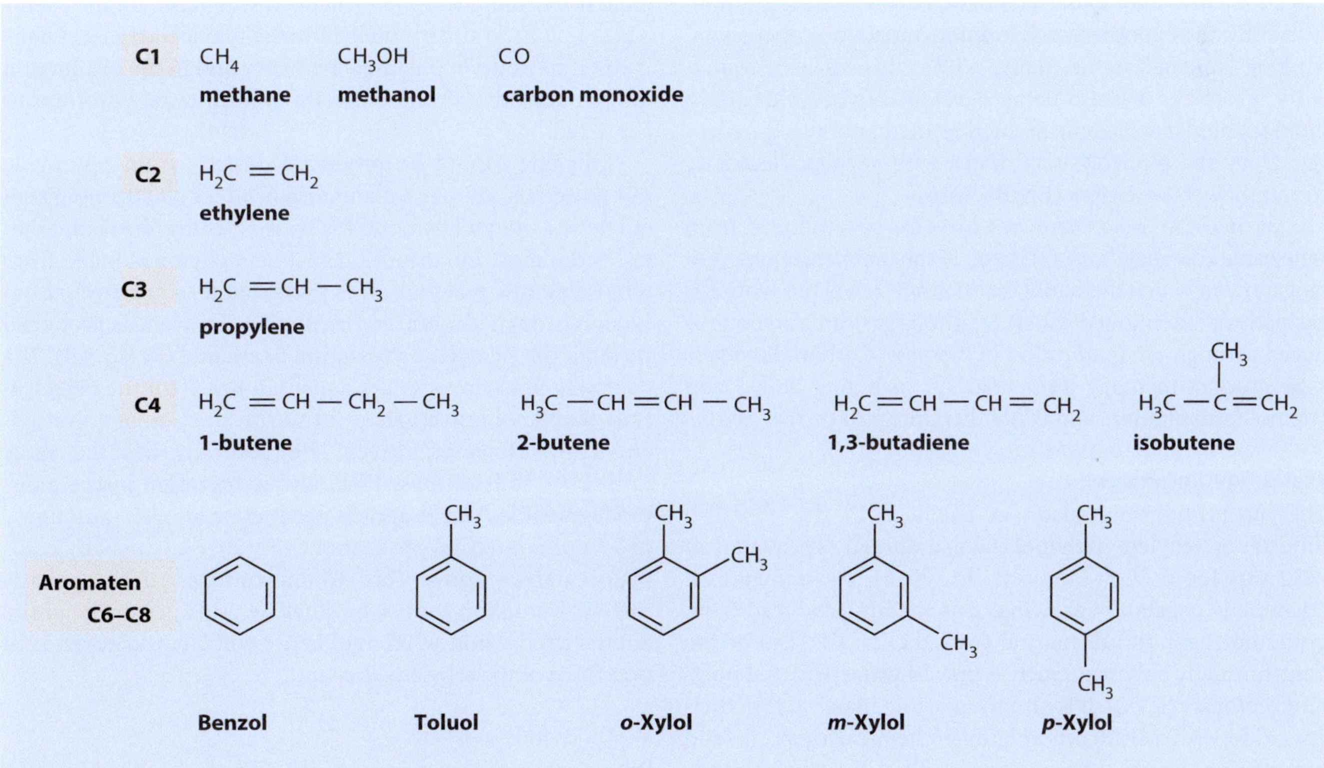

◘ Fig. 4.7 Examples of important (fossil) platform chemicals in the chemical industry. (FCI 2009), modified

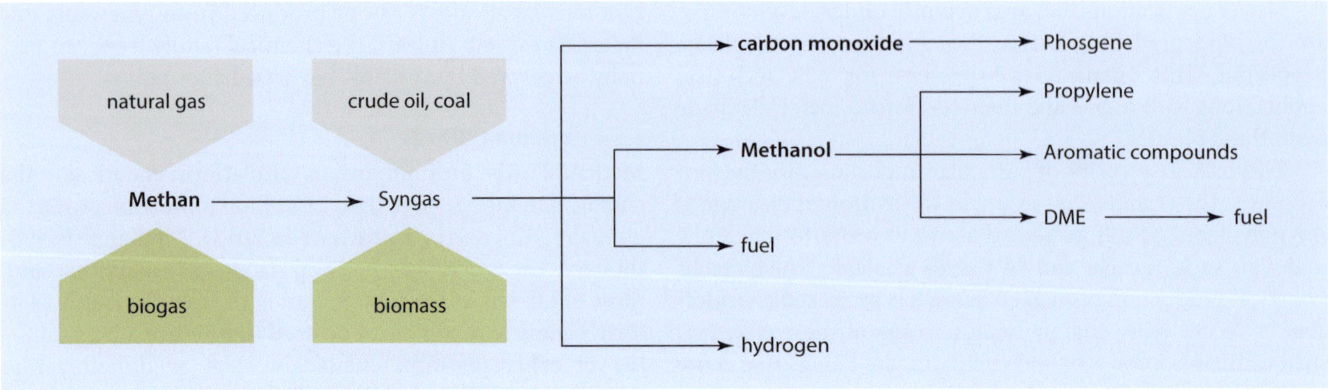

Fig. 4.8 Important pathways of the value chain for C1 building blocks in the chemical industry

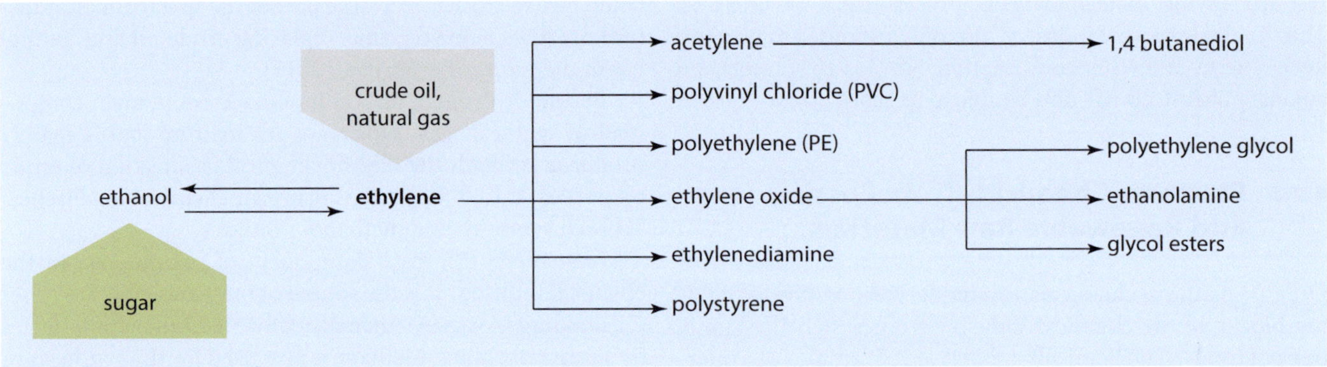

Fig. 4.9 Important pathways of the value chain for C2 building blocks (ethylene) in the chemical industry

Carbon dioxide as a component of industrial exhaust gases may also play a role as a basic chemical in the future. Currently, the major research funding initiative "Kopernikus" is being launched in Germany. Within this research framework, processes are also being developed to recycle carbon dioxide into basic chemicals such as methanol using hydrogen from the electrolysis of water with surplus electricity from renewable energies (BMBF 2016).

All of these basic chemicals can also be produced from renewable raw materials: methane is the main component of biogas, from which methanol is also available via the synthesis gas pathway mentioned above. Synthesis gas can also be produced through the gasification of biomass. Carbon dioxide is a by-product in many fermentations, including biogas and ethanol fermentation, and could therefore also be recycled.

C2 Building Blocks

The most important platform chemical in the chemical industry is ethylene (ethene). Global annual production in 2012 was 156 million tonnes (CIEC 2016). The demand for ethylene is constantly growing. It is mainly produced from crude oil (naphtha) or natural gas (ethane). Ethylene is the basis for many polymers, such as polyethylene (PE) and polyvinyl chloride (PVC). It is also the starting material for ethylene oxide, which, in turn, is a platform chemical for mono- and polyethylene glycol. Approximately half of the ethylene produced annually is used for polyethylene. Acetylene (ethine), which was obtained from coal until the 1950s and was of great importance as a platform chemical in the Reppe chemical reactions, is also produced from ethylene today. Acetylene is still important in polymer chemistry and in the production of 1,4-butanediol, from which the solvent tetrahydrofuran is extracted.

Ethylene cannot be produced directly from renewable raw materials, although plants also produce small amounts of ethylene as phytohormone for fruit ripening. Instead, ethanol is the most important C2 platform chemical made from renewable raw materials. It is produced fermentatively from sugars through alcoholic fermentation. Ethylene is produced through the catalytic dehydration of ethanol (Fig. 4.9). The company Braskem operates a plant in Brazil for the dehydration of ethanol into ethylene for the production of polyethylene (PE). However, "green" PE naturally has the same properties as fossil-based PE, and is therefore just as non-biodegradable. World annual production in 2015 amounted to 77 million tonnes of ethanol (RFA 2016), which, by complete catalytic conversion, would correspond to approximately 47 million tonnes of ethylene. Thus, even the entire annual production of ethanol is not sufficient to cover even one-third of the ethylene demand.

C3 Building Blocks

Propylene is the most important C3 platform chemical. Like ethylene, it is produced by thermal cracking from natural gas

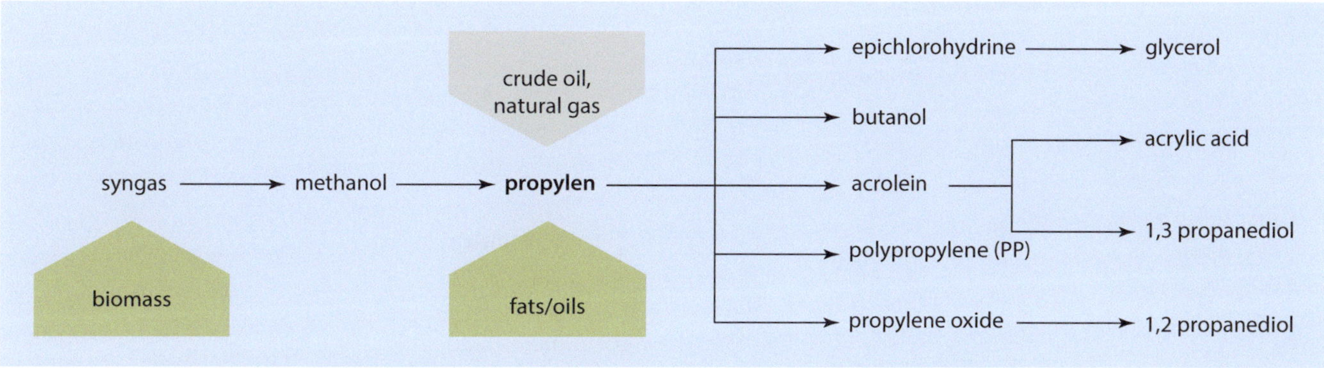

■ **Fig. 4.10** Important pathways of the value chain for C3 building blocks (propylene) in the chemical industry

or naphtha. About two thirds of the production is used for the synthesis of polypropylene. Other product lines are propylene oxide (educt for 1,2-propanediol), acrolein, acrylic acid (educt for superabsorbers, paints and plastics), 1,3-propanediol and 1-butanol (solvent). Epichlorohydrin is used in epoxy resins and, until a few years ago, had also been used for the synthesis of glycerol. In addition to the already described synthesis pathway via the gasification of biomass, there are other ways to produce bio-based propylene, e.g., by the catalytic cracking of fats and oils (■ Fig. 4.10).

In addition, other bio-based platform chemicals are available, which often lead to the same end products, but for which different synthesis pathways have to be used. Glycerol is the by-product of biodiesel production. From the transesterification process, 100 kg of glycerol accrue from each tonne of biodiesel produced. In 2014, three million tonnes of biodiesel were produced in Germany alone, and thus also 300,000 tonnes of glycerol (FNR 2016). However, this bio-based glycerol can be used to produce many of the chemicals that are currently synthesized from propylene (■ Fig. 4.11). Even epichlorohydrin is now made from glycerin, by performing the reverse reaction of the above-mentioned glycerol synthesis. Acrylic acid and 1,3-propanediol can also be directly produced biotechnologically from glycerol (not shown).

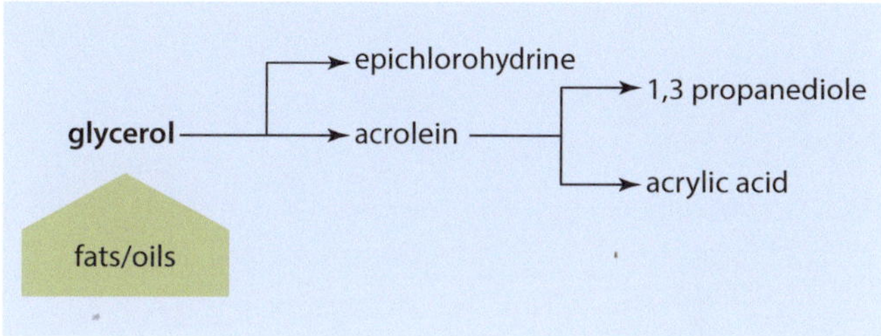

■ **Fig. 4.11** Important pathways of the value chain for C3 building blocks (glycerin) in the chemical industry

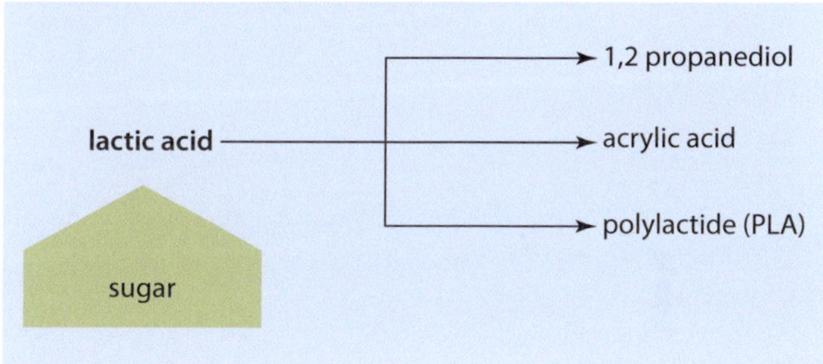

■ **Fig. 4.12** Important pathways of the value chain for C3 building blocks (lactic acid) in the chemical industry

Lactic acid, which is obtained biotechnologically from sugar, can also be converted into acrylic acid or 1,2-propanediol. Enantiomerically pure lactic acid can also be polymerized to polylactide (PLA). To this purpose, lactic acid bacteria are used, which preferably produce L-(+)- or D-(−)-lactic acid (Idler et al. 2015). This bio-based polymer has similar properties to polypropylene (PP) or polyethylene terephthalate (PET) (■ Fig. 4.12).

C4 Building Blocks

In the petrochemical industry, the C4 platform chemicals are obtained as the so-called C4 cut in the cracking process of naphtha. The C4 cut essentially consists of the four butenes (■ Fig. 4.7) and *n*-butane. More than 90% of 1,3-butadiene is processed into synthetic rubber (styrene-butadiene rubber), e.g., for the tire industry. 1-butene is used as a co-monomer in the production of polyethylene or polypropylene to improve plastic properties or it is dimerized to 1-octene, which, in turn, serves as a platform chemical. 1-butene can easily be isomerized to 2-butene. The fuel components MTBE and ETBE are produced with isobutene. They serve as an antiknock agent in gasoline for better fuel combustion. Furthermore, polymers and synthetic rubber are produced from isobutene (■ Fig. 4.13). Maleic anhydride is obtained from *n*-butane on a large scale and the platform chemical succinate is derived from it.

Due to the aforementioned fracking boom, the availability of these C4 platform chemicals has declined, at least in the USA. This is because natural gas contains only very small amounts of C4-olefins. For this reason, alternative – as well as bio-based – methods for the production of these platform chemicals have been increasingly researched in recent years. At the very least, pilot plants exist for the subsequent selection of biotechnological processes:

- Bio-based 1,3-butadiene can be produced from ethanol through dimerisation. Another possibility is syngas

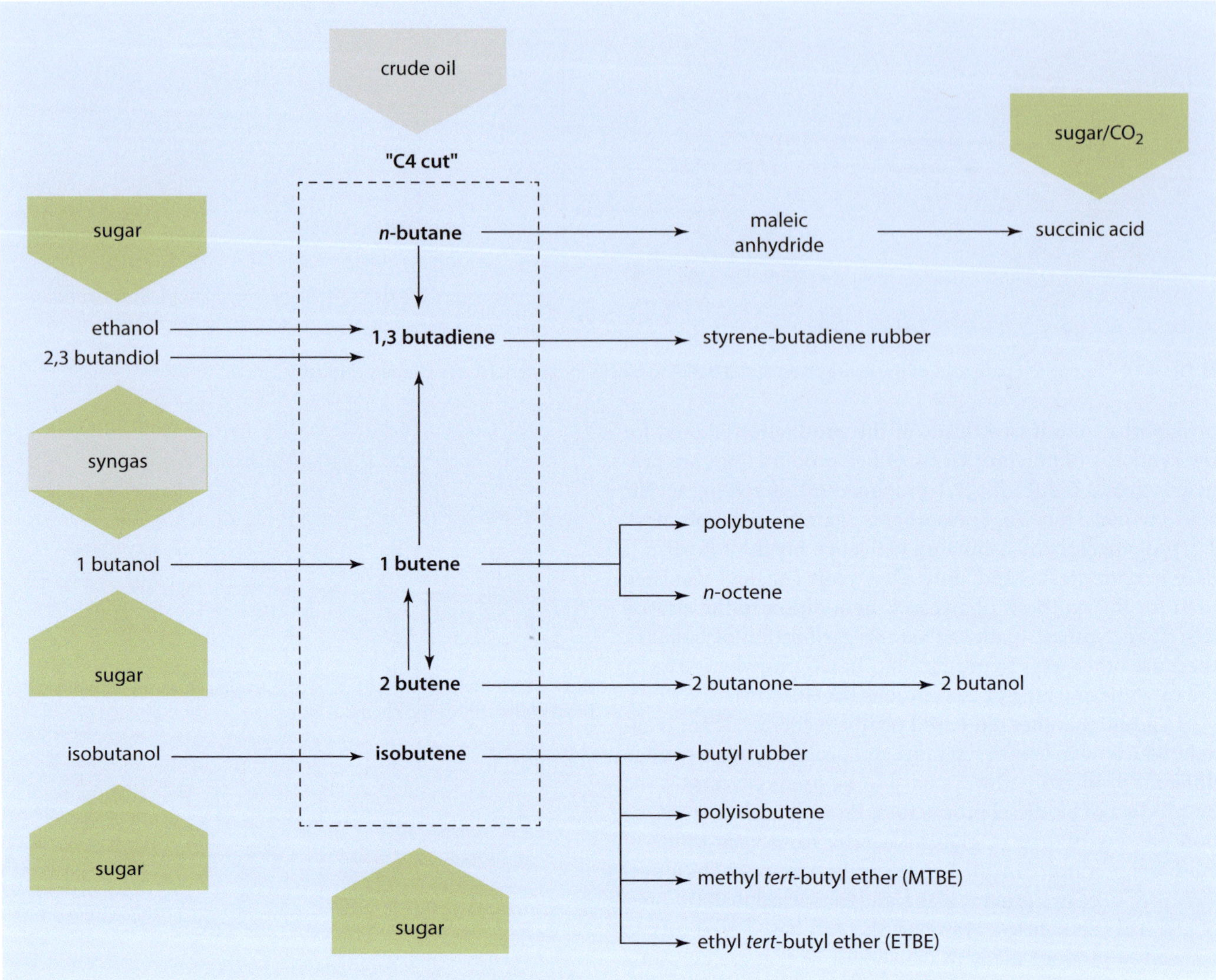

Fig. 4.13 Important pathways of the value chain for C4 building blocks (C4 section) in the chemical industry. The syngas described here comes from power plant flue gases. It can be biotechnologically converted into ethanol and 2,3-butanediol or butanol

fermentation into 2,3-butanediol, which can, among other routes, be dehydrated into butadiene. Direct fermentation into butadiene is also possible using genetically modified microorganisms.

- Bio-based 1-butene is produced through the dehydration of 1-butanol, which is obtained in the classic acetone-butanol-ethanol (ABE) fermentation process or by syngas fermentation with flue gas from, e.g., power plants or steel mills.
- Bio-based isobutene is obtained by the dehydration of isobutanol, which can be produced through fermentation. Using genetically modified bacteria, isobutene can also be obtained directly by biotechnological processes.

Another significant C4 platform chemical is succinic acid. It can be accessed via fossil routes (from *n*-butane, see above) as well as fermentatively. Succinic acid is therefore predicted to have great potential in the bioeconomy. It had already been ranked among the most promising top 12 bio-based platform chemicals by 2004 (Aden et al. 2004), which was confirmed in 2010 (Bozell and Petersen 2010): Succinic acid is the source of quite important fine chemicals (Fig. 4.14). One of these downstream chemicals is 1,4-butanediol, which has already been mentioned in the ethylene family tree. 1,4-butanediol can be converted into γ-butyrolactone, which is a widely used solvent in the chemical industry, but also forms the basis for many pharmaceuticals and polymers. Other products manufactured from 1,4-butanediol include tetrahydrofuran, which is also an important solvent, and the polyesters of terephthalic acid (polybutylene terephthalate). Other downstream chemicals are fumaric acid and maleic acid, which – like succinic acid – are used as dicarboxylic acids in polyester chemistry. The family tree of succinic acid is shown in de Jong et al. 2011.

Aromatic Building Blocks

Aromatic synthesis building blocks only occur in small quantities in crude oil. Benzene, toluene, the xylenes and higher aromatics are therefore produced today in refineries by the

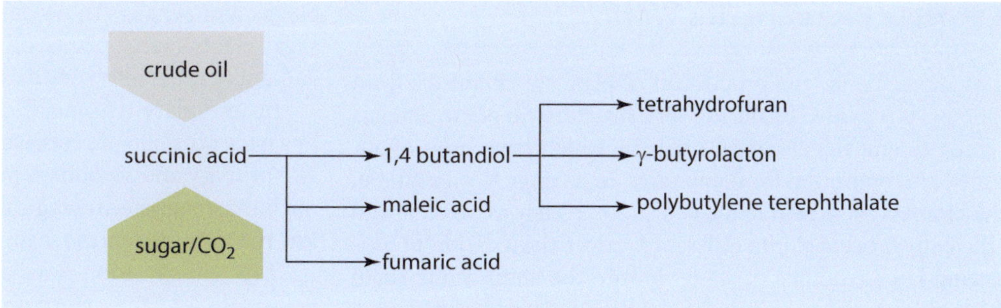

Fig. 4.14 Important pathways of the value chain for C4 building blocks (succinic acid) in the chemical industry

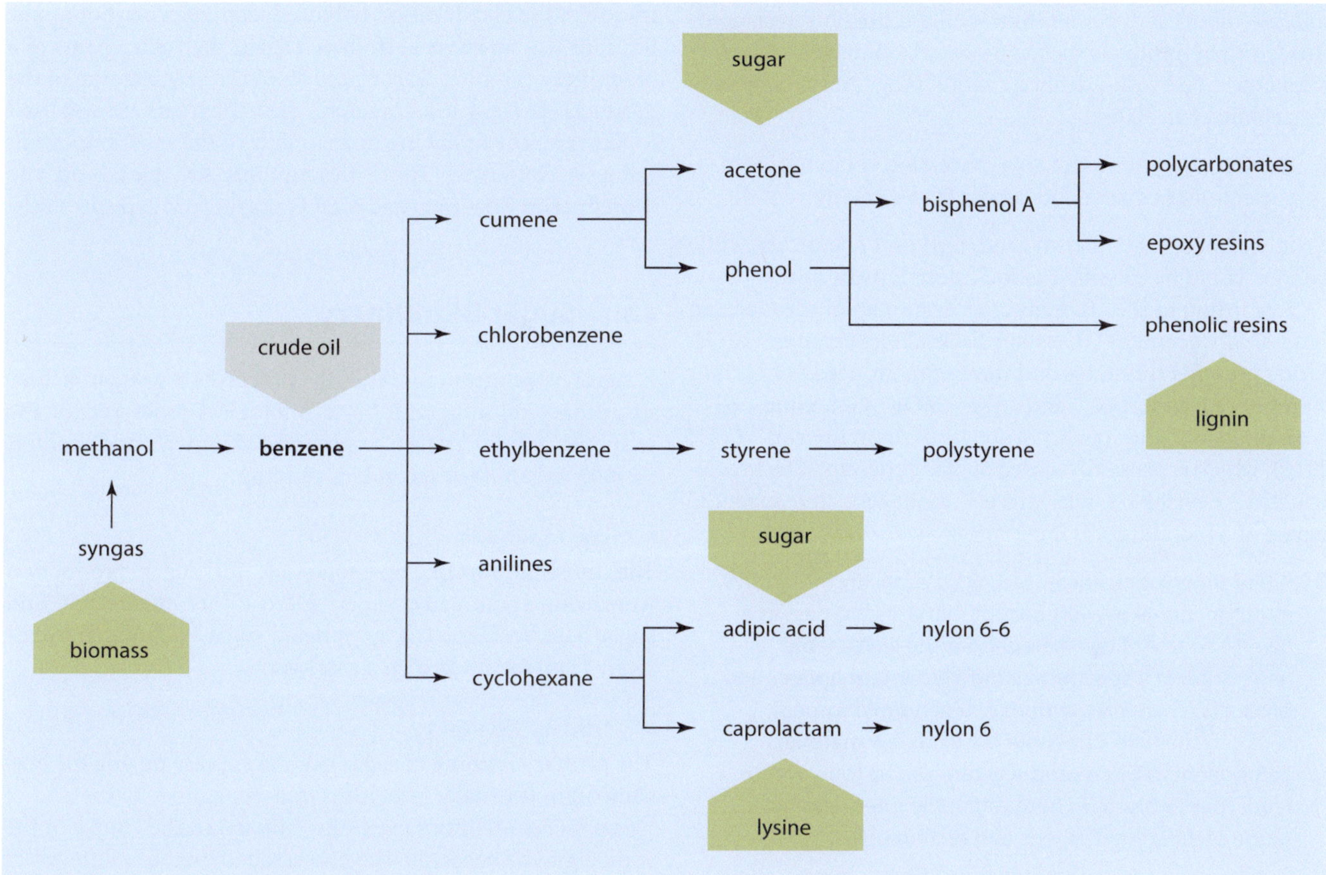

Fig. 4.15 Important pathways of the benzene value chain as a representative of aromatics in the chemical industry

catalytic reforming of naphtha (fractions). Substituted aromatics (phenols, nitrobenzenes, aminobenzenes) are produced from these synthesis components. Aromatics are widely used in the chemical industry as solvents and as platform chemicals for the synthesis of plastics and specialty chemicals (e.g., aniline, styrene, nylon, synthetic rubber, detergents, insecticides, dyes, explosives, pharmaceuticals). In an exemplary manner, the main pathways of the benzene value chain are depicted in Fig. 4.15. The chemical industry annually uses around 50 million tonnes of these synthetic building blocks worldwide.

Despite this high added value of aromatics in the chemical industry, there are hardly any aromatic platform chemicals based on renewable raw materials, although many natural substances also contain aromatic structures. The most important natural source of aromatics is lignin, which accounts for 20 to 30% of the dry matter of woody plants. Since it is not yet possible to selectively depolymerize the lignin polymer into defined monomers, lignins are rarely used in the chemical industry. One of the few applications is the partial replacement of phenol with lignin in phenol-formaldehyde resins (Fig. 4.15). These resins are used for bonding both chipboard and plywood panels.

From Fig. 4.15, it is also evident that there are indeed bio-based starting materials from which building blocks of the benzene value chain can be derived. Bio-based acetone can be produced by the ABE fermentation mentioned above. Adipic acid, which is mainly used for the production of nylon 6–6, can be produced biotechnologically and biotechnologically-chemically from sugar. Caprolactam is accessible through the chemical conversion of lysine, while lysine is nowadays produced fermentatively from sugar.

4.3 How Biorefineries Work

The archetype for the production of platform chemicals from biomass is provided by the network structures in petrochemical refineries and the chemical industry, which have been established and optimized for many years. In analogy to oil refining, the biomass must be fractionated as completely as possible and efficiently processed into different product classes without producing large amounts of waste. Ideally, the entire value chain from biomass production to the end product must be integrated.

A biorefinery can simultaneously produce fuels, energy and raw chemical materials from biomass, possibly even combined with by-products such as food and animal feed. The *IEA Bioenergy Task 42* defined biorefining very generally (Cherubini et al. 2009):

> Biorefining is the sustainable processing of biomass into a spectrum of marketable products and energy

'Products' here may refer to food, feed and chemicals, while 'energy' could be classified as fuel, electricity or heat.

According to the "Biorefineries Roadmap of the German Federal Government" (German Federal Government 2012), which presents the status and development needs of various biorefinery concepts, and the VDI Guideline 6310 "Classification and quality criteria of biorefineries" (VDI 2016), which is primarily aimed at plant operators, technical-scientific developers and project managers, the essential aspect of a biorefinery is integration:

> A biorefinery is characterised by an explicitly integrative, multifunctional overall concept that uses biomass as a diverse source of raw materials for the sustainable generation of a spectrum of different intermediates and products (chemicals, materials, bioenergy/biofuels), allowing the fullest possible use of all raw material components. The co-products can also be food and/or feed. These objectives necessitate the integration of a range of different methods and technologies.

According to this definition, industrial plants that primarily produce food or feed, or plants that do not separate components, are not biorefineries. Even plants that are only part of an integrated biorefinery process chain are not, per se, biorefineries. However, these plants can be integrated into biorefineries or upgraded to biorefineries. This bottom-up approach to the development of biorefineries has already been successfully implemented in the past, for example, when a starch factory integrates the production of ethanol and animal feeds or a biodiesel plant is integrated into an oil mill, which then further processes the accumulated glycerine fraction.

The classification of biorefineries used today was established by the *IEA Bioenergy Task 42 Biorefining* (IEA) (IEA Bioenergy 2016). It classifies biorefineries according to four main characteristics (Cherubini et al. 2009):

The name-giving element in the system is the **platform**, since various **raw materials**, after their processing within the primary refining, lead to these intermediates. The main platforms are sugar, starch, lignocellulose, fats and oils, fibres, biogas and syngas. These determine the raw materials used: sugar plants, starch plants, oil plants, wood or woody biomass, and grasses, but also residual and waste materials and algae.

In secondary refining, the platforms are further converted by a wide portfolio of **processes** to the **products** of a biorefinery. The primary and secondary processes used in a biorefinery are not limited to biotechnological processes, but also include physical, mechanical, thermochemical and chemical processes.

This classification was also used in the "Biorefineries Roadmap of the German Federal Government." VDI Guideline 6310 also makes use of it (VDI 2016). For the illustration of the classification of biorefinery concepts, both publications use flowcharts. In these charts, the value chains of a biorefinery can be lucidly traced from the raw material to the products (▫ Fig. 4.16). Therefore, these diagrams are also used to illustrate the following descriptions of the most important refinery platforms. These descriptions are based on the Biorefineries Roadmap and VDI Guideline 6310, respectively.

4.3.1 Sugar Biorefinery

A sugar biorefinery combines the processes of a sugar factory in primary refining with biotechnological processes for the fermentative production of platform chemicals and biofuels (usually ethanol) in secondary refining.

Raw Materials

The most important raw materials for sugar production worldwide (granulated sugar, sucrose) are sugar cane and sugar beet. In Germany, granulated sugar is obtained exclusively from locally grown sugar beet.

Primary Refining

The primary refining of sugar beet takes place during the beet campaign. It usually lasts from mid-September to the end of December, sometimes even into January. In the course of the campaign, the beets must be processed to the storable levels of thick juice or granulated sugar. The sugar beets are washed and crushed after harvesting. The sugar beet cuttings are then extracted. In addition to sugar, the raw juice obtained by extraction also contains non-sugar substances, some of which are separated in the subsequent juice purification process. The purification of the raw juice results in a clear, thin, light yellow juice with a sugar content of about 16%. In a multi-stage evaporation station, the thin juice is thickened into a dry matter content of 70–75% sugar. By further concentration and multiple crystallization of this thick juice, granulated sugar is obtained. The effluent syrup of the final crystallization stage is molasses, which contains the sugar that cannot be crystallized and the non-sugar substances originating from the sugar beet.

Secondary Refining

For further processing into bio-based products and/or for the production of bioenergy, thick juice and granulated sugar are available as raw products (platforms) and sugar-containing molasses and extracted cuttings as by-products:

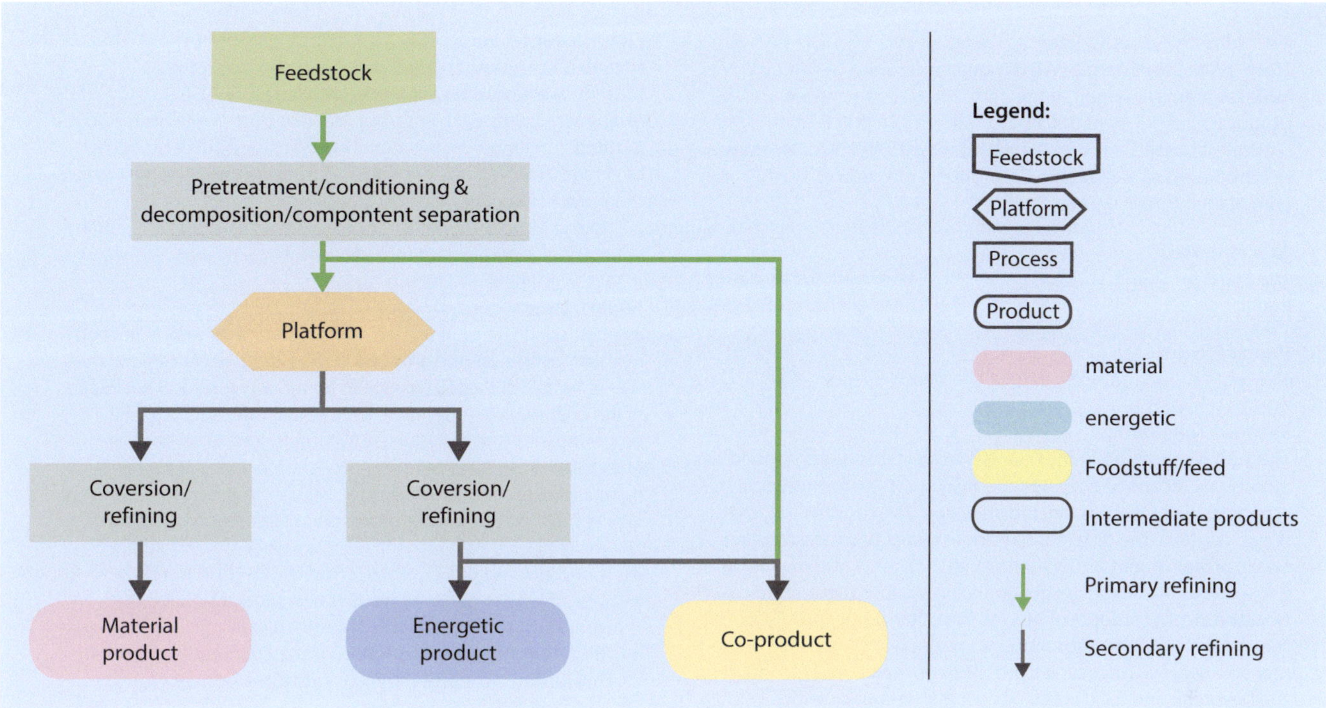

Fig. 4.16 Schematic graphic representation of a biorefinery. (After the German Federal Government, 2012)

- thick juice is usually processed into granulated sugar as described above, but can also be used directly as a raw material for fermentation.
- granulated sugar is also used as a raw material for fermentation in the chemical-technical sector, and can, in addition, be used as a raw material for chemical intermediate or finished products (e.g., sugar surfactants).
- molasses is also usually used as a raw material for fermentation (e.g., for feed yeast, bioethanol, and chemicals) or for the production of animal feed. In addition to sugar (e.g., sucrose), molasses contains other ingredients, like organic acids, betaine, vitamins, and inorganic salts, that can be extracted and further processed.
- The extracted cuttings are dried and then mostly pelletized and mixed with molasses. These molassed beet cuttings are a sought-after animal feed.

4.3.2 Starch Biorefinery

In principle, a modern starch factory can already be considered as a biorefinery. It produces starch on its own during primary refining, can also isolate proteins and oil from germs and produces technical starches, modified starches and various glucose/fructose syrups during secondary refining. The latter, in particular, are ideal starting points for new biotechnological value chains in secondary refining.

Raw Materials

The most important crops for starch production worldwide are cereals (corn, wheat, rice), potatoes and manioc. In Germany, cereals (maize, wheat) and potatoes are the main crops used.

Primary Refining

The first steps in the processing of cereals are the soaking and swelling of the grains (where appropriate, after previous grinding), separation of the germs, grinding and sieving. Potatoes are mashed in after cleaning. Further processing is then similar to that of grain: starch is extracted, fibres and protein are separated. The starch suspension is cleaned and dried to obtain pure starch.

Secondary Refining

For further processing into bio-based products and/or for the production of bioenergy, native starch is available as a raw product and the fractionated proteins and fibre residues as by-products:

- Native starch is either processed directly or converted into either modified starch or starch saccharification products:
  - Chemical and technical processing takes place in the production of paper and cardboard, in that of chemical materials such as adhesives and in the manufacture of finished products such as tyres.
  - Native starch is chemically or physically modified to produce modified starch. The resulting modified starch (e.g., starch esters, starch ethers, dextrins) and starch mixtures are then further processed.
  - Starch saccharification products (e.g., glucose) are formed by enzymatic hydrolysis of the polymer starch. The monomers are either used as fermentation substrates or further converted chemically (e.g., into sugar alcohols) or biochemically (e.g., into fructose).
- The separated proteins are used for animal feed. The cereal protein gluten is used as a binder and adhesive in the food industry and in the chemical-technical sector.
- The fibre residues are usually used for animal feed.

4

Cereal Crops-Based Starch Biorefinery

Since 2005, CropEnergies AG has been operating Germany's largest bioethanol plant at the Zeitz site, with an annual production capacity of 360,000 m^3 (about 280,000 t) of bioethanol based on cereals and sugar syrup (thick juice) from a neighbouring sugar factory. Co-products are carbon dioxide and animal feed (CropEnergies 2016) (Fig. 4.17).

Raw Materials
The plant processes various cereals, such as wheat, maize, barley and triticale.

Primary Refining
The grain is dry ground to extract the starch.

Secondary Refining
The starch contained in the flour is macerated and saccharified to glucose by means of enzymes and heat and then fermented into ethanol. Finally, the ethanol is distilled off. The residue is stillage, which contains the distillation residues (yeast biomass, residual sugar, proteins, salts and fibre residues). For each litre of alcohol, 8 to 10 litres of stillage are generated. In a subsequent rectification, the alcohol is concentrated to 96%. For biofuel production, the ethanol is dried or absoluted via molecular sieves. Part of the raw ethanol is also processed into high-purity neutral alcohol for the beverage and food industries, for the cosmetics and pharmaceutical industries and for technical applications. The stillage is evaporated, dried, pelletized and marketed as protein- and fibre-rich animal feed called DDGS *(Distillers Dried Grains with Solubles)*. The carbon dioxide from the fermentation is purified and liquefied. It is mainly used as carbonic acid in the beverage industry, as a cooling and freezing agent, as well as for the production of dry ice and as a cleaning agent.

In the biorefinery, thick juice from the neighbouring sugar factory is also fermented into ethanol.

Another large sugar and starch biorefinery cluster has been established in France (IEB 2016). Various industrial companies and research institutions are grouped in the extensive cluster known as "Les Sohettes" of the IEB (European Biorefinery Institute) near Reims in the Champagne-Ardenne region. As raw materials, wheat is processed into starch or glucose (Chamtor) and sugar beet into granulated sugar (Crystal Union). Both companies supply raw sugar materials for ethanol production (cristanol). In the cluster, the following methods for the further processing of intermediate products and residual materials, respectively, take place, among others:

- Conditioning of CO_2 from fermentation (AirLiquide)
- Use of wheat germ oil from starch production for the production of detergents (Wheatoleo)
- Biotechnological production of the cosmetic components hyaluronic acid and dihydroxyacetone (Soliance)

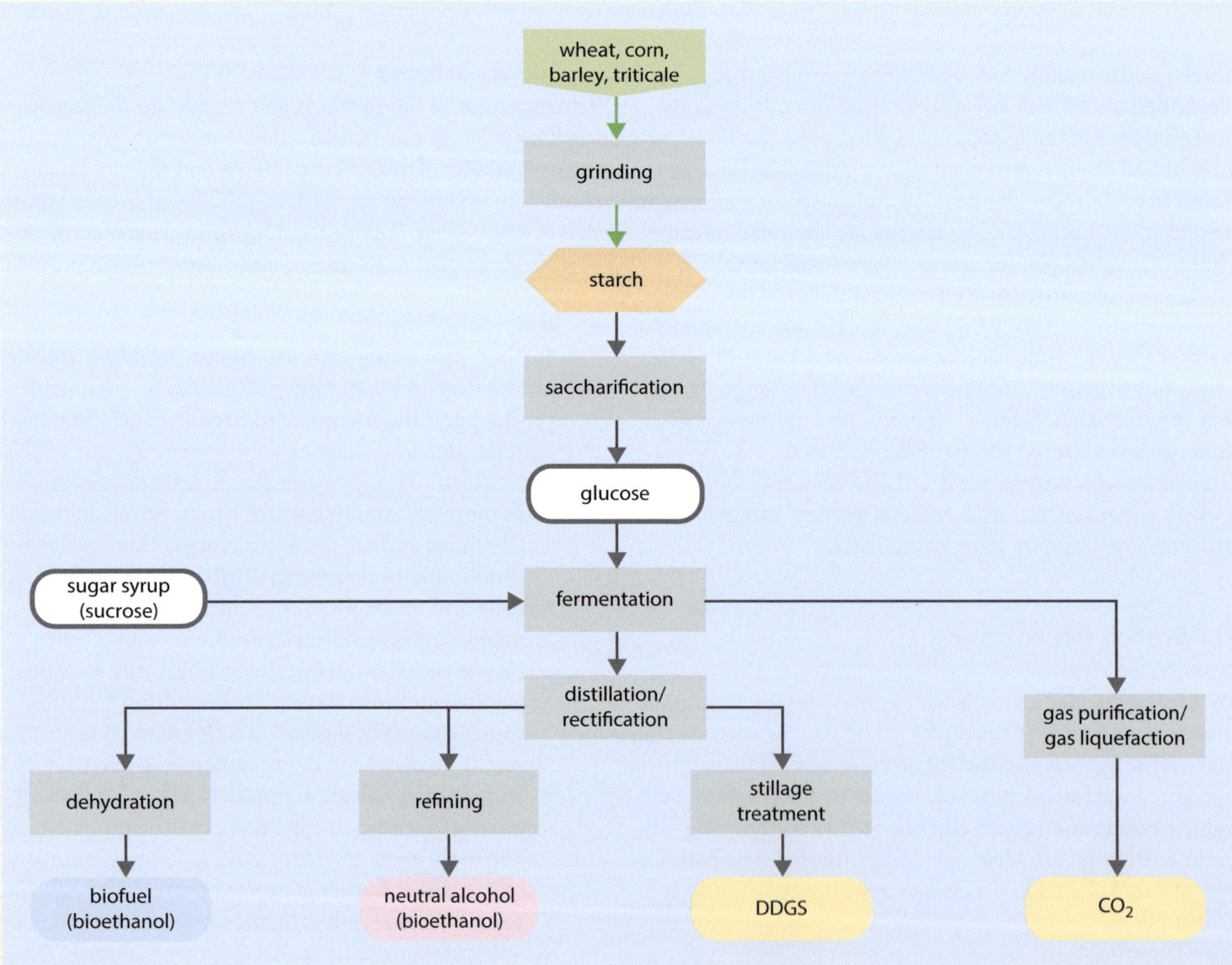

Fig. 4.17 Starch biorefinery pathway of CropEnergies Bioethanol in Zeitz, Germany

- Demonstration plant for the production of succinate (Biodemo)
- Pilot plant for the production of cellulose ethanol (Futerol)

The cluster also includes the private research centre Agro-Industry, Research & Development (ARD), the Open Innovation Platform B.R.I., and training and education facilities, which are bundled in the *Excellence Center in White Biotechnology* (ECWB). All companies share energy, heat and steam from the energy-related recycling of residual materials and the process water, which is used and treated in a closed circuit. However, the interaction among the companies, as well as the individual refining steps and the product portfolio of the cluster, is much more extensive than outlined here.

Maize-Based Integrated Starch Biorefinery

In Blair, Nebraska, USA, Cargill now operates a so-called biorefinery campus (Thielen 2010). In 1995, Cargill opened a corn-based starch factory there for the production of bioethanol, animal feed and corn syrup with a high fruit sugar content. The plant was then initially expanded so as also to produce maize germ oil. Since the year 2000, a large biorefinery campus has developed on this area, which resembles the integrated system of a chemical park: several companies have settled there that purchase raw materials, energy and operating supplies from the maize-based starch factory (▪ Fig. 4.18).

Raw Materials

Only maize is processed in the plant.

Primary Refining

The corn is wet milled. The husks, fibres, insoluble gluten and germ are separated from the starch. The germ oil is squeezed off from the germ. The husks, fibres and proteins are further processed into animal feed.

Secondary Refining

The starch is saccharified and then processed into either glucose or corn syrup with a high fructose content. Glucose is fermented in the factory into bioethanol and processed into biofuel (capacity: 800,000 m^3 or 628,000 t).
In addition, chemically modified starch is produced.
On the biorefinery campus, the following production lines of the companies located there are supplied, directly or indirectly, with glucose from the corn starch factory:

- Biotechnological production of erythritol, a sugar substitute derived from glucose (Cargill Polyols, a joint venture of Cargill and Mitsubishi Chemical), capacity: 20,000 t per year
- Chemical production of maltitol, a sugar substitute derived from maltose (Cargill Polyols)
- Biotechnological production of lactic acid from glucose for pharmaceutical and cosmetic purposes (Cargill/PURAC-PGLA-1), capacity: 35,000 t per year
- Biotechnological production of lactic acid from glucose and chemical processing into polylactide (PLA), a bio-based plastic made from lactic acid (joint venture of Cargill and Dow, now NatureWorks), capacity: 140,000 t per year
- Biotechnological production of L-lysine, an amino acid for animal feed (joint venture Cargill/Evonik), capacity: 280,000 t per year
- Biotechnological production of enzymes for biofuels and applications in industrial biotechnology (Novozymes)

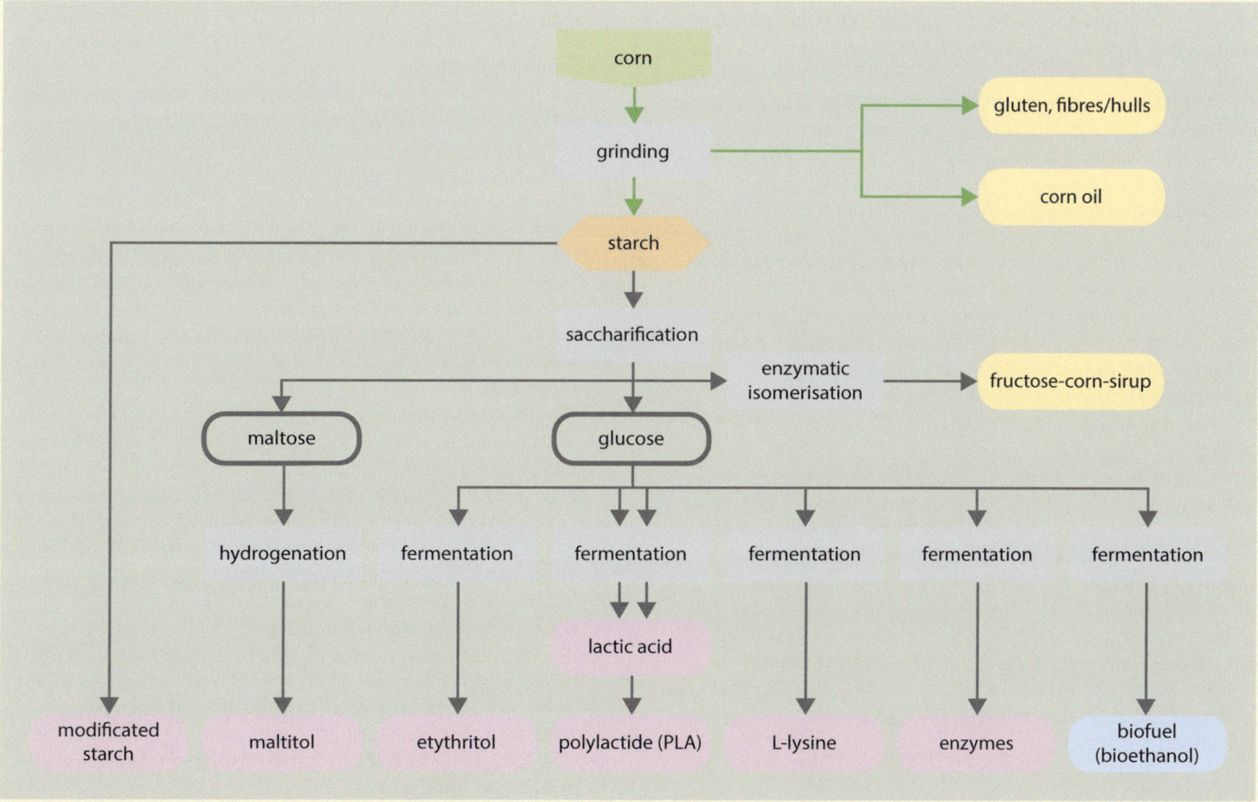

▪ **Fig. 4.18** Maize-based starch biorefinery pathway of the Cargill Biorefinery Campus in Blair, Nebraska, USA

4.3.3 Vegetable Oil Biorefinery

A vegetable oil biorefinery combines the processes of an oil mill for oil extraction in primary refining with oleochemical applications in secondary refining to produce platform chemicals or biofuels (biodiesel).

Raw Materials

The raw materials for the production of vegetable oil are oilseeds and oleaginous fruits. The most important oilseeds worldwide are rapeseed, soybeans, sunflowerseed, cottonseed and peanuts. The most important global oleaginous fruits are oil palm fruits and oil palm kernels, coconuts and olives. Rapeseed is by far the most important oil plant in Germany.

Depending on the origin of the oilseeds and oleaginous fruits, the oils obtained differ in their fatty acid spectra, which is important to consider for subsequent use. Short chain, saturated fatty acids are mostly found in fats and fatty oils from tropical countries of origin (palm oil, palm kernel oil, coconut fat). Indigenous oils, contrastingly, contain long-chain and (poly)unsaturated fatty acids (rapeseed oil, sunflower oil, linseed oil, soybean oil).

Primary Refining

The oilseeds or oleaginous fruits are cleaned, crushed and pre-swollen. The vegetable oils are then extracted by pressing. In large oil mills, extraction of the press cake with hexane further increases the oil yield. Native vegetable oil is usually purified of accompanying substances before use or further processing.

Secondary Refining

For further processing, native vegetable oil (fats and fatty oils) is available as a raw product. Extraction meal from oilseeds such as rape and soya or press cakes occur as by-products, which are marketed as protein-rich animal feed.

In the energy sector, vegetable oil is used either as a fuel after transesterification (biodiesel) or as a native vegetable oil to generate electricity and heat, for example, in combined heat and power (CHP) plants.

In the chemical-technical sector, vegetable oil is either used directly, e.g., as a solvent, or is initially split into fatty acids and glycerol. Fatty acids are the starting material for a whole range of chemical products and, after refinement, can be found in, e.g., cosmetics, surfactants, paints and varnishes. Glycerol also has many applications. Further processing results in pharmaceutical glycerol or other chemical intermediates and products. Glycerol can also be used as a raw material for fermentation.

Thistle-Based Integrated Vegetable Oil Biorefinery

The Italian companies Novamont, one of the largest producers of bio-based plastics, and Versalis, the chemicals division of the Italian mineral oil company Eni, have set up a joint venture called Matrìca to transform a discontinued chemical site in Porto Torres, Sardinia, into a vegetable oil biorefinery based on thistles. They are thus implementing the idea of a biorefinery integrated into the regional community: On the one hand, agriculture benefits from the cultivation of previously abandoned land that is not suitable for growing food or animal feed, as well as from the development of new agricultural machinery for growing, harvesting and processing plants. On the other hand, universities and research institutes are also involved in order to optimise processing and develop new value chains. Since 2015, the Matrìca biorefinery has therefore been a so-called lighthouse demonstration project of the EU funding initiative *Bio-based Industries Initiative* (BBI), whose objectives are the promotion of investment and the creation of a competitive market for bio-based products and materials *made in Europe* (FIRST2RUN 2015).

The main raw material source of the biorefinery is an indigenous thistle species, which is undemanding and grows on barren, unirrigated soils. The thistle oil is processed in an environmentally-friendly way into bio-based, biodegradable plastics, lubricants, additives and other products. The production capacity is currently 35,000 tonnes per year. Part of the vegetable oil and the protein fraction from the thistles are used as animal feed. Residual materials containing lignocellulose (stems, leaves, husks) are processed into pellets and are used in the biorefinery as fuel to provide energy and heat (◘ Fig. 4.19).

Raw Materials

The plant mainly processes artichoke thistle, which has a fatty acid spectrum similar to that of sunflower.

Primary Refining

The seeds of the thistle are separated from the other parts of the plant and pre-ground. The oil is squeezed out and purified.

Secondary Refining

Native thistle oil is available as a raw product. A press cake made of protein-rich meal and vegetable oil, which is used as animal feed, and the lignocellulose-containing plant parts, which are processed into pellets, are the by-products.

The thistle oil is chemically processed into various products: In a first step, short-chain, saturated fatty acids are separated from the oil. This is achieved by hydroxylation of double bonds in the oil and subsequent oxidative cleavage. As thistle oil contains a lot of oleic acid, the saturated C9 fatty acid pelargonic acid is mainly formed during this splitting process. The acid is separated from the other cleaved saturated fatty acids (called the "fatty acid mix," C5-C9) and the remaining oil.

The remaining oil is then hydrolyzed to glycerol and fatty acids and fractionated. In addition to unsaturated long-chain fatty acids (palmitic acid, C16, and stearic acid, C18), the fatty acid fraction also contains dicarboxylic acids, which were formed during the cleavage of the double bonds mentioned above. Since most unsaturated fatty acids in thistle oil carry the first double bond on the ninth carbon atom (omega-9), this is mainly azealic acid, the C9 dicarboxylic acid. The aforementioned products are either marketed or further processed within the biorefinery complex:

- Pelargonic acid has many applications: It is esterified into biolubricants, but can also be used in the cosmetics and food industries. Moreover, pelargonic acid is a natural pesticide.
- The fatty acid mix (C5-C9) is also processed into biolubricants for various applications.
- Glycerol is purified and used in the pharmaceutical and cosmetics industries. It is also a component of antifreeze agents.

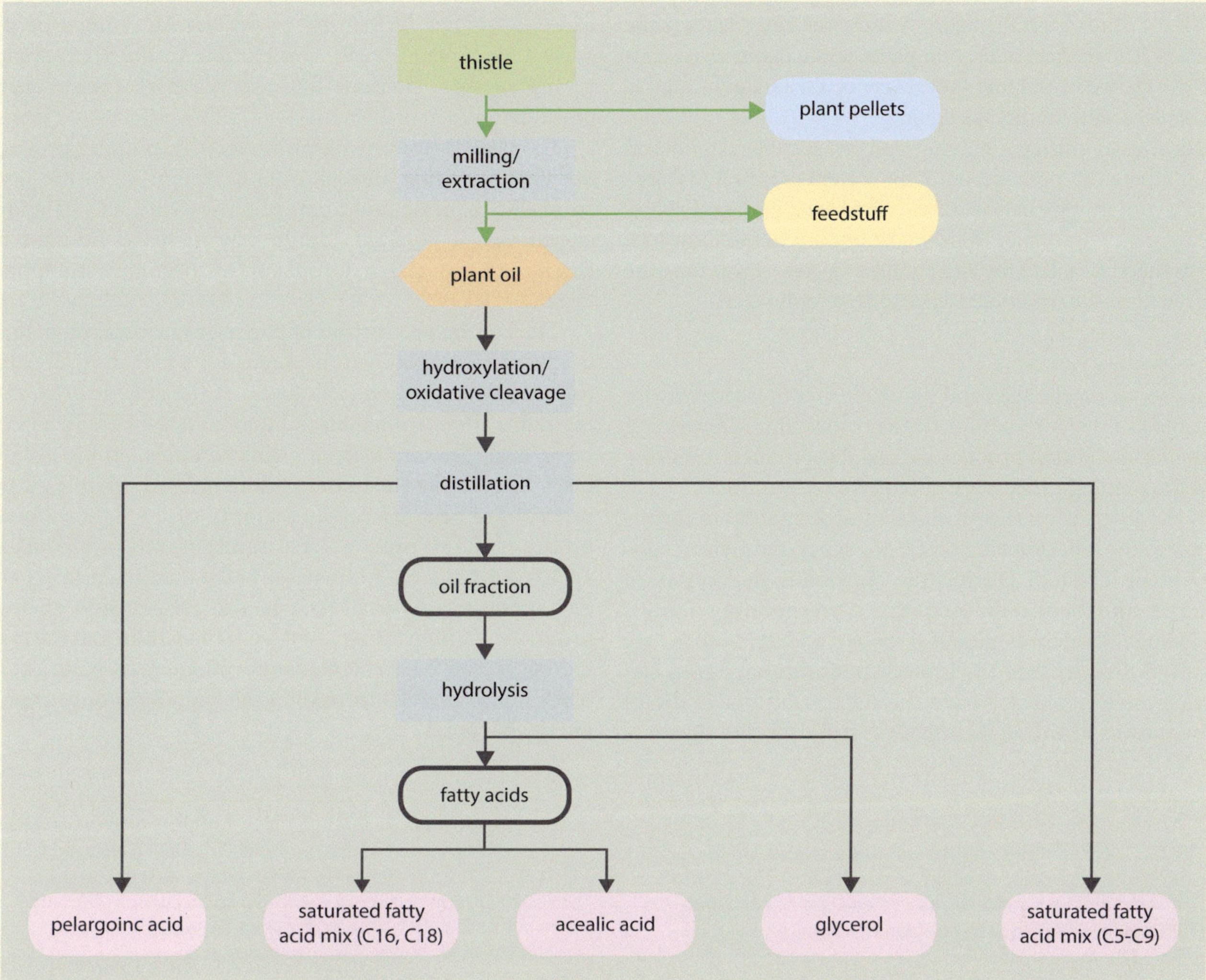

Fig. 4.19 Thistle oil biorefinery pathway of the Matrìca biorefinery (monomer production) in Porto Torres, Sardinia, Italy

- The mixture of unsaturated long-chain fatty acids forms a wax-like substance at room temperature. It is used in the production of waxes, soaps and candles, and as a vulcanization accelerator in rubber production.
- Finally, azealic acid is a major component of bio-based plastics as a dicarboxylic acid. It is also used for the production of lubricants. In PVC plastics, it can replace carcinogenic phthalates as plasticizers. Azealic acid is also used as an active component in the treatment of acne.

The thistle oil can, of course, also be processed into biodiesel, which can then be used in agriculture for the climate-friendly cultivation of thistles. Another direct application of the oil is that of a plasticizer oil in the tire industry.

4.3.4 Lignocellulosic Biorefinery

Raw Materials

A lignocellulosic biorefinery uses agricultural residues (e.g., straw, bagasse, husks and pods, corncobs) and wood and woody biomass, but also (woody) annual and perennial grasses, as its sources of raw material. The term lignocellulose represents the three platforms – cellulose, hemicelluloses and lignin – of which lignified biomass mainly consists.

Primary Refining

After a mechanical pre-treatment step (shredding, grinding), the lignocellulose is physico-chemically treated by pressure and temperature and, if necessary, with the aid of organic solvents and reagents (pulping process). Depending on the process and application, fractionation into raw products takes place within the course of primary or secondary refining.

The ratios of the three fractions of lignocellulose differ depending on the type and origin of the biomass, wherein – roughly speaking – cellulose makes up half and hemicelluloses and lignin a quarter each (▶ Sect. 4.2.1). The type of biomass also determines the fibre length of the cellulose and the chemical composition of the hemicelluloses and lignin in particular. This explains why, depending on the raw material source and production target, there are already many variants of primary refining. A paper mill, for example, uses processes that primarily protect the cellulose fibres and their length, while a plant that produces glucose via enzymatic hydrolysis optimises the separation of the fibres from the lignin because the latter hinders the enzymes. The desired utilization paths of the lignin also influence the choice of pulping process, since it can chemically, as well as physically, modify the lignin.

Thus, each primary refining and its variable parameters (e.g., temperature, pressure, duration, solvent use, catalysts) represent a compromise between the desired exploitations. Process development *(Upscaling)* from the laboratory to the industrial scale is therefore a lengthy and complex process. Each time the scale changes, the parameters have to be re-optimized.

Secondary Refining

Cellulose, hemicelluloses and lignin, or mixtures thereof, are available as raw materials for further processing. Depending on the kind of main product desired, by-products resulting from the pulping process are obtained as co-products.

If the focus is on the production of fermentable carbohydrates, the fractions are often not separated before subsequent processing. During the enzymatic hydrolysis of cellulose and hemicelluloses, the corresponding monomeric sugars, such as glucose, xylose and arabinose, are dissolved directly into the broth. The insoluble lignin can then be easily separated from the sugar solution. In direct fermentation of the sugar solution, however, the lignin is often separated from the stillage only after it has been fermented. If, on the other hand, the sugar solution is not directly further fermented or processed, it must be preserved by concentration, comparable to the processing in a sugar factory. Thereby, lignin serves as a source of energy and heat.

In order to use the lignin materially, pulping processes are necessary that allow for good separation of the lignin from the cellulose and hemicelluloses without changing it chemically, if possible. Lignin can currently be used primarily as a material in duroplastic or thermoplastic applications.

During the production of paper or chemical pulp, lignin is chemically modified in such a way that it dissolves, together with the hemicelluloses, in the aqueous treatment medium. The so-called black liquor can then easily be separated from the undissolved cellulose fibres. So far, only the black liquor from the rather seldom utilised sulphite pulping process has been used. The majority of the lignosulfonates produced in this process serve as dispersants or binders and are used as concrete additives or in the manufacture of paints and varnishes. A small part is also refined into chemical products: Vanillin is extracted from the lignin, and the residual sugar can be fermented into ethanol. In contrast, the black liquor derived from the kraft pulping is only used for energy purposes.

Straw-Based Lignocellulosic Biorefinery

The specialty chemicals manufacturer Clariant Produkte Deutschland GmbH operates a demonstration plant in Straubing for the production of ethanol from straw. The integrated approach of the plant is represented in the fact that the enzymes and microorganisms required for the fermentation process are produced *on-site* straight from partial flows of the process (◘ Fig. 4.20). The energy required for the process is completely provided by lignin combustion. This means that CO_2-savings of up to 95% compared to conventional fuels are possible. The demonstration plant has a production capacity of 1,000 t of ethanol per year (Clariant 2014). The ethanol is marketed under the name Sunliquid® and is used as biofuel and in cleaning agents (Werner & Mertz 2016). As part of the EU project SUNLIQUID (2014–2018), the construction and operation of a large-scale plant in Romania will demonstrate the economical production of cellulose ethanol using this technology on a commercial scale (SUNLIQUID 2015).

Raw Materials

In addition to wheat straw, the plant could process other agricultural residues, such as bagasse and maize straw.

Primary Refining

The raw materials are shredded and treated in a mechanical and thermal pulping process without fractionation of the straw components cellulose, hemicelluloses and lignin.

Secondary Refining

A part of the product stream of the primary refinery is used as substrate for the biotechnological production of enzymes (cellulases, hemicellulases) for saccharification. However, enzyme production is optimised for the raw material used in each case. The enzyme cocktail produced in this way is used to saccharify the cellulose and hemicelluloses platforms. Following saccharification, in which the fermentable sugars have been dissolved, the insoluble lignin is separated and dried. It is used to generate process energy (electricity, steam, heat) by combustion.

A partial stream of fermentable C5 and C6 sugars is used to produce precultures for inoculation of the actual ethanol fermentation (yeast propagation). The specialized yeast strains used are able to ferment C6 sugars (glucose), as well as the C5 sugars (xylose, arabinose), from hemicelluloses into ethanol simultaneously. In Straubing, the main product ethanol can be separated either by distillation or by using an adsorber process from the stillage (vinasse). For its application as biofuel, ethanol is also dehydrated. The stillage (vinasse) is utilized in a nearby biogas plant or applied directly to the fields as a liquid fertilizer. As is true with all second-generation ethanol plants, the vinasse cannot be used as animal feed.

Beech Wood-Based Lignocellulose Biorefinery

Located at the Fraunhofer Center for Chemical-Biotechnological Processes (CBP) in Leuna, the pilot plant of a lignocellulose biorefinery was constructed and commissioned in 2012 with the support of a publicly funded research project (◘ Fig. 4.21). The aim of the plant is to optimize the pulping of hardwood and the fractionation of lignocellulose in such a way that not only will the saccharification of cellulose and hemicelluloses be possible, but so will the coincident processing of the lignin for material use. The applica-

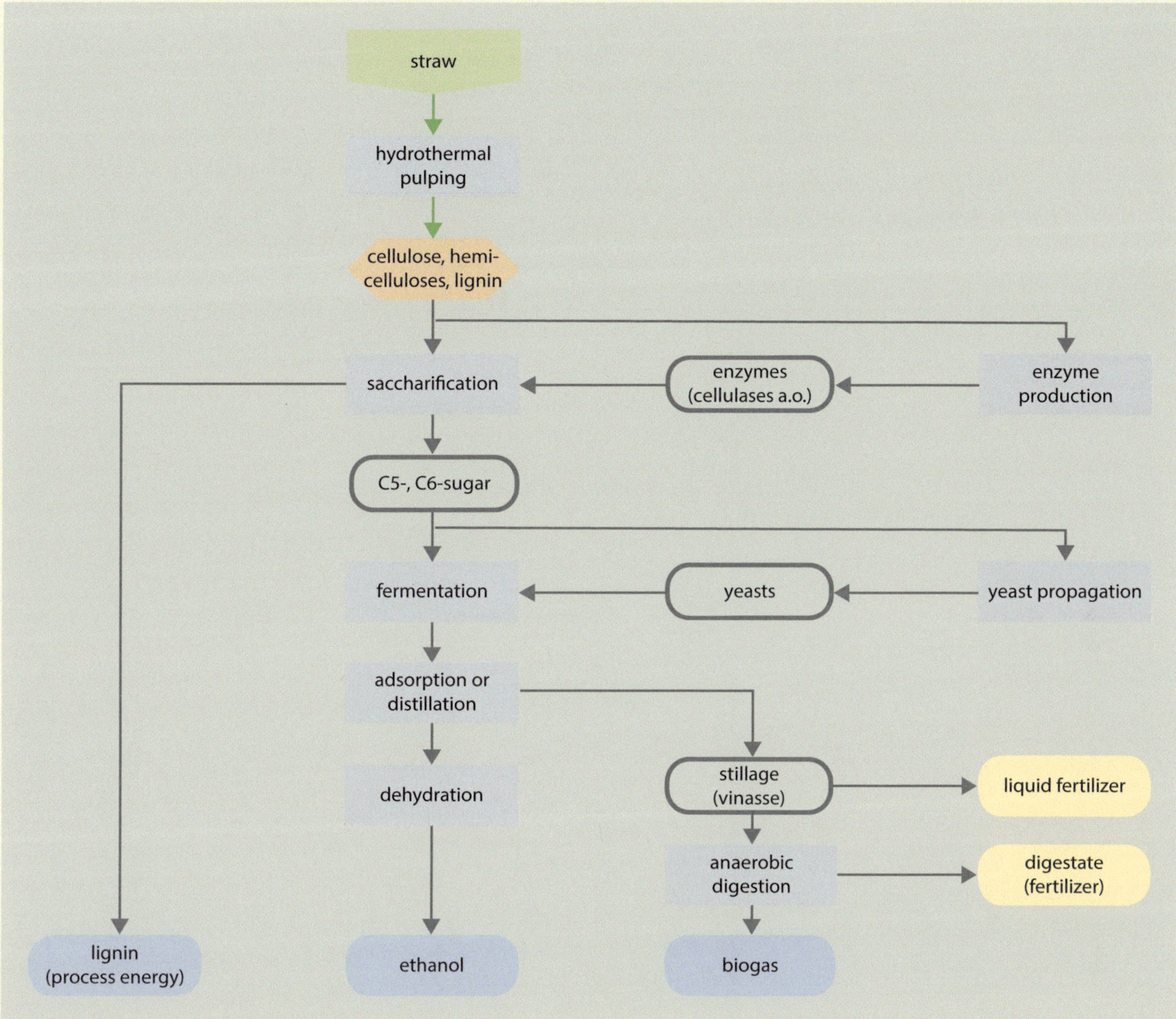

Fig. 4.20 Lignocellulose biorefinery pathway at the Clariant Sunliquid® demonstration plant in Straubing, Germany

tion spectrum of the sugar solutions obtained has been tested for various chemical and biotechnological processes. The suitability of the lignin fraction in various material and chemical utilization pathways has also been investigated. In the pilot plant, about 1 t of beech wood chips per week can be treated and fractionated (Laure et al. 2014).

Raw Materials
In the pilot plant, beech and poplar wood chips are processed.

Primary Refining
The beech wood chips are treated using the so-called Organosolv process. This hydrothermal pulping takes place in a solvent mixture of water and ethanol. As a result, a large fraction of the lignin is dissolved, together with the hemicelluloses, in the course of the pulping process and can easily be separated from the undissolved cellulose fibres. Thereafter, the lignin can be precipitated from the mother liquor by reducing the ethanol:water ratio – either by adding water or by distilling the ethanol. Thus, after primary refining, the three platforms cellulose, hemicelluloses and lignin are separated.

Secondary Refining
The three platforms are processed individually. The cellulose fraction is hydrolyzed enzymatically into glucose (i.e., saccharified). Any residual lignin still adhering to the fibre is separated as a solid after hydrolysis and added to the main lignin fraction. For preservation, the glucose solution is thereafter concentrated to a sugar content of more than 60%, similar to the preparation of thick juice in the sugar industry.

The lignin is merely washed and dried after precipitation.

The hemicellulose fraction not only contains hemicelluloses, but also sugar monomers, which have already formed during pulping. In addition, the fraction also includes soluble co-products from beech wood, like acetic acid, or by-products formed in the pulping process, like furfural and hydroxymetylfurfural, as well as soluble lignin components. Before C5 sugar can be used as raw material for fermentation, co- and by-products must be separated.

4

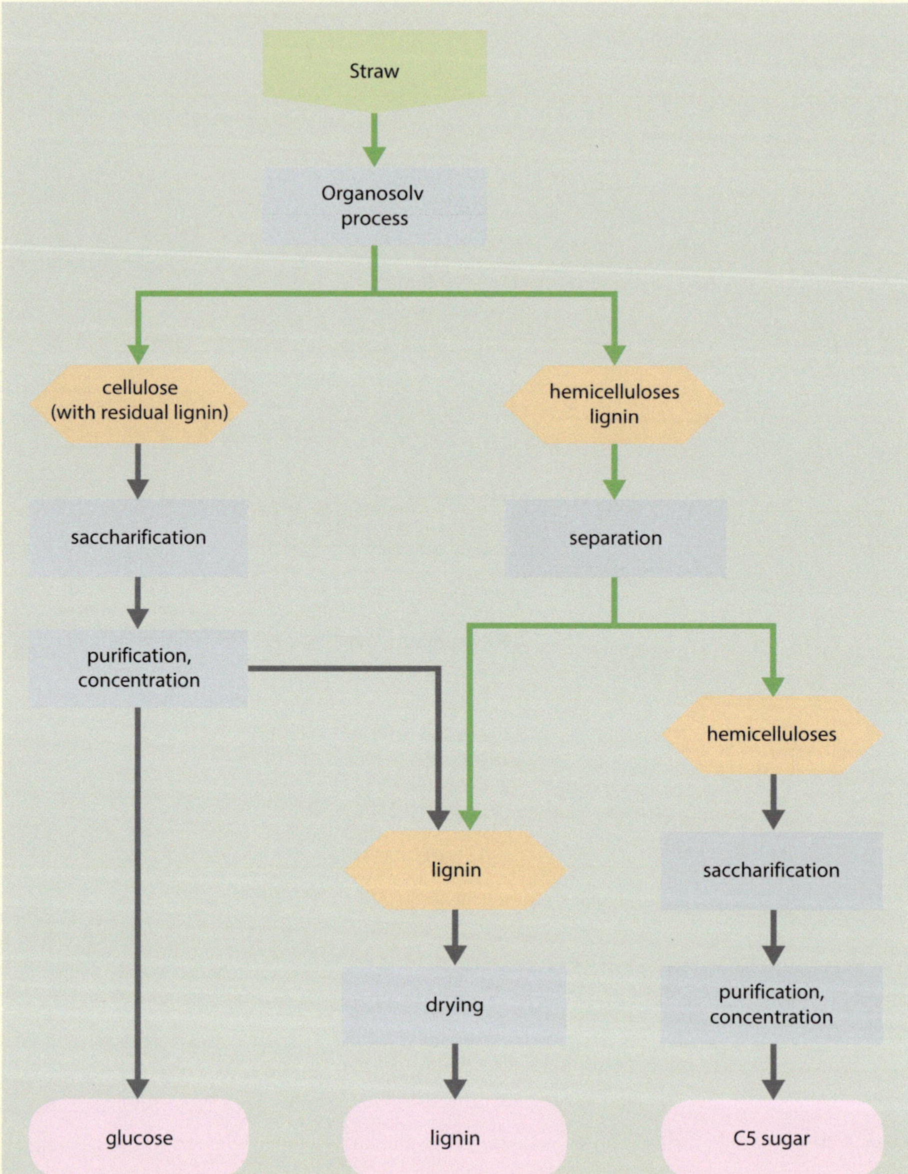

Fig. 4.21 Lignocellulose biorefinery pathway of the pilot plant at CBP in Leuna

Further refinement of the products does not take place in the pilot plant. Within the scope of the research project, however, the following applications were investigated:

- Fermentations with glucose as a substrate to ethanol, acetone-butanol-ethanol (ABE), succinic acid, itaconic acid, lactic acid and acetic acid as platform chemicals
- Reduction of glucose to sorbitol as a sweetener or as a platform chemical for polyurethane production
- Use of lignin as thermoplastic in blends and composites
- Use of lignin for partial replacement of phenol in phenol-formaldehyde resins
- Use of lignin as a polyol component in polyurethane (PU) and PU foams

The soluble lignin components formed during pulping were also tested as mediators in enzymatic pulp bleaching. Various cleavage reactions were also tested to obtain phenolic monomers from the lignin fraction.

Within the scope of further research work, new fields of application for lignin in particular are currently being developed. A much larger demonstration plant is currently in the planning stage.

4.3.5 Green Biorefinery

Raw Materials

In a green biorefinery, wet biomasses such as annual or perennial grasses and grains in green or silaged form are used as raw materials. The proper platforms are the press juice and the plant fibres in the press cake. However, the term 'green biorefinery' has become established for this type of plant.

Primary Refining

The green or ensilaged biomass is cleaned and shredded. The liquid components are subsequently separated by pressing. All soluble components can be found in the press juice. The fractions remaining are the press juice and the press cake.

Secondary Refining

The press juice and press cake are available as raw materials for further processing.

Green press juice contains, among other things, sugar, amino acids and proteins. Silaged press juice contains organic acids instead of sugar, mainly lactic acid. Press juice is either used directly in a biogas plant or its constituents are separated. Green press juice can also be used as a raw material or supplement, for example, as a nitrogen source for fermentations.

The press cake is processed directly into animal feed or serves as raw material for fibre-based products (insulating materials, cellulose fibres, fibre-reinforced plastics). After hydrolytic cleavage (saccharification), the fibre fraction can also be used as a raw material for fermentation.

Residuals from the press juice, press cake and fibre processing can be used as co-substrates in a biogas plant.

Examples of Further Green Biorefinery Concepts Still in the Research Stage

While, in Brensbach, the focus is on the material use of the fibre, other biorefinery concepts focus more on the utilization of the press juice. After ensilaging, it contains a high proportion of organic acids, in particular, lactic acid, acetic acid and amino acids, as well as proteins. At the "Grüne Bioraffinerie Utzenaich" demonstration plant in the Austrian state of Upper Austria, lactic acid and amino acids are obtained, while the fibre fraction and other residues are fermented in a biogas plant (BMVIT 2009).

In the context of a research project at the TU Kaiserslautern, Germany, it was investigated whether both the fiber fraction and the press juice are suitable for the production of platform chemicals. Lactic acid and acetic acid were isolated from the press juice. Lactic acid could be further fermented into lysine. The fibre fraction was pulped in a way similar to that of a lignocellulose biorefinery, hydrolyzed enzymatically into glucose and fermented. The special feature of this process is that saccharification and fermentation take place simultaneously in a reactor (SSF). Fermentations into ethanol, 1,2-propanediol, succinic acid and itaconic acid were tested (Sieker et al. 2011).

Meadow Grass-Based Green Biorefinery

In the grass refinement plant of Biowert Industrie GmbH in Brensbach, Germany, grass from permanent grassland from regional cultivation within a maximum radius of 30 km is processed (◘ Fig. 4.22). The throughput capacity is 5,000 t of grass dry matter per year, which corresponds to a silage volume of 16,000–18,000 t. While the grass juice is co-fermented with regional biowastes into biogas and fertilisers (AgriFer) in the company's own biogas plant, the dried grass fibres are further processed into insulating material (AgriCell) or used in composite materials with plastics (AgriPlast).

Raw Materials

Grass cuttings are used as raw material. In addition, regional biowaste (food waste and cattle manure) is used for the co-fermentation of the grass juice in the biogas plant.

Primary Refining

The grass cuttings that are slightly dried on the field are first ensilaged on site for conservation and preconditioning in bunker silos under exclusion of air. Before mechanical treatment, the grass silage is slurried. Fibres and the liquid phase (press juice) are ultimately fractionated by pressing.

Secondary Refining

The plant fibres are cleaned and dried. For the production of the insulating material (AgriCell), the fibres are impregnated with a fire retardant (borax/boric acid). The composite material (AgriPlast) is produced by mixing and granulating the grass fibres with conventional (polypropylene, PP) or bio-based (polylactide, PLA) plastic, depending on the application. The granulate is suitable for injection moulding or extrusion.

The pressed juice is fermented, together with regional biowaste, into biogas in the company's own biogas plant. The fermentation residue is separated into liquid and solid phases and marketed regionally as a fertilizer (AgriFer) in order to close the material cycles. The liquid phase is concentrated through reverse osmosis.

The plant is autarkic. The source of all of the plant's process energy (electricity, heat) is self-generated biogas, which is provided via combined heat and power (CHP). The waste heat is used for slurrying and drying, as well as for the hygienisation of biowaste. The plant also produces net electricity. The water from reverse osmosis is also reused as process water for slurrying the silage.

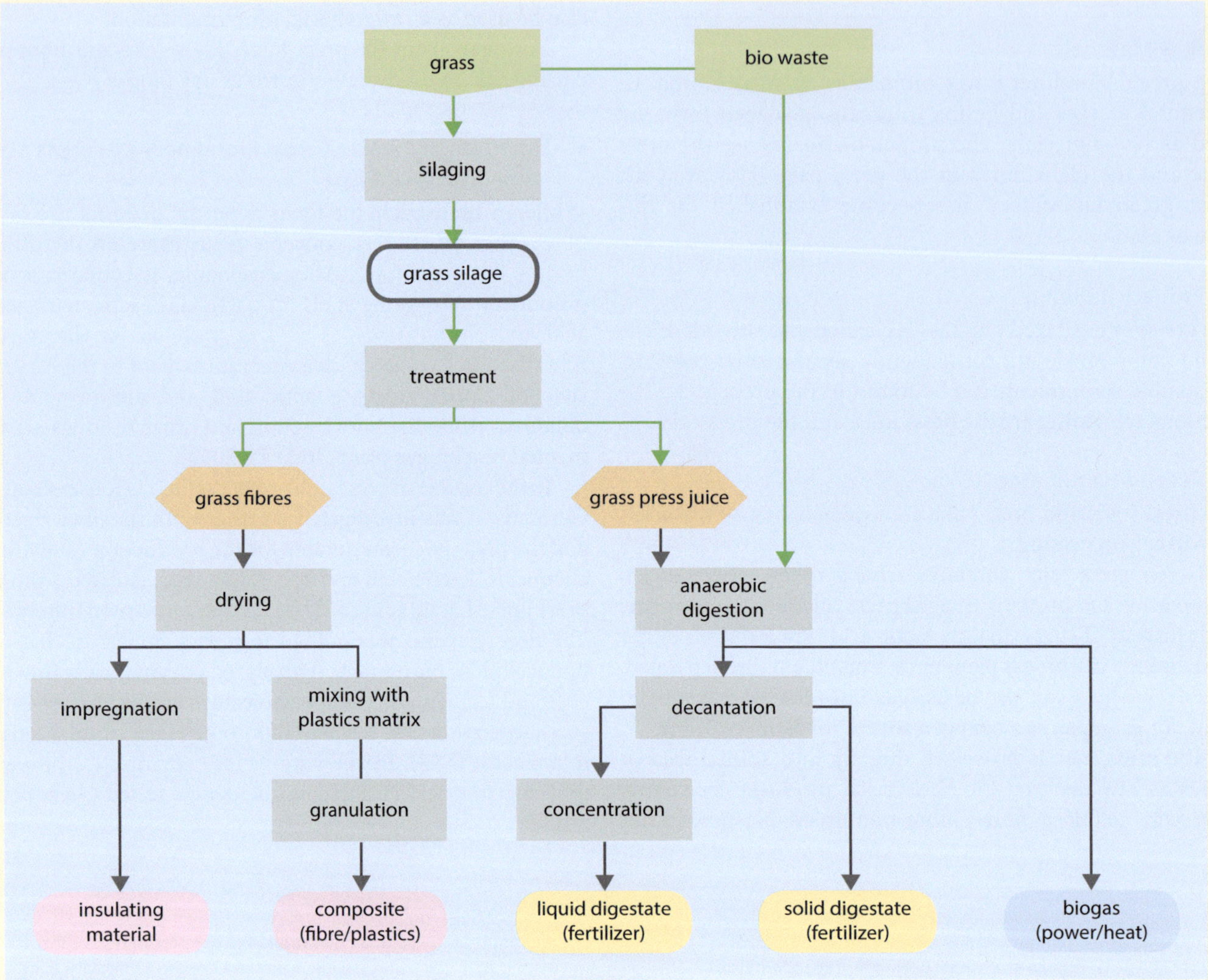

Fig. 4.22 Green biorefinery pathway of Biowert Industrie GmbH in Brensbach, Germany

4.3.6 Synthesis Gas Biorefinery

In this kind of biorefinery, biomass is completely gasified. Platform chemicals and biofuels can then be produced from the synthesis gas obtained via methanol synthesis or Fischer-Tropsch synthesis. The special feature of this biorefinery type is the complete defunctionalization of the biomass into synthesis gas (carbon monoxide and hydrogen) in primary refining and the total synthesis of chemical products from synthesis gas in secondary refining. Although this means that nature's advance synthesis performance is completely lost, it has the advantage that secondary refining can take place independently of the raw material in question. Originally, the process was developed within the course of coal gasification and liquefaction. But natural gas can also be converted into synthesis gas by steam reforming. As part of the energy transition, processes are now also being investigated into the use of excess energy from renewable energies in order to generate synthesis gas from CO_2 and water for the production of liquid fuels. According to their source of raw materials, fuels from these processes are called CtL – *Coal to Liquid*, GtL – *Gas to Liquid*, PtL – *Power to Liquid* or just BtL – *Biomass to Liquid*. The latter refers to the technology underlying a synthesis gas biorefinery.

Raw Materials

Dry biomass, in particular, annual and perennial grasses, agricultural residues (e.g., straw, bagasse, husks and pods, corncobs), wood and woody biomass, and biogenic residues (e.g., waste paper and lignin), can be considered as a source of raw materials.

Primary Refining

The first step in primary refining is biomass pre-treatment. The biomass is either dried in a short-term drying process at 200–300 °C with oxygen excluded (torrefaction) with subsequent grounding or it is thermally decomposed at higher temperatures (400–600 °C) with oxygen excluded (pyrolysis). In the latter process, pyrolysis coke, oil, and gas, as well as water, are produced in different ratios, depending on the process conditions. The products of this pre-treatment are ultimately gasified at much higher temperatures (800–1200 °C) by sub-stoichiometrically adding an oxidizing agent (oxygen, air, steam), which converts the carbon and hydrogen chemically bound in the biomass into carbon monoxide and elementary hydrogen. Afterwards, the synthesis gas must be purified of accompanying gases, such as CO_2, CS_2, HCl, NH_3, HCN and, if required, N_2, in the case in which air is used as the oxidizing agent.

Secondary Refining

The raw synthesis gas is available for further processing. For the subsequent syntheses, the CO:H_2 ratio has to be set specifically. The water-gas shift reaction, which allows for the adjustment of any ratio of the two gases, serves this purpose.

In the next step, the so-called synthesis step, synthesis gas is processed into chemical platforms (e.g., methanol or dimethyl ether, hydrocarbons) or fuels. Using the water-gas shift reaction, it is also possible to provide H_2 as the sole product. Synthesis of methane (*Synthetic Natural Gas*, SNG) is also possible through steam reforming. Alternatively, the synthesis gas can be fermented into alcohols, as mentioned above.

A disadvantage of the process is that remaining ashes and other solid components generally cannot be utilized, due to their properties and legal frameworks (e.g., fertilizer and waste management legislation), and must therefore be disposed of.

Straw-Based Synthesis Gas Biorefinery

The Karlsruhe Institute of Technology (KIT), Germany, has constructed a pilot plant for a synthesis gas biorefinery that can produce fuels from straw or wood using the so-called Bioliq process. The special feature of this concept is that the pyrolysis step in the course of primary refining is carried out decentrally. Useful for transportation, energy-compacted slurry is produced from straw, which is then gasified in the central Bioliq pilot plant. The target of the synthesis is dimethyl ether (DME), which is further processed as a platform chemical into fuels (gasoline and diesel fuels), among others (Fig. 4.23). 1 t of fuel can be produced from about 7 t of straw; the energy required is generated in the process itself. The following process steps can also be found in the data sheet on the Bioliq process (KIT 2016).

Raw Materials

The main raw material used for the process is straw.

Primary Refining

The dry, shredded biomass is pyrolysed with hot sand in a twin-screw mixing reactor within a few seconds at around 500 °C. The result is liquid pyrolysis oil and pyrolysis coke. What remains is a pyrolysis gas. Its combustion can be used to heat the sand or to dry the biomass. Pyrolysis coke and pyrolysis oil are mixed to form a suspension (biosyncrude).

The production of the biosyncrude is planned to be carried out on a large scale in a decentralised manner in the vicinity of straw occurence, while the gasification and the further process steps take place in a central plant.

The biosyncrude is atomized with technical oxygen in a high-pressure entrained-flow gasifier and converted at over 1,200 °C into a tar-free, low-methane raw synthesis gas. The type of gasifier used is particularly suitable for ash-rich biomass.

The resulting raw synthesis gas is purified of particles, alkali salts and gases by hot gas cleaning, in order to prevent catalyst poisoning during fuel synthesis.

Secondary Refining

For the methanol/DME synthesis, an adaptation of the CO:H_2-ratio from the gasification of biomass is not required. The synthesis gas is directly converted into dimethyl ether via methanol in a single-stage process. Fuel synthesis is then carried out by converting DME into long-chain alkanes via ethene and propene.

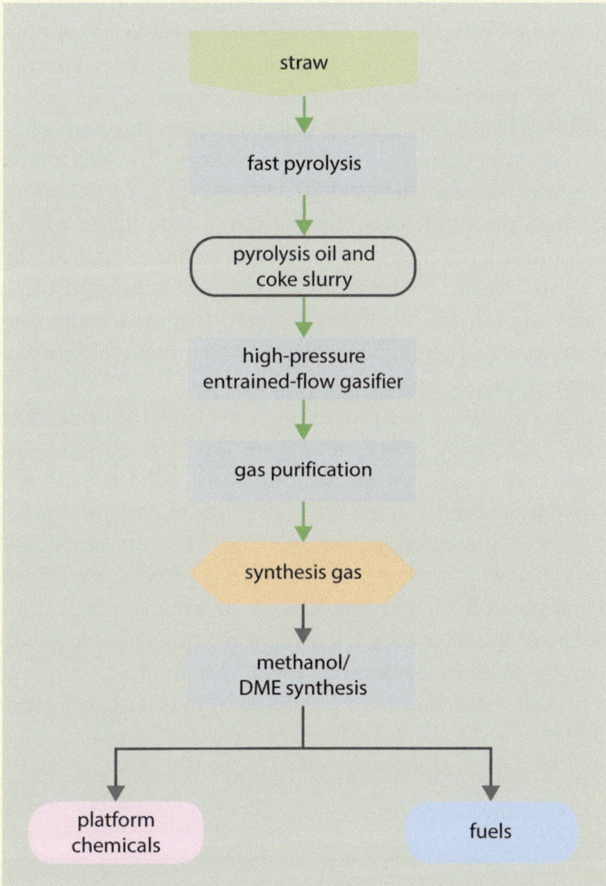

Fig. 4.23 Synthesis gas biorefinery pathway of the Bioliq process at the KIT in Karlsruhe, Germany

4

4.3.7 Biorefinery Concepts Based on Algae

Biorefinery concepts based on microalgae are special, inasmuch as the cultivation of algal biomass is an integral part of the entire plant concept. The advantages of algal cultivation are that the algae perform with a better efficiency of photosynthesis than land plants and that it is not dependent on agricultural land. Microalgae biomass is particularly suitable for processing in a biorefinery, because, unlike land plants, microalgae do not contain lignocellulose. Therefore, they are easy to disrupt and can be processed as a whole biomass or in their individual components without problems. Certain types of microalgae can have an oil content of up to 50% of their dry matter, which makes them an interesting source of biofuels. In addition, various microalgae species contain a variety of ingredients that are suitable for the production of food, cosmetics and pharmaceuticals. Their exploitation is a prerequisite for making an algae biorefinery profitable (DECHEMA 2016).

The various utilization opportunities of microalgal biomass in a biorefinery are not yet fully foreseeable, as very few algae species have been used industrially to date. For this reason, no practical example of such a type of biorefinery should be given here. Theoretically, however, the path of an algae biorefinery can be outlined as follows: The high-value polyunsaturated fatty acids (PUFA) from the oil fraction are used for human nutrition. The saturated fatty acids, on the other hand, are processed for technical purposes, such as biodiesel production. Proteins for animal nutrition and other valuable ingredients are then extracted from the remaining biomass and marketed, while the resulting residual biomass is used to generate energy in a biogas plant.

Algae cultivation and processing can also be integrated into existing biorefinery concepts if large quantities of CO_2 are formed. Such a concept has recently been tested as part of a research project at the starch biorefinery site in Zeitz, which has already been presented above: The CO_2 from the bioethanol plant is used to grow algae, so as to isolate proteins for animal feed and starch from the algae biomass for ethanol production, while the residual biomass is used in the biogas plant to generate energy (combined heat and power). The final report on the joint project has already been published by the German Agency for Renewable Resources (Schmid-Staiger et al. 2016).

References

acatech (2012) Biotechnologische Energieumwandlung in Deutschland – Stand, Kontext, Perspektiven. acatech – Deutsche Akademie der Technikwissenschaften. http://www.acatech.de/de/publikationen/stellungnahmen/acatech/detail/artikel/biotechnologische-energieumwandlung-in-deutschland-stand-kontext-perspektiven.html. Accessed: 30.05.2016

Aden A, Bozell J, Holladay J, White J, Manheim A (2004) Top value added chemicals from biomass: volume I – results of screening for potential candidates from sugars and synthesis gas. http://www.pnl.gov/main/publications/external/technical_reports/PNNL-14808.pdf. Accessed: 31.05.2016

Bertau M, Räuchle K, Offermanns H (2015) Methanol – die Basischemikalie. Chemie in unserer Zeit 49(5):312–329. https://doi.org/10.1002/ciuz.201500689. Accessed: 31.05.2016

BGR (2012). Energiestudie 2012 – Reserven, Ressourcen und Verfügbarkeit von Energierohstoffen. Bundesanstalt für Geowissenschaften und Rohstoffe. http://www.bgr.bund.de/DE/Themen/Energie/Produkte/energiestudie2012_node.html. Accessed: 31.07.2016

BMBF (2016) Sicher, bezahlbar und sauber: 230 Partner starten größte Forschungsinitiative zur Energiewende / Wanka: "Erneuerbare Energieversorgung ohne Wohlstandsverlust ist machbar". Bundesministerium für Bildung und Forschung. Pressemitteilung 033/2016. https://www.bmbf.de/de/sicher-bezahlbar-und-sauber-2624.html. Accessed: 29.07.2016

BMVIT (2009) Internetseite: Demonstrationsanlage Grüne Bioraffinerie Utzenaich. http://www.fabrikderzukunft.at/results.html/id5502. Accessed: 25.05.2016

Bozell JJ, Petersen GR (2010) Technology development for the production of bio-based products from biorefinery carbohydrates-the US Department of Energy's "Top 10" revisited. Green Chem 12:539–554

BP (2016) Internetseite: BP statistical review of world energy. http://www.bp.com/en/global/corporate/energy-economics/statistical-review-of-world-energy/downloads.html. Accessed: 31.07.2016

Cherubini F, Jungmeier G, Wellisch M, Willke T, Skiadas I, Van Ree R, De Jong E (2009) Toward a common classification approach for biorefinery systems. Biofuels Bioprod Biorefin 3(5):534–546. https://doi.org/10.1002/bbb.172. Accessed: 31.05.2016

CIEC (2016). Internetseite: the essential chemical industry – online: ethene. http://www.essentialchemicalindustry.org/chemicals/ethene.html. Accessed: 01.05.2016

Clariant (2014) Internetseite: Sunliquid® – cellulosic ethanol from agricultural residues. https://www.clariant.com/en/Solutions/Products/2014/10/16/16/16/sunliquid. Accessed: 25.05.2016

CropEnergies (2016) Internetseite: CropEnergies AG. http://www.cropenergies.com/de/Home/. Accessed: 18.05.2016

De Jong E, Higson A, Walsh P, Wellisch M (2011) Bio-based chemicals: value added products from biorefineries. IEA Bioenergy Task42 Biorefinery. http://www.ieabioenergy.com/publications/bio-based-chemicals-value-added-products-from-biorefineries/. Accessed: 31.07.2016

DECHEMA (2016) Positionspapier: Mikroalgen-Biotechnologie – Gegenwärtiger Stand, Herausforderungen, Ziele. DECHEMA e.V. http://dechema.de/dechema_media/PP_Algenbio_2016_ezl-p-20001550.pdf. Accessed: 31.07.2016

EIA (2016) Internetseite: Europe Brent Spot Price FOB (Dollars per Barrel). U.S. Energy Information Administration. http://www.eia.gov/dnav/pet/hist/LeafHandler.ashx?n=PET&s=rbrte&f=D. Accessed: 31.07.2016

Faix O (2008) Chemie des Holzes. In: Wagenführ A, Scholz F (eds) Taschenbuch der Holztechnik. Fachbuchverlag im Carl Hanser Verlag, München, pp 127–259

FAO (2015) Pulp and paper capacities, 2014–2019. Food and Agriculture Organization of the United Nations. http://www.fao.org/forestry/statistics/81757/en/. Accessed: 31.05.2016

FCI (2009) Informationsserie "Nachwachsende Rohstoffe". Fonds der Chemischen Industrie im Verband der Chemischen Industrie e.V. https://www.vci.de/fonds/schulpartnerschaft/unterrichtsmaterialien/detailpage-84.jsp?fsID=30789. Accessed: 31.07.2016

FIRST2RUN (2015) Internetseite: flagship demonstration of an integrated biorefinery for dry crops sustainable exploitation towards bio-based materials production. http://www.first2run.eu/. Accessed: 25.05.2016

FNR (2014) Marktanalyse Nachwachsende Rohstoffe. Fachagentur Nachwachsende Rohstoffe e.V.. http://fnr.de/marktanalyse/marktanalyse.pdf. Accessed: 31.07.2016

FNR (2016) Basisdaten biobasierte Produkte. Fachagentur Nachwachsende Rohstoffe e.V.. http://www.fnr.de/fileadmin/allgemein/pdf/broschueren/Basisdaten_biobasierte_Produkte-2016_web.pdf. Accessed: 31.07.2016

German Federal Government (2012). Biorefineries Roadmap. Federal Ministry of Food, Agriculture and Consumer Protection; Federal Ministry of Education and Research; Federal Ministry for the Environment, Nature Conservation and Nuclear Safety; Federal Ministry of Economics and Technology. https://www.bmbf.de/pub/Roadmap_Biorefineries_eng.pdf. Accessed: 28.01.2019

Gosselink RJA (2011) Lignin as a renewable aromatic resource for the chemical industry. Dissertation, Wageningen University. http://edepot.wur.nl/186285. Accessed: 31.05.2016

Idler C, Venus J, Kamm B (2015) Microorganisms for the production of lactic acid and organic lactates. In: Kamm B (ed) Microorganisms in biorefineries. Springer, Berlin Heidelberg, pp 225–273. https://doi.org/10.1007/978-3-662-45209-7_9. Accessed: 31.05.2016

IEA Bioenergy (2016) Internetseite: IEA Bioenergy Task 42 Biorefining. http://www.iea-bioenergy.task42-biorefineries.com. Accessed: 29.07.2016

IEB (2016) Internetseite: Institut Européen de la Bioraffinerie Reims Champagne-Ardenne. http://www.institut-europeen-de-la-bioraffinerie.fr/. Accessed: 19.05.2016

Karhumaa K, Wiedemann B, Hahn-Hagerdal B, Boles E, Gorwa-Grauslund M-F (2006) Co-utilization of L-arabinose and D-xylose by laboratory and industrial *Saccharomyces cerevisiae* strains. Microb Cell Factories 5:18

Khirsariya P, Mewada RK (2013) Single step oxidation of methane to methanol – towards better understanding. Proc Eng 51:409–415. http://www.sciencedirect.com/science/article/pii/S1877705813000581. Accessed: 31.05.2016

KIT (2016) Das Karlsruher bioliq®-Verfahren: Forschung und Entwicklung von Biokraftstoffen der 2. Generation aus trockener Restbiomasse. Karlsruher Institut für Technologie. http://www.bioliq.de/downloads/Flyer%20als%20Datenblatt_DE_2016.pdf. Accessed: 30.05.2016

Laure S, Leschinsky M, Frohling M, Schultmann F, Unkelbach G (2014) Assessment of an Organosolv lignocellulose biorefinery concept based on a material flow analysis of a pilot plant. Cellul Chem Technol 48(9–10):793–798

RFA (2016) Internetseite: renewable fuels association – industry statistics. http://www.ethanolrfa.org/resources/industry/statistics/. Accessed: 01.05.2016

Schmid-Staiger U, Holdmann C, Ripplinger P, Bergmann P, Wach W, Klingeberg M, Haußmann H (2016) Bioraffinerie auf Basis kohlenhydratreicher Algenbiomasse, Nutzung von Stärke und Protein. Fachagentur Nachwachsende Rohstoffe e. V. (FNR). http://www.fnr.de/index.php?id=11151&fkz=22403211. Accessed: 30.07.2016

Sieker T, Neuner A, Dimitrova D, Tippkötter N, Muffler K, Bart H-J, Heinzle E, Ulber R (2011) Ethanol production from grass silage by simultaneous pretreatment, saccharification and fermentation: first steps in the process development. Eng Life Sci 11(4):436–442. https://doi.org/10.1002/elsc.201000160. Accessed: 31.05.2016

SUNLIQUID (2015) Internetseite: FP7-Project: Sunliquid® large scale demonstration plant for the production of cellulosic ethanol. http://sunliquid-project-fp7.eu. Accessed:25.05.2016

Thielen M (2010) Biorefinery Blair, Nebraska. Bioplastics Magazine 5(3):22–24. http://www.bioplasticsmagazine.com/en/online-archive/issues2010/201003.php. Accessed: 31.05.2016

UNFCCC (2016a) Report of the Conference of the Parties on its twenty-first session, held in Paris from 30 November to 13 December 2015, Part one: Proceedings. United Nations Office at Geneva (Switzerland). http://unfccc.int/resource/docs/2015/cop21/eng/10.pdf. Accessed: 31.05.2016

UNFCCC (2016b) Report of the Conference of the Parties on its twenty-first session, held in Paris from 30 November to 13 December 2015, Addendum. Part two: Action taken by the Conference of the Parties at its twenty-first session. United Nations Office at Geneva (Switzerland). http://unfccc.int/resource/docs/2015/cop21/eng/10a01.pdf. Accessed: 31.05.2016

Vanholme R, Demedts B, Morreel K, Ralph J, Boerjan W (2010) Lignin biosynthesis and structure. Plant Physiol 153(3):895–905. http://www.plantphysiol.org/content/153/3/895.short. Accessed: 31.05.2016

VCI (2013) VCI-Prognos-Studie: Die deutsche chemische Industrie 2030. Verband der Chemischen Industrie e.V. https://www.vci.de/vci/downloads-vci/publikation/langfassung-prognos-studie-30-01-2013.pdf. Accessed: 27.04.2016

VCI (2015) VCI-Positionspapier: Chancen und Grenzen des Einsatzes nachwachsender Rohstoffe in der chemischen Industrie. Verband der Chemischen Industrie e.V. https://www.vci.de/langfassungen-pdf/chancen-und-grenzen-des-einsatzes-nachwachsender-rohstoffe-in-der-chemischen-industrie.pdf. Accessed: 21.05.2016

VDI (2016) Richtlinie VDI 6310 Blatt 1: Klassifikation und Gütekriterien von Bioraffinerien. VDI-Gesellschaft Technologies of Life Sciences (VDI-TLS), Fachbereich Biotechnologie. https://www.vdi.de/technik/fachthemen/technologies-of-life-sciences/fachbereiche/biotechnologie/gremien-ausschuesse/bioraffinerie/. Accessed: 31.07.2016

Werner & Mertz (2016) Internetseite: Stroh zu Gold machen – Werner & Mertz setzt auf neuartigen Bio-Spiritus und kooperiert mit dem Spezialchemie-Unternehmen Clariant. Werner & Mertz GmbH. http://werner-mertz.de/Pressecenter/Pressetexte/Detail_3008.html. Accessed: 25.05.2016

The Importance of Biotechnology for the Bioeconomy

Manfred Kircher, Michael Bott, and Jan Marienhagen

Enzyme fermentation in a laboratory of evoxx technologies. Biotechnologically produced or optimized enzymes increasingly enhance the environmental compatibility and efficiency of chemical syntheses as biocatalysts. (© evoxx technologies)

J. Pietzsch (ed.), *Bioeconomy for Beginners*, https://doi.org/10.1007/978-3-662-60390-1_5

5.1 Biotechnology as a Production Process

Manfred Kircher

Biotechnology is the basis for numerous processes for the production of food and feed, pharmaceuticals, chemical products and energy sources. It is also the technology that prepares raw biological materials and systems (cells and their components) for use in such processes. Although humans have been using microorganisms, animals, plants and enzymes for millennia, it is only modern biotechnology that has made it possible to optimise them specifically for defined processes.

5.1.1 Microorganisms

Because they decompose plant and animal biomass, microorganisms enable the natural material cycle. However, this also means that food, feed and other organic substances must be protected from microbial degradation. For this purpose, in addition to cooling, drying and salting food, traditional biotechnology has proven its worth for thousands of years. For a long time, the ability of microorganisms to influence environmental conditions was unknowingly used in their favour. Some microorganisms inhibit the competing microflora by excreting organic acids, alcohols or antibiotics. Since preindustrial times, humankind has learned to make these metabolic services useful to itself by applying their use for the conservation of plant biomass and, nowadays, also for the industrial production of organic acids, alcohols, amino acids, antibiotics and other intermediates.

Lactic acid bacteria and other acid-forming bacteria provide a classic example of traditional biotechnology. They acidify vegetable biomass into silage, which is used for the preservation of animal feed. Fresh grass, maize, beet leaves and other vegetable biomass with a dry matter content of between 25% and 50% are packed in airtight silos or under plastic film. The aerobic bacteria and fungi present in the natural microbial flora of biomass excrete enzymes such as cellulases, proteases and lipases in order to hydrolyse carbohydrates, proteins and lipids of biomass into mono- and disaccharides, amino acids, glycerol and fatty acids. This allows these microorganisms to multiply by decomposing the biomass. With increasing oxygen consumption, acid-forming species such as *lactobacillus*, *leuconostoc* and *streptococcus* prevail. Under anaerobic conditions, they secrete, in particular, acetic acid and lactic acid. Other bacteria such as clostridia release propionic acid and butyric acid. The resulting acidification to a pH value between 4 and 4.5, in combination with the exclusion of oxygen, prevents further microbial degradation. The silage is thus preserved.

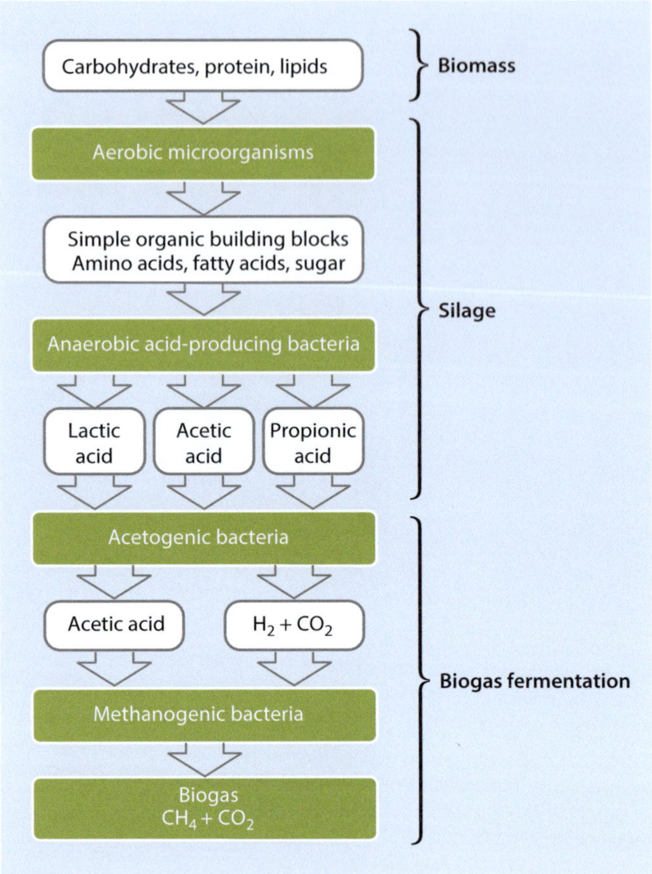

Fig. 5.1 From biomass via silage to biogas

For the energy-related utilization of the silage for biogas, the biotechnological conversion is continued by acetogenic bacteria, which transform the acids present in the silage to acetic acid, carbon dioxide and hydrogen (Fig. 5.1). These are the substrates that are ultimately converted by methanogenic bacteria into methane. Under strictly anaerobic conditions, most bacteria metabolise acetic acid, while others convert carbon dioxide and hydrogen into methane:

Biomass > hydrolysis > acidification > acetic acid:

$$CH_3COOH = CH_4 + CO_2$$

$$CO_2 + 4H_2 = CH_4 + 2H_2O$$

Biogas is a mixture of about two thirds methane and one third carbon dioxide (Rec-energy 2016).

Biomethane (natural biogas) is mainly used today as an energy carrier. However, methane can also be used to produce methanol, the most important and most versatile C1-platform chemical in the chemical industry, from which a broad family tree of chemical products can be produced (▶ Chap. 4). This potential for the material utilization of methane must not be overlooked.

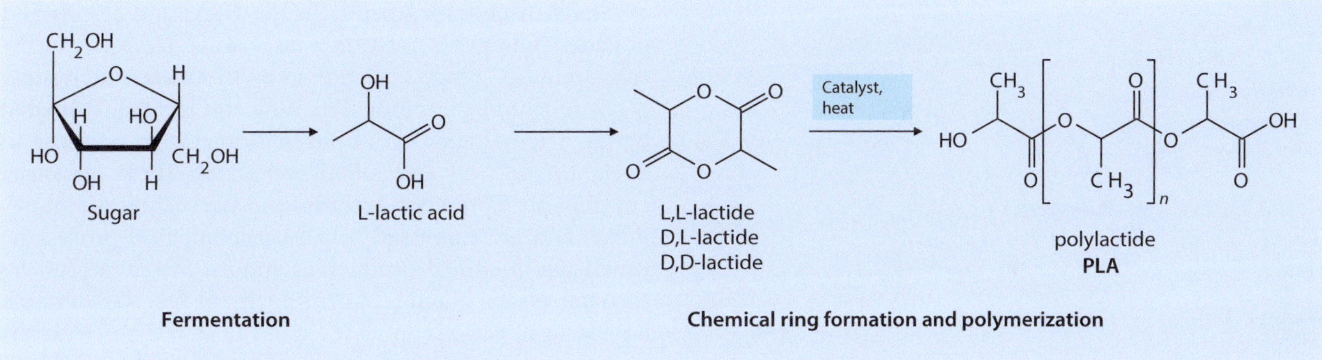

Fig. 5.2 Preparation of polylactide (PLA) from lactic acid

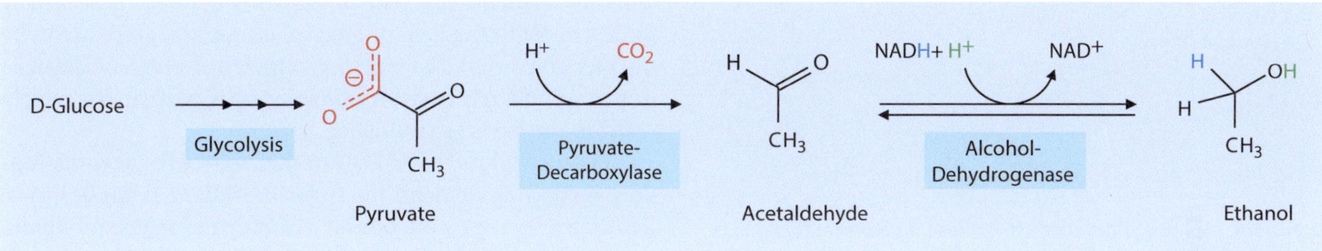

Fig. 5.3 Ethanol fermentation in yeast (*Saccharomyces cerevisiae*)

Silage offers further options for the material utilization of biomass. For example, silage produced from grass is used as a raw material for bio-based construction materials, such as insulating materials and terrace planks, or as a filler for bio-based plastics in various applications. In this case, the silage process not only serves to preserve the biomass, but also opens it up in such a way that grass fibres can be separated for further processing.

Lactic acid has also gained singular industrial significance. Produced on an industrial scale, it is the microbial intermediate that led to one of the first modern biopolymers, namely, polylactides (*polylactic acid*, PLA). These can be processed into fibres, films or bottles. PLA is the aliphatic polyester of lactic acid (Fig. 5.2). Lactic acid is formed by fermentation from sugar. The following steps from lactic acid to lactide, and on to the polymer, are based on chemical synthesis. This is an example of the way in which biotechnology and chemistry complement each other in the bioeconomy.

Because the technical properties of PLA are similar to those of the large-volume fossil-based aromatic polyesters polyethylene terephthalate (PET) and polybutylene terephthalate (PBT), a strongly growing market potential is expected for polylactides. A disadvantage, however, is their relatively low temperature resistance. The higher the proportion of left-turning lactic acid, the greater it is. Although lactic acid bacteria naturally synthesize the racemic mixture of D- and L-lactic acid, it has become possible, through strain development, for modern production strains exclusively to produce L-lactic acid.

Other organic acids that are produced microbially on an industrial scale include acetic acid (for food applications), adipic acid (pilot scale), succinic acid, citric acid and D-gluconic acid. Their production organisms are bacteria or fungi. For example, the fungus *Aspergillus niger* is established in citric acid production and the bacterium *Acetobacter acetii* in acetic acid fermentation.

Eucaryotic **baker's yeast** (*Saccharomyces cerevisiae*) is another example of a microbial species that, coming from the traditional food culture, also plays an important role in the modern bioeconomy. In sugar-containing solutions such as fruit juices, acid-forming bacteria initially propagate. After lowering the pH value and the oxygen content, the yeast prevails: It anaerobically converts sugar into alcohol (Fig. 5.3), excretes it, and thereby inhibits the growth of other microorganisms.

The fruit juice is thereby preserved as wine. Since ancient times, this process has been used to obtain hygienic (wine) drinking water by means of alcohol, i.e., to "conserve" water. An additional biotechnological step is necessary in beer production. Barley does not itself contain fermentable sugar, but rather the vegetable storage molecule starch. Starch is a polysaccharide that must be split into the disaccharide maltose in order to be absorbed by yeast. For this reason, barley is first germinated at 15–17 °C by adding water in order to initiate the expression of enzymes that dismantle starch and other storage molecules of the seedling, such as proteins and the phosphate storage polymer phytin. These enzymes are mainly cytases (hemicellulases), which dissolve cell walls, proteinases, which cleave proteins, phosphatases,

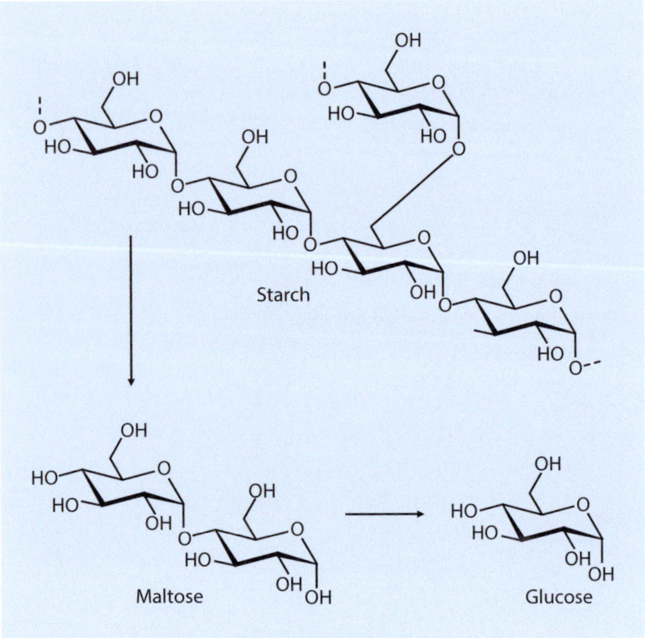

Fig. 5.4 Splitting from starch to maltose, and further to glucose

which release phosphate, and amylase, which cleave starch into dextrins (short starch units) and the disaccharide maltose. Germination is broken off through drying (kilning) at 80–120 °C. The maltose of the malt thus produced can now be absorbed by yeast cells. Using the enzyme maltase, they split maltose into the monosaccharide glucose (Fig. 5.4), which is further metabolized into ethanol under anaerobic conditions.

Alcohol fermentation has long since outgrown the food industry, and is involved today in the production of bioethanol, thus serving as the most important industrial and volumetric fermentation process for the fuel sector. Bioethanol can be converted chemically by splitting off water into bioethylene. Oil-based Ethylene is presently the most significant platform chemical, with an annual production of 156 million tonnes (2012) (▸ Chap. 4).

Sugar is the basis of a large number of applications in the food and fermentation industry. Depending on market requirements, other sources of sugar have been developed based on the large demand and different costs. Sucrose, a disaccharide of glucose and fructose extracted from sugar beet and cane sugar, can be used both for food and directly as a carbon source for fermentation. In contrast, starch and lignocellulose must be degraded. In both procedures, biotechnology plays a decisive role.

Starch from corn (mainly in the USA) and cereals and potatoes (Europe) is used on a large scale. However, in the cost-optimised processes required on this scale, starch sugar is not released by traditional malting, but is rather degraded by the external addition of enzymes analogous to those in Fig. 5.4 into di- and monosaccharides. These industrial enzymes are optimized according to the technical requirements and are expressed in a biotechnological process by genetically modified production strains. Starch processing then follows the example of malting: In the first step of starch liquefaction, various amylases split the starch into dextrins (short starch units) and the disaccharide maltose. This is followed by starch saccharification using glucoamylase and pullulanase. The result is a glucose syrup that is used in bioethanol fermentation and in the confectionery and bakery industry. With the help of glucose isomerase, glucose can be further converted into fructose, which has a higher sweetening power. Starch-based fructose syrup is particularly widely used in the beverage industry.

Lignocellulose is used as another source to meet the rapidly increasing demand for sugar in industrial applications. However, its sugars are bound in a polymer matrix of lignin, hemicellulose and cellulose that is difficult to break down. It is for this reason that wood is relatively difficult to degrade compared to other biomass. Industrially produced enzymes also play a key role in the saccharification of woody biomass such as straw.

Corynebacterium glutamicum is a species that has no pre-industrial history of use, but rather stands exclusively for modern biotechnology. *Corynebacterium glutamicum* is the result of a scientific screening in which microorganisms were specifically sought for the fermentative production of the amino acid L-glutamic acid. *Corynebacterium glutamicum* naturally excretes L-glutamic acid and has, since its discovery, established itself as a production organism for many industrially relevant amino acids. In addition to L-glutamic acid, which is marketed as a seasoning, L-lysine is mainly used for supplementing animal feed, for supplementing special diets and as an essential amino acid in infusion solutions.

Amino acid fermentation is, in two respects, an impressive example of the role of biotechnology in the bioeconomy. On the one hand, biotechnology contributes to nutrition and reduces both the amount of land required for meat production and its environmental impact (▸ Chap. 8). On the other hand, biotechnology in this field is the key to the development of efficient production strains (▸ Excursus 5.1).

Excursus 5.1 From the Strain Optimization to the Systems Biology

Industrial amino acid fermentation does not use the wild type of *corynebacterium glutamicum*. That's because the synthesis path to L-lysine, like all other biosynthetic pathways, is designed for nutrient-poor conditions in nature (Fig. 5.5). It only supplies for the bacterium's own needs and stops as soon as it is satisfied. This is ensured by the strictly regulated key enzyme aspartokinase, which is inhibited by threonine and lysine. This is where biotechnology comes into play as a method of strain development (Eggeling and Bott 2015). In a first step, on the way to a production strain that produces lysine beyond its own needs, it introduces the mutated gene of a non-inhibitable aspartokinase.

In the course of further optimisation, the uptake of carbon and nitrogen sources (e.g., sugar and ammonia), the excretion of L-lysine from the cell and other metabolic pathways may prove to be limiting. They will also then be the target of genetic engineering measures. The material flow of the microbial biosynthesis will be pushed and expanded as comprehensively as possible towards the target product L-lysine. Similar optimizations are possible for other microorganisms or cells. The prerequisite for this is a precise knowledge of the cellular physiology, as well as of the substance flows and their regulation. In addition to other methods, the most complete possible inventory of cellular metabolites and their exchange with the culture medium under different culture conditions and growth states contributes to this. These metabolites are chemically very different. They can include sugars, fats, amino acids, proteins or nucleic acids. They are analysed using a variety of sophisticated methods, the combined result of which provides a snapshot of the metabolic state of the bacterium. Mass spectrometry (MS), in combination with liquid chromatography (LC-MS) or gas chromatography (GC-MS), is used to identify the metabolites on the basis of their mass and their decay products occurring during chemical analysis. LC, enzymatic analytical methods and NMR spectroscopy can be used to determine the uptake of substrate from the culture medium into the cell or to detect excreted metabolites. In addition, transcription factor-based metabolite sensors enable the measurement of defined metabolites in individual cells. In this way, the status of metabolites (metabolome) and proteins (proteome) is determined. The expression state of the genome is observed using DNA microarrays or RNA sequencing, which measure the overall status of the m-RNA, i.e., the transcriptome. On the basis of this data, bottlenecks can be identified, and their overcoming can be planned. For more in-depth optimization strategies, *in-silico metabolic design* is a valuable tool, which creates algorithms for metabolic models, and thus calculates various metabolic states. Thus, not only are metabolic bottlenecks in the biosynthesis pathway leading directly to the target product identified, but so are those of the complex metabolic background, for example, in the cellular energy balance. The cell is thus understood as a system of finely regulated, coordinated, complex and interdependent reaction chains; this is an approach that includes all cellular processes, and is therefore referred to as *systems biology*. Within the framework of *metabolic engineering*, the identified genes are optimized using genetic engineering methods. For example, a gene for a regulating intermediate can be switched off. If the intermediate is missing in the cell, the enzyme in question is no longer inhibited. Alternatively, the regulatory properties of the key enzyme of a metabolic pathway can be directly influenced in its structural gene. High performance lysine strains developed with such sophisticated technologies accumulate, within 48 hours, up to 170 g/l L-lysine with a yield of 40–45 g of amino acid per 100 g of sugar (Sahm and Eggeling 2009). With this arsenal of methods, both biosynthetic pathways that occur naturally in a strain and foreign and synthetic reaction chains can be processed (► Sect. 5.2).

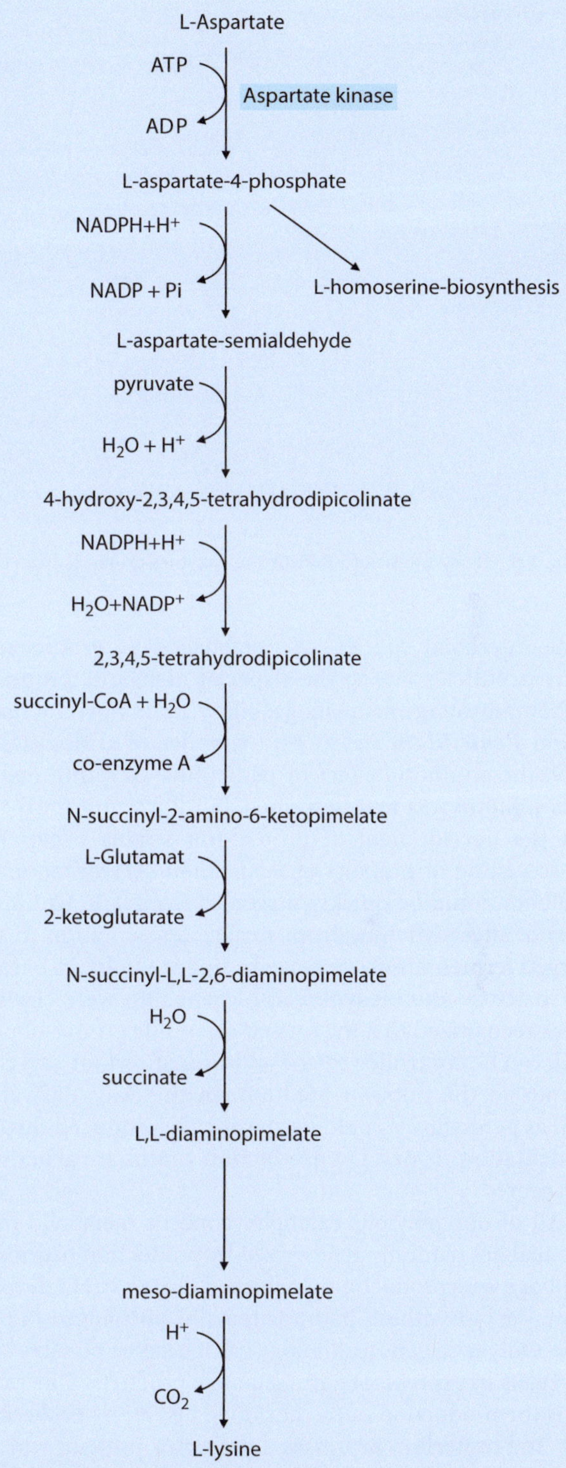

Fig. 5.5 Biosynthesis into L-lysine in wild type and in production strains of *Corynebacterium glutamicum*

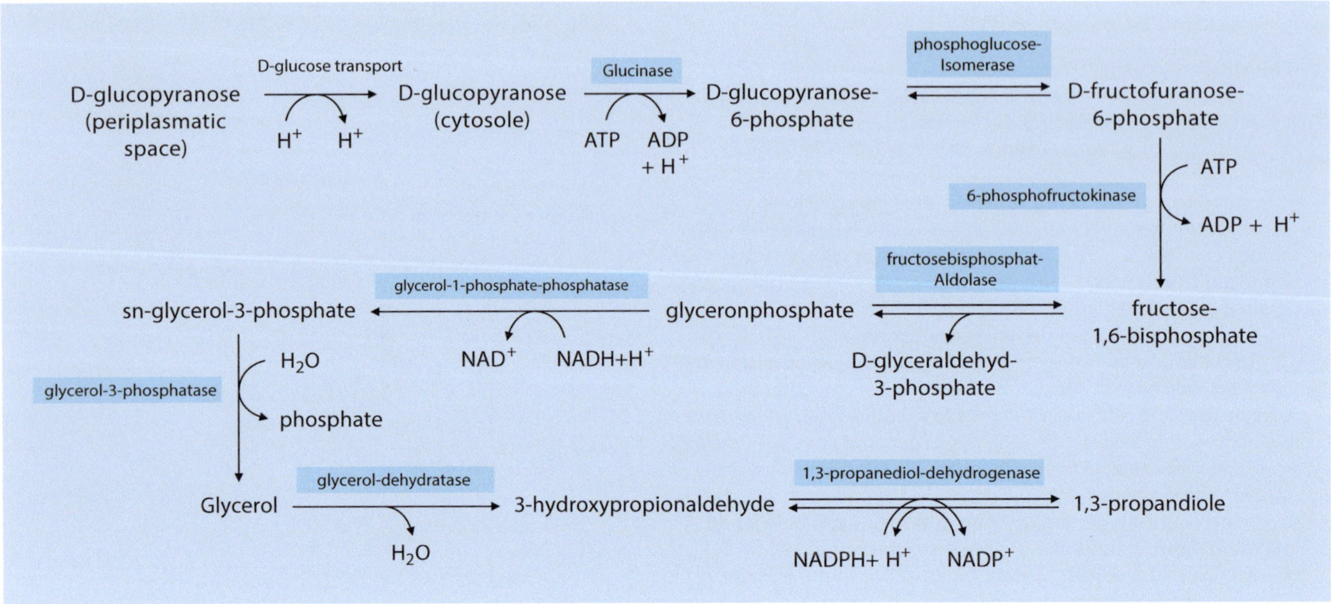

Fig. 5.6 Biosynthesis of 1,3-PDO; the enzymes originate from *E. coli* (Ec); *Saccharomyces cerevisiae* (Sc) and *Klebsiella pneumoniae* (Kp)

Some microbial species are established as producers of pharmaceuticals due to their special metabolic properties or their advantageous manageability. That's how the fungal mould ***Penicillium*** serves as a supplier of antibiotics. In 1929, the antibiotic effect of penicillins excreting *Penicillium notatum* was first described, but it was not until 1941 that the development of production strains began with the screening of mutants of *Penicillium chrysogenum*. The efficiency could be quickly increased from 0.06 to 1.8 mg/ml and, after switching from surface fermentation to submerged fermentation, increased a hundredfold. In parallel, the structure and biosynthesis of penicillin were clarified. It was recognized that the formation of side groups of penicillin can be controlled by the addition of certain acyl compounds to the nutrient medium. In this way, derivatives such as penicillin V could be specifically produced through fermentation. Today, the production strains are genetically engineered.

All of our previous examples concern metabolic products that are naturally synthesized by production organisms: Although overproduction is achieved during strain development, the biosynthesis pathway remains unchanged in principle. Only with gene technology has it become possible since the 1970s to express foreign metabolic products. One example is the production of the hormone insulin by ***Escherichia coli***. The bacterium expresses a precursor protein, which is converted into humanized insulin by chemical and enzymatic post-processing. The process has revolutionised the production of insulin by replacing the previous isolation of the hormone from the pancreas of pigs and cattle. These products were not human-identical and led to side effects in long-term use.

The production of 1,3-propanediol (PDO) goes one step further with the help of *E.coli* – because PDO is a molecule that only enterobacteria produce from glycerol under anaerobic conditions. By combining biosynthesis sequences from *Klebsiella pneumoniae* and *Saccharomyces cerevisiae* in *E. coli*, it is, however, possible for the bacterium to transform sugar into PDO (Fig. 5.6). This fermentation process is used industrially and is competitive with the chemical synthesis of PDO from fossil-based acrolein. PDO is processed into textile fibres and floor coverings, among other things.

Microalgae can photosynthetically fix carbon dioxide. They are therefore of interest for bioeconomic applications (▶ Sects. 2.3 and 4.3). However, they still form more of a niche application, e.g., as a protein source for diet food (*chlorella*) and as a supplier of ß-carotenoids and polyunsaturated fatty acids (PUFA). Microalgae have a high fatty acid content and multiply relatively quickly. With a high productivity per unit area, it is therefore possible, in principle, to produce biodiesel on a large scale. However, algae broth has only low light transmission, which requires the cultivation of algae in thin layers a few centimetres deep in reactors made of light-transmitting materials. In sunny locations, they can be cultivated inexpensively in open basins. However, this only works with wild type algae. In addition, contamination by the surrounding microbial outdoor flora must be expected. Microalgae have therefore not yet achieved an industrial breakthrough. However, the increasing pressure on the bioeconomy to recycle carbon may increase the attractiveness of this photosynthetic system in the coming years.

Clostridia was mainly of academic interest until the turn of the century, although research had been carried out for many decades. Only in acetone-butanol-ethanol fermentation did they play an earlier role on an industrial scale. *Clostridia* are anaerobic acetogens. *C. ljungdahlii* and *C. autoethanogenum* are able to convert carbon monoxide and/or carbon dioxide and hydrogen into acetic acid, ethanol and other metabolites. Carbon monoxide is the main component of synthesis gas (CO, CO_2, H_2), which can be produced from any organic (waste) material, whether it comes from biomass or a fossil carbon source. The bacterium *Clostridium ljungdahlii* naturally excretes ethanol synthesized on the basis of carbon monoxide. The fact that this can be used for industrial production can already be shown on a demonstration scale. The way in which acetogens utilize CO and CO_2/H_2 via the Wood-Ljungdahl Trail, and fix the gaseous carbon in the process, can be seen in ◘ Fig. 5.7.

The synthesis of the central intermediate acetyl-CoA can be catalyzed by the enzyme CO-dehydrogenase/acetyl-CoA-synthase in two different ways – either solely by consumption of carbon monoxide and water or by the additional consumption of carbon dioxide with hydrogen as the energy source.

With carbon monoxide as the sole source of carbon and energy, ethanol is produced according to the following equation:

$$6\,CO + 3\,H_2O \rightarrow CH_3CH_2OH + 4\,CO_2$$

If hydrogen is also available as an energy source, carbon dioxide is emitted:

$$3\,CO + 3\,H_2 \rightarrow CH_3CH_2OH + CO_2$$

$$2\,CO + 4\,H_2 \rightarrow CH_3CH_2OH + H_2O$$

This can be converted into ethanol in the same way as externally fed carbon dioxide:

$$2\,CO_2 + 6\,H_2 \rightarrow CH_3CH_2OH + 3\,H_2O$$

Because *clostridium* can recycle the by-product CO_2, high carbon yields can be expected from the metabolic pathways described. This makes acetogens very attractive for industrial processes (► Chaps. 4 and 7), because, for a *bulk*-chemical like ethanol, raw materials account for the largest share of production costs. Even with wild type strains of Clostridia isolated from nature, 48 g of ethanol per litre could be produced on a laboratory scale, with a yield of 76% based on carbon monoxide. It is foreseeable that this efficiency and the variety of products could be significantly increased by the described methods of strain optimization.

Acetone can also be produced fermentatively with *Clostridium ljungdahlii*. For this purpose, the carbon sources sugar and synthesis gas are combined, a process that is known as mixotrophic fermentation. Since the energy source hydrogen can be used to utilize both the carbon dioxide produced in the bacterium's metabolism and the carbon from the synthesis gas, carbon yields of more than 100% are achieved in relation to the sugar used:

- Sugar: 65% yield
- Sugar + hydrogen: 72–83% yield
- Sugar + synthesis gas: 86–195% yield

Fermentation is therefore carried out with the highest yield and without carbon dioxide emissions when synthesis gas is fed into the system. On a laboratory scale, 23 g of acetone per litre are enriched with a productivity of 5 g/l/h (Jones 2016).

A technical hurdle of this so-called gas fermentation with clostridia is the problem of introducing sufficient gaseous carbon sources into the aqueous medium of a fermenter, which is not easy to solve from a process engineering point of view, because the solubility of carbon monoxide and carbon dioxide in water is only low under culture conditions (Daniell et al. 2012).

5.1.2 Enzymes

Like the use of microorganisms, the use of isolated enzymes has a millennia-long tradition. From time immemorial, the **rennin enzyme** (chymosin, pepsin) has been used in the production of cheese for the coagulation of the milk protein casein. In the past, it was extracted from the abomasum of lactating ruminants. Today, the enzymes used in the food industry are biotechnologically produced through the use of genetically modified *E. coli*-bacteria, yeasts or moulds. In pharmaceutical research, protein biosynthesis is also carried out cell-free (► Excursus 5.2).

The first industrial enzyme was launched in 1907 by Otto Röhm in Darmstadt. His product, a **protease**, revolutionized leather tanning by replacing the use of dog excrement in leather stains.

Amylases are among the most important industrial enzymes, because they enable the use of starch as a source of sugar. Alpha-amylase first breaks down starch into short-chain fragments, which are then split by glucoamylase into glucose. This so-called starch saccharification is the prerequisite for the production of bioethanol based on maize and cereal starch (► Sect. 5.1.1). The beverage industry also uses starch-based sugar to a large extent. Since the saccharification process requires high temperatures, the enzymes must be adapted to these technical conditions. Here, too, biotechnology offers a solution. With the procedures of *protein design*, *directed evolution* and *protein engineering*, performance parameters are specifically developed. At first, in addition to the amino acid sequence of the enzyme, its three-dimensional structure is investigated. Of particular interest here is the active center in which the substrate fits and where the enzymatic reaction is catalyzed (◘ Fig. 5.8). Other structural elements determine, for example, the temperature or pH stability of the enzyme.

5

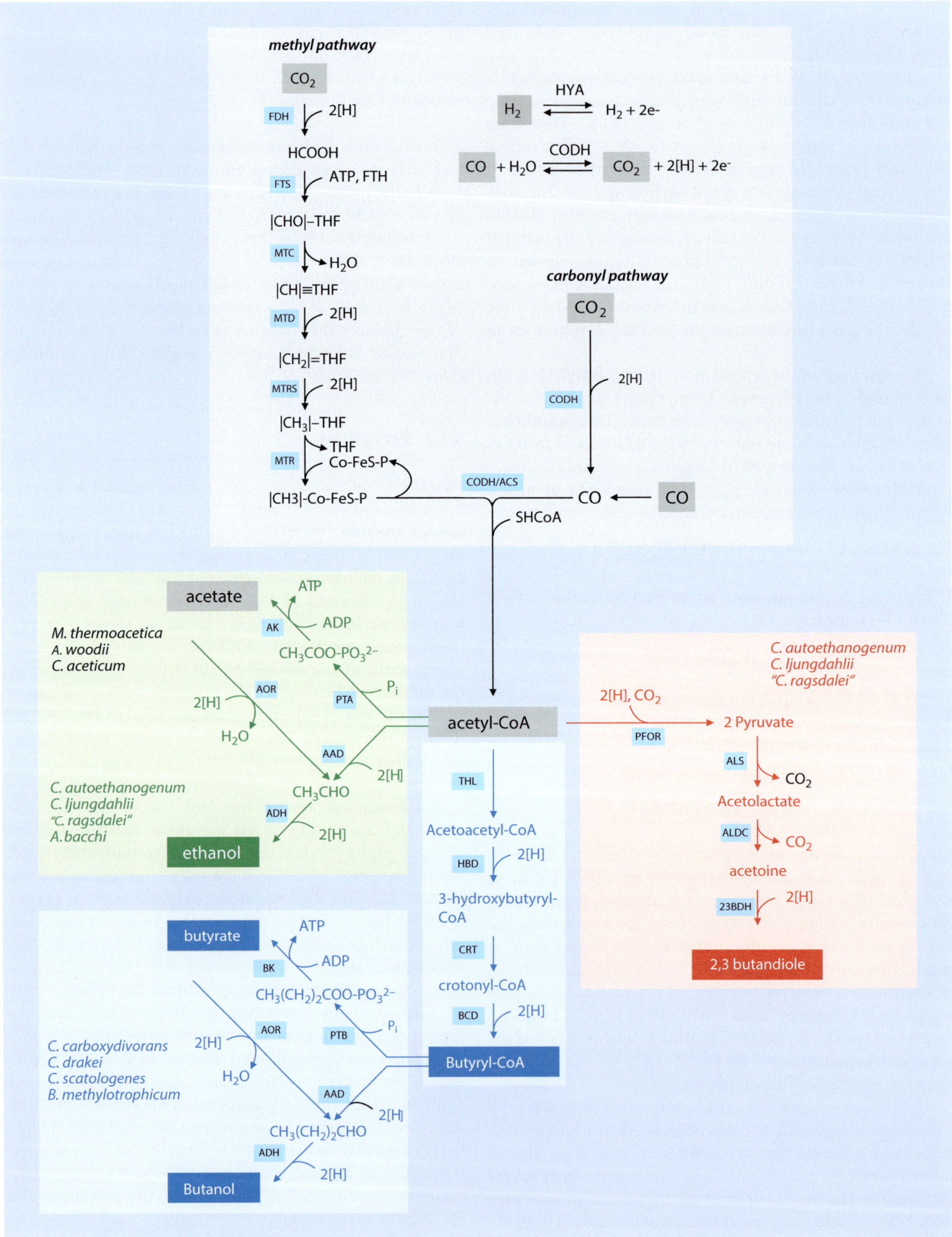

Fig. 5.7 CO/CO_2-metabolism in Clostridia

In the next step after the structural analysis, which is enzyme design, bioinformatics plays a decisive role. It can be used to calculate the effects of amino acid sequence changes on enzyme properties. Thus, an optimal target sequence is determined and then realized by means of *protein engineering*. Through site-specific mutagenesis or DNA synthesis, the DNA corresponding to the sought-after enzyme protein is produced and expressed in a suitable host organism, frequently in *E.coli*. In practice, the genetic information for an enzyme is also deliberately altered by chance (*gene shuffling*) to produce a whole set of differently modified enzymes. For the production of such a genetic library (*gene library*), a faulty DNA polymerase chain reaction (PCR) may be used. The resulting enzymes are tested in specific assays. The best candidates are then varied and optimized in repeated cycles – a procedure that has become known as *directed evolution* (■ Fig. 5.9).

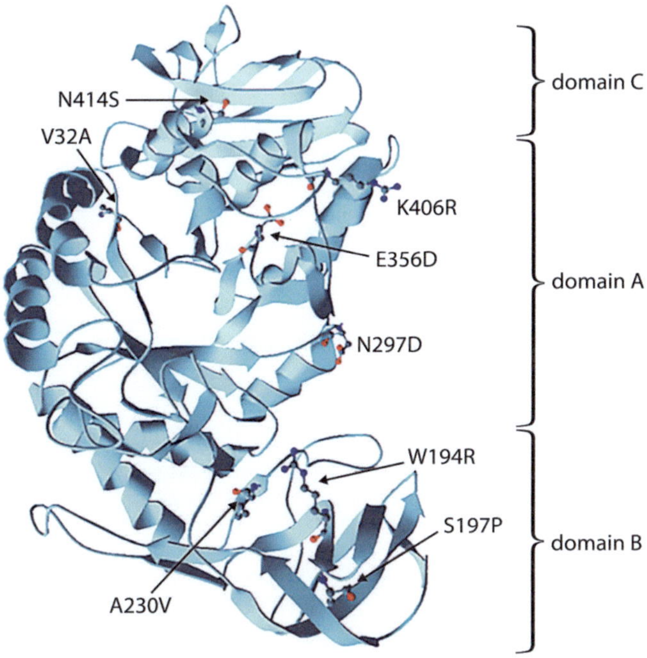

■ **Fig. 5.8** Structure of the alpha-amylase of *Bacillus amyloliquefaciens*. Domain A contains the N-terminus and the active center. Domain B has the highest variability between different amylases. Domain C ends with the C-terminus. (Bessler et al. 2003)

The alpha-amylase from *Bacillus amyloliquefaciens* shown in ■ Fig. 5.8 has been increased by one unit to pH 7 and by a factor of 9 in its specific activity to make it applicable in detergents. Even at a pH of 10, its activity is 5 times higher than that of the initial enzyme (Bessler et al. 2003).

Phytase is an enzyme that hydrolytically degrades the vegetable phosphate storage molecule phytic acid and releases phosphate at the same time (■ Fig. 5.10).

Phytase is added to vegetable animal feed to make the phosphate accessible for the animal (▶ Chap. 8). The phytase supplementation of pig and poultry feed reduces the external administration of phosphate and, at the same time, the phosphate load of the liquid manure. The phytase is added to the animal feed in pressed form as a non-dusting pellet, and must therefore be optimised for temperature resistance. This is because pelleting takes place at temperatures of 50–60 °C, which must not inactivate the enzyme. However, the optimum temperature of phytase should be within the limits of the body temperature of the livestock. For the supplementation of pig feed, 38–39 °C, and for poultry, 40–42 °C should

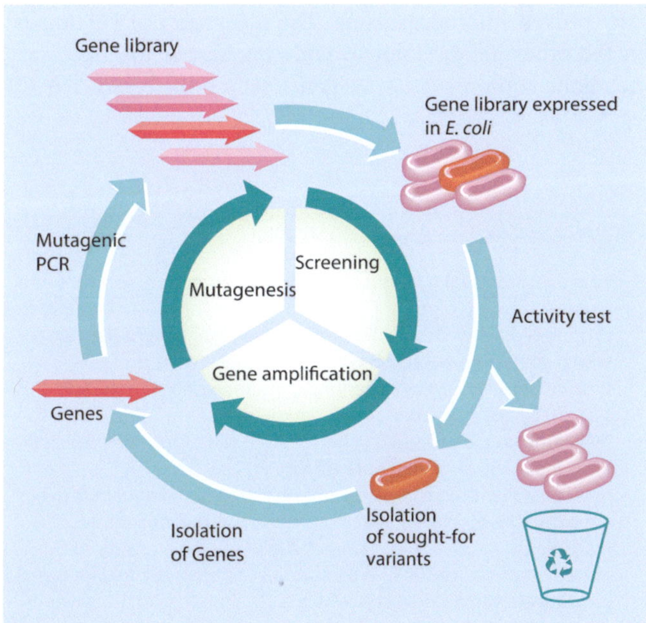

■ **Fig. 5.9** The principle of the procedure of *directed evolution*

■ **Fig. 5.10** Hydrolysis of phytic acid

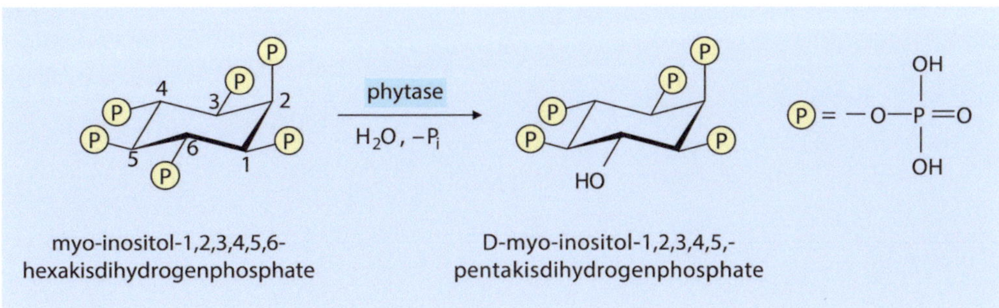

therefore be aimed for. Phytase is also applied for the production of bioenergy (biogas, ethanol fermentation) when biomass containing phytic acid is used (wheat, barley, rye, rice, maize or soya). The enzyme then contributes to the efficient use of all components of the complex biogenic raw materials.

Raw material efficiency is also at stake when **cellulases**, **xylanases** and **glucanases** are used. A large part of the vegetable biomass has so far not been accessible for either nutrition or for material conversion into chemical products. This is lignocellulose, a wood-like biomass found in large quantities in straw, maize and rice husks and palm oil fruit waste (► Chap. 4). It consists of cellulose, hemicellulose and lignin. Hemicellulose is a sugar polymer consisting of different polysaccharides, depending on their origin, made of C6 sugar (glucose, mannose, galactose) and C5 sugar (xylose, arabinose). Cellulose is composed of several hundred to ten thousand glucose units. Lignin is a phenolic macromolecule. Overall, lignocellulose accounts for by far the largest share of plant biomass. If it were possible to unlock this very large source of sugar, it would be accessible for industrial use. The C6 sugars of lignocellulose can be used by all biotechnologically proven microorganisms. The utilization of C5 sugars, on the other hand, is not so widespread, but the necessary metabolic equipment can be genetically transferred. The C5 sugar utilisation of the yeast *Pichia stipitis* was thus successfully transferred into the baker's yeast *saccharomyces cerevisiae* that is established in ethanol fermentation. However, the rigid structure of lignocellulose makes the release of both sugar polymers difficult. In the production of bioethanol, this release is achieved by first breaking up lignocellulose using acids or heat. Depending on the process, the C6 sugars of the cellulose are released by enzymatic and/or acid hydrolysis and the C5 sugars of the hemicellulose by cellulase enzymes. A mix of three enzymes is used: endoglucanase, which splits the cellulose polymer into irregular fragments; cellulase, which releases glucose dimers; and xylanase, which separates the C5 sugars. These enzymes are produced separately by fermentation and added to the prepared lignocellulose broth for digestion. Ethanol fermentation is then carried out using C5 and C6 sugar-using strains.

An alternative process for the production of ethanol from lignocellulose has reached the technical scale. The enzymes are not fermented separately. The microbial enzyme producer is cultivated directly on the lignocellulose substrate.

The development of lignocellulose as a carbon source by microorganisms and enzymes will considerably relieve the burden on agriculture as an industrial supplier of raw materials. It is therefore a key to raw material efficiency in the bioeconomy (► Chap. 4).

Excursus 5.2 Cell-Free Protein Biosynthesis

Enzymes are usually produced by *in vivo* expression, i.e., obtained from the protein biosynthesis of living cells. The method of cell-free protein biosynthesis transfers this process into a technical reactor, which contains isolated ribosomes and the sites of protein biosynthesis, as well as initiation, elongation and termination factors. By adding genetic information (messenger RNA), transfer RNA, amino acids and ATP, as well as GTP, the *in vitro* synthesis of proteins is also possible there (◘ Fig. 5.11).

The advantage compared to expression in living cells is the considerably shorter processing time. It could, for example, increase the productivity of *directed evolution*. According to the current status of this *in vivo* technique, each genetic variant is first introduced into a single cell, which is then multiplied for several days, expressing the desired enzyme. The cell culture is then lysed, and the desired product purified, if necessary, and finally tested. *In vitro*, the steps of cell proliferation and lysis are completely omitted. Pharmaceutical research, in particular, makes use of cell-free protein biosynthesis in order to have proteins at its disposal that cannot be produced sufficiently *in vivo*. These include, for example, those membrane proteins whose overexpression destabilizes the cellular membrane to such an extent that it inactivates the producing cell. This technology also has potential for enzymes relevant to the bioeconomy. The hydrogenase involved in hydrogen cleavage in Clostridia is sensitive to oxygen, which makes it difficult to express it in the usual aerobic systems. For the development of such enzymes, cell-free protein synthesis can become the method of choice. However, the previous processes have not yet exceeded volumes of 100 l.

Fig. 5.11 Comparison of in-vivo and in-vitro protein expression. (Carlson et al. 2012)

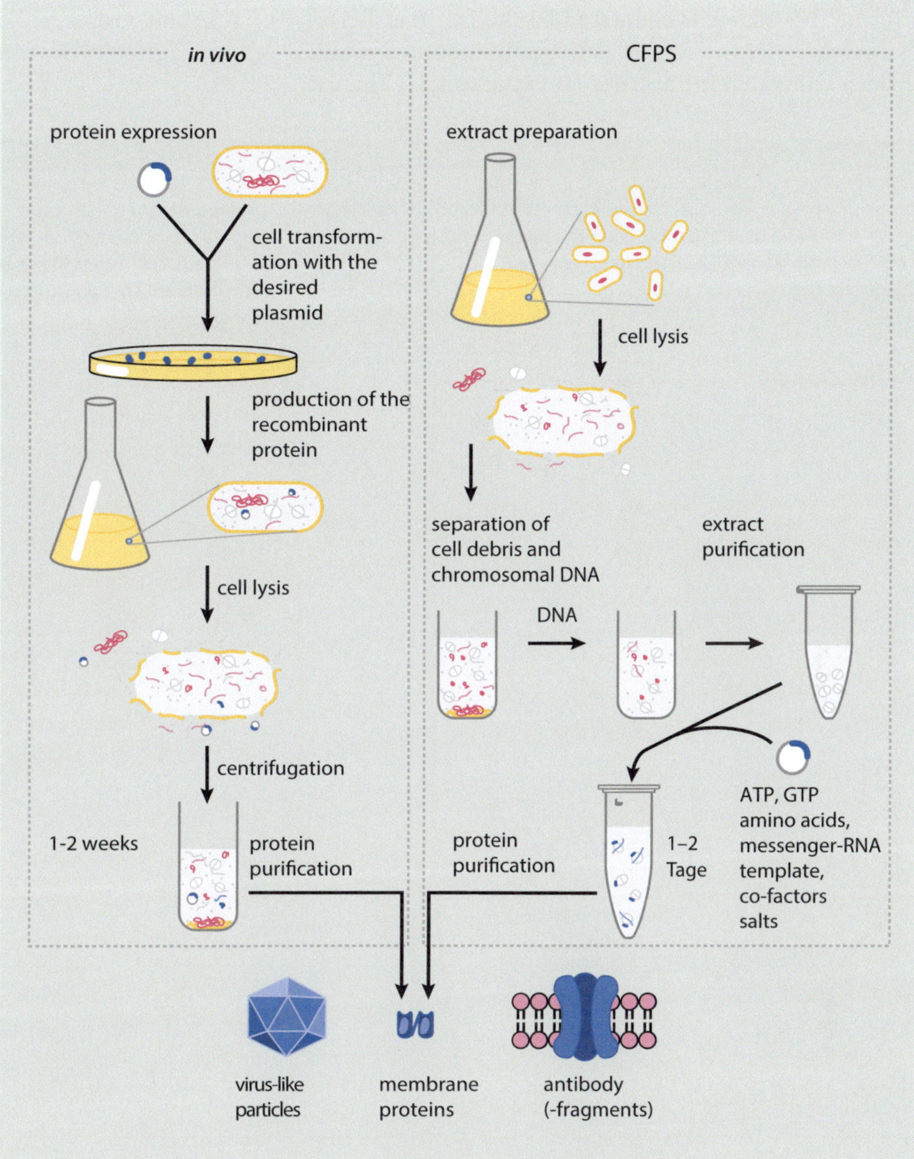

5.1.3 Plants

Biotechnology simplifies the propagation of plants and enables their breeding adaptation to growing conditions (*input traits*), as well as the optimization of plant biomass for industrial requirements (*output traits*).

For *in vitro* reproduction, the meristem culture was established. Meristem tissue in shoot, root and axillary buds consists of undifferentiated cells. These cells are isolated from selected plants that meet certain quality requirements and are propagated in culture media so that they can be regenerated into complete plants with the addition of phytohormones. A source plant can provide thousands of clones in this way. This process is state-of-the-art for potatoes, maize, sugar cane, oil palms and ornamental plants.

Desirable *input traits* are, e.g., resistance to drought and resistance to pests and herbicides, such as to the herbicide glyphosate. This herbicide inhibits an enzyme that only plants and certain bacteria have, but not humans or animals, specifically, enolpyruvylshikimate phosphate synthase (EPSP synthase), which catalyses a step in the biosynthesis of the aromatic amino acids phenylalanine, tyrosine and tryptophan. However, even the exchange of only one amino acid in this enzyme cancels the inhibitory effect of glyphosate. Glyphosate-resistant crop plants are therefore genetically engineered with an appropriately modified gene for EPSP synthase. This gene is selected in bacteria because they can be screened in much larger numbers than plants.

Glyphosate is the subject of much critical discussion. While experts agree that it is harmless under its conditions of application, given its many years of use, some organizations have classified it as potentially carcinogenic. Moreover, many consumers reject the use of herbicides and the genetic modification of plants in principle or in intensive form. This stance is countered by the fact that herbicides increase yields and that genetically modified plants have been cultivated for decades (▶ Excursus 5.3). The use of a total herbicide such as glyphosate and the cultivation of resistant crops makes ploughing unnecessary, which reduces agricultural greenhouse gas emissions (Staropoli 2016).

Excursus 5.3 From Plant Breeding to Green Gene Technology

Humanity began to domesticate wild plants thousands of years ago by selecting the plants that best met their yield, resistance or quality requirements (◘ Fig. 5.12). The selected characteristics of these plants are to be found in the next generation (hereditability). The prerequisite for making such a selection is genetic diversity in the population of plants from which it is selected. Humans began to increase genetic diversity at an early age by interbreeding plants with different characteristics. Inheritance follows from what are

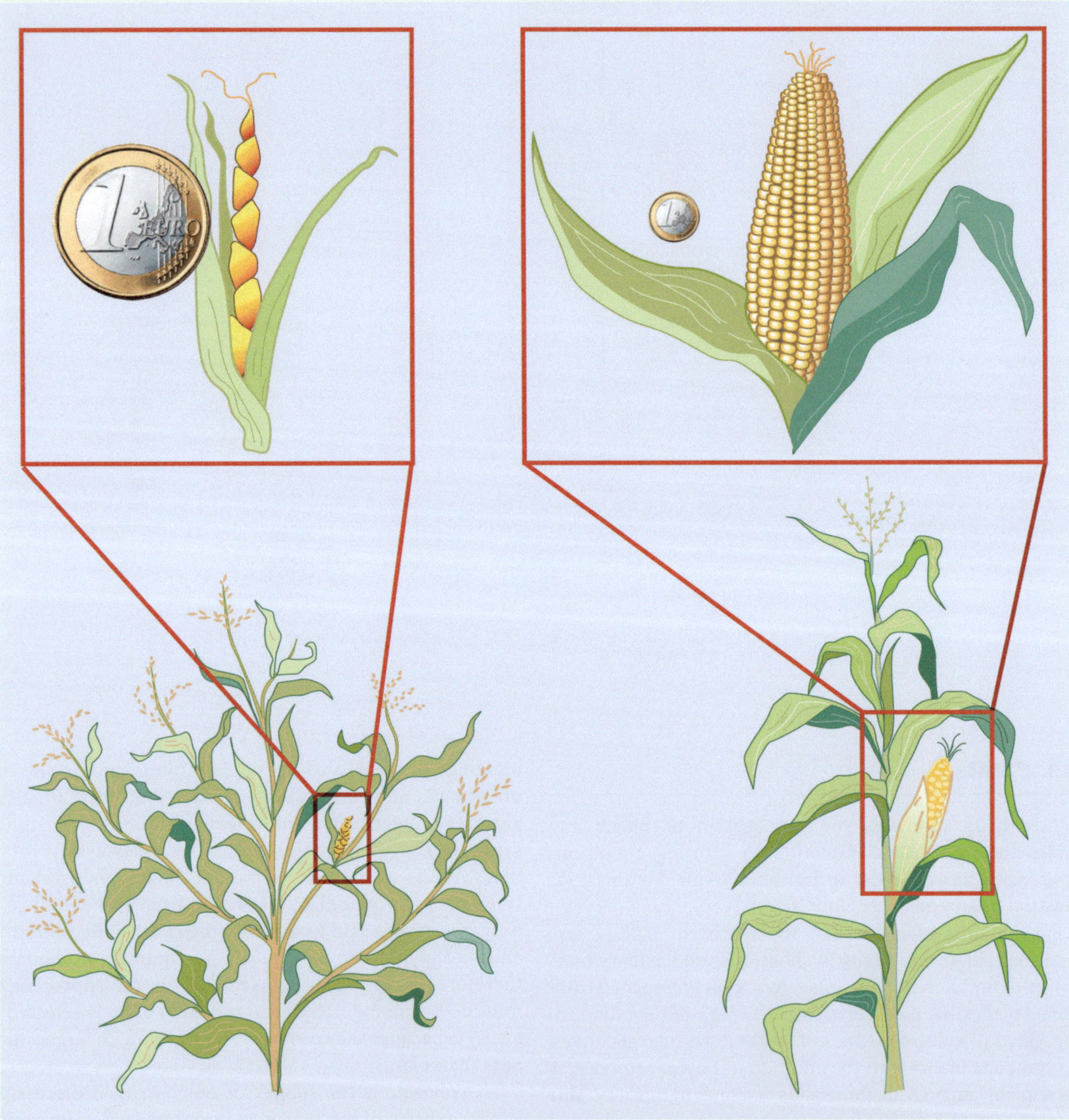

◘ **Fig. 5.12** Comparison of wild grass teosinte and modern maize. The difference becomes particularly clear with the fruit stands ("corncobs"). (Based on a template by Nicolle Rage Fuller, National Science Foundation, USA)

called Mendel's rules, discovered by the monk Gregor Mendel. Genetic diversity can also increase due to errors in DNA replication, repair mechanisms or "jumping genes" (transposons). However, mutations can also be induced by radiation or chemicals. The task of breeding is thus to identify desired hereditary traits and match them with the variety of other desired traits in cultivars.

Besides this unspecific induction of mutations, which are statistically distributed throughout the genome, genetic engineering has developed methods over the past three decades through which the genetic properties of plants are specifically modified. This includes procedures in which genes or regulatory elements from other organisms (transgenes) or closely related species/varieties (cisgenes) are incorporated into the genome of the target plant. For a few years now, methods have also been available for plants that can be used without the need for gene transfer to introduce specific mutations into certain parts of the genome (genome editing, ► Sect. 5.2). The methods used are similar to those used in microbial biotechnology.

In Europe, the cultivation of genetically modified plants is currently totally or partially prohibited at the level of the member states of the European Union. Currently, only transgenic maize is cultivated, in Spain in particular, over a very small area of about 130,000 ha, which corresponds to about 1.3% of the maize cultivation area in Europe. This is the variety MON810, which, due to a transgenic alteration, contains the toxin *Bacillus thuringiensis* (Bt toxin), which protects the plant against the pest corn borer, a butterfly. While practically no transgenic plants are cultivated in Europe, the proportion of land with transgenic plants worldwide rose to about 140 million ha in the period from 1996 to 2015. More than 50% of these areas are planted with soya beans, 30% with maize, 13% with cotton and 5% with rapeseed. This means that around 80% of soya production, 30% of maize production, 68% of cotton production and 25% of rapeseed production worldwide are based on genetically modified plants.

Insects can also cause high crop losses. In order to make crops biotechnologically resistant to insects, they can be genetically modified with a gene of *Bacillus thuringiensis*, which is only toxic to certain insects. As a result of the insect resistance achieved, the use of chemical pesticides can be reduced. However, there are also fundamental concerns in Europe about this genetic modification of plants.

An example of desirable *output traits* can be found in the *Golden Rice* rice variety, which has been genetically enriched with the biosynthesis pathway to ß-carotene, and therefore provides more provitamin A than would occur naturally. This rice could help protect the population in regions that suffer from a lack of food containing provitamin A from blindness due to vitamin deficiency. Because of the fundamental resistance against genetically modified plants, *Golden Rice* has yet to achieve breakthrough success (► Chap. 3).

An example of an *output trait*, which is still in the developmental stage, is the production of chicken egg white in plants (Bobo 2015). Because of its amino acid profile and its digestibility for humans, chicken egg white is particularly nutritionally valuable. The transformation of a plant with the gene for chicken egg white is genetically possible. Suitable production systems are plants that are naturally rich in proteins and have the corresponding organelles for protein storage. These include, for example, soya and pea. Biotechnology could therefore be used to transfer the production of animal food to a plant system.

A similar shift is being worked on in the field of biopolymers. Plants are transformed with genes for the biosynthesis of biopolymers or their monomers. They then accumulate these products in defined tissues. For the model plant established in research, *Arabidopsis thaliana* (thale cress), the feasibility of this principle has already been demonstrated using polyhydroxybutyric acid as an example.

Another field of activity of plant biotechnology is the optimization of lignocellulose. It is an important raw material in the production of cellulose and paper, and demand is growing for other industrial applications. While solutions for lock-up for cellulose and hemicelluloses have been found or are in the process of being found, this is not yet the case for the lignin content of lignocellulose. A reduction in the lignin content would therefore meet the requirements of industrial processes. In fact, it has already been possible to influence the lignin content of various tree species (poplar, eucalyptus, aspen) through breeding. The biosynthesis of lignin starts with the Shikimat pathway, which, besides tyrosine, provides the amino acid phenylalanine. A key enzyme for the further steps towards the biosynthesis of[?] lignin is phenylalanine ammonia lyase. Gymnosperms and angiosperms differ in the final steps of lignin biosynthesis, but it could be shown for both that f by 50% by processing the biosynthesis genes, while, at the same time, recording an increase in cellulose of 30% (Castellanos-Hernandez et al. 2011) (◘ Fig. 5.13).

5.1.4 Animals

Biotechnology also plays a role in the use of animals. For example, horses have traditionally been used as "bioreactors" to manufacture snake venom antisera. For this purpose, poisonous snakes are first "milked." In diluted form, this poison is injected into horses. In contrast, they develop antibodies that are taken from them with the serum of their blood. This horse serum is used to save people bitten by poisonous snakes.

In regards to livestock farming, biotechnology opens up new options. In 2015, a genetically modified type of salmon containing the growth hormone of chinook salmon was approved for marketing as a food in the USA. This salmon species grows particularly fast and reaches its catch weight after 16–18 months instead of 32–36. It may only be kept on fish farms (AquaBounty 2016).

1996 saw a milestone in the breeding of mammals with the birth of the sheep Dolly. Dolly was the first cloned mammal. She was created from a differentiated body cell of a sheep, whose nucleus was implanted into an egg cell. This ovum was carried by a surrogate sheep and grew into a healthy lamb. This technology is still up to-date. For example, horn-

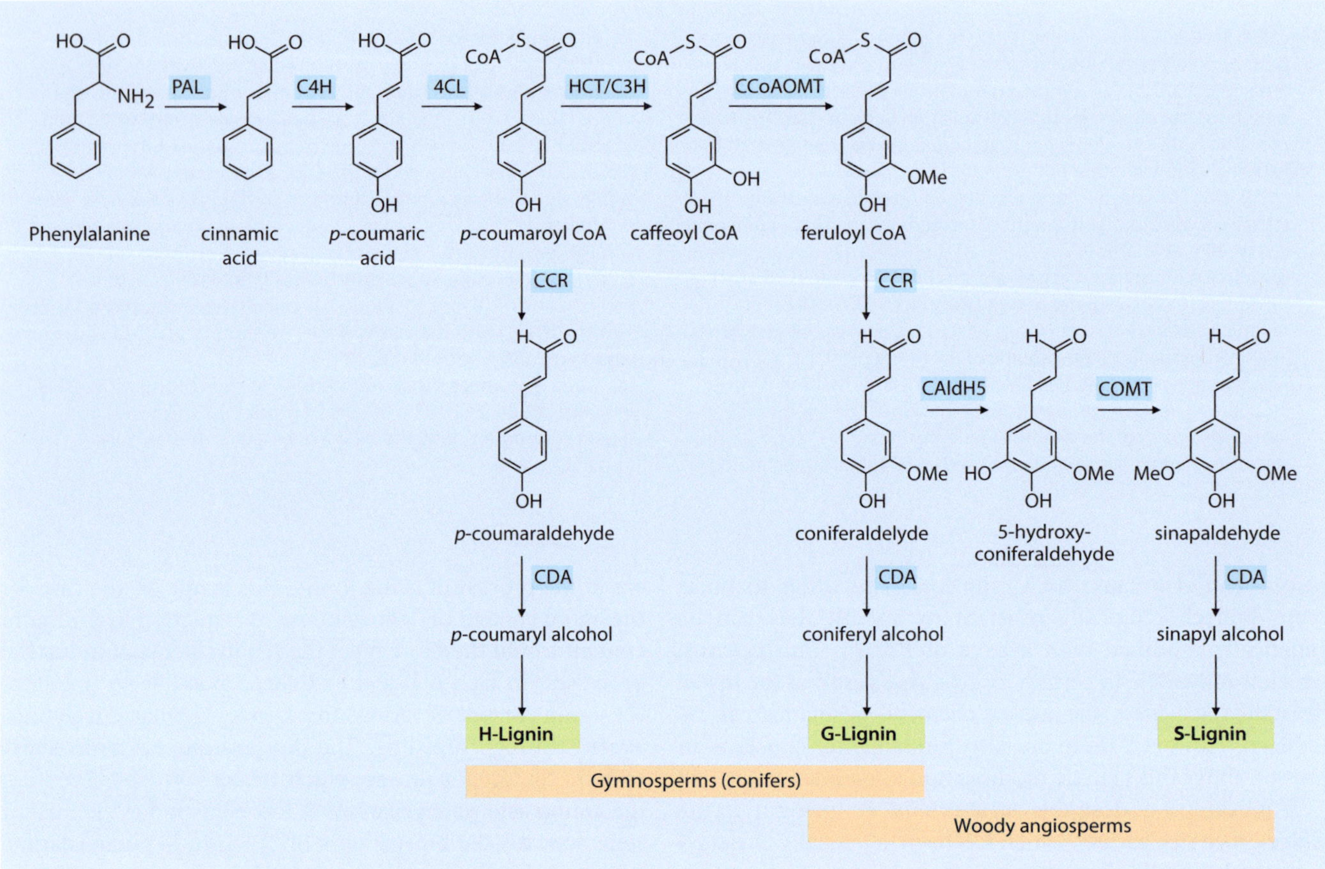

Fig. 5.13 Lignin biosynthesis in forest trees. (Castellanos-Hernandez et al. 2011)

less cattle can be bred by genetically transferring genes for hornlessness into bovine embryos, which are carried to term by cow-borrowing mothers analogous to Dolly (Carlson et al. 2016). Such breeding lines are useful and sought after because they eliminate the risk of horn injuries in the herds of animal breeding facilities.

Crispr/Cas9, the new method of genome editing (▶ Sect. 5.2), is also already used to modify animals. In China, for example, it is tested in the breeding of Koi carp.

5.2 The Perspectives of Synthetic Biology

Michael Bott and Jan Marienhagen

Biotechnology today makes extensive use of the natural metabolic capacities of living systems and adapts its production processes to their needs. Synthetic biology aims to construct completely new biological functional units and systems with desired properties that do not occur in nature or seeks to fundamentally redesign existing biological systems for new tasks following engineering principles. Firstly, synthetic biology wants to gain new insights into the fundamentals of life, in keeping with the statement of the physicist Richard Feynman: "What I cannot create, I do not understand." Secondly, it aims to develop solutions for the future bioeconomy in which biological systems are already adapted to technical requirements at the design stage. For the bioeconomy, synthetic biology opens up new possibilities. It brings within reach products that cannot be produced chemically in an economic manner or for which no natural pathways allowing for their biosynthesis are known. Alternative biotechnological processes based on renewable raw materials are opening up for chemical products that are currently only accessible from raw fossil materials such as crude oil. Raw materials that have not yet been biologically convertible can also be used in synthetic biotechnology processes. Synthetic biology is thus becoming a key element of the hoped-for future bioeconomy. A number of companies, particularly in the USA, are already developing microbial production strains rigorously using synthetic biology methods.

5.2.1 BioBricks

The constructive character of synthetic biology as an applied life science is reflected, in particular, in the concept of *BioBricks*. *BioBricks* is the name given to standardised genetic building blocks with compatible connecting sites, each of which fulfils a defined task and can be freely combined in a

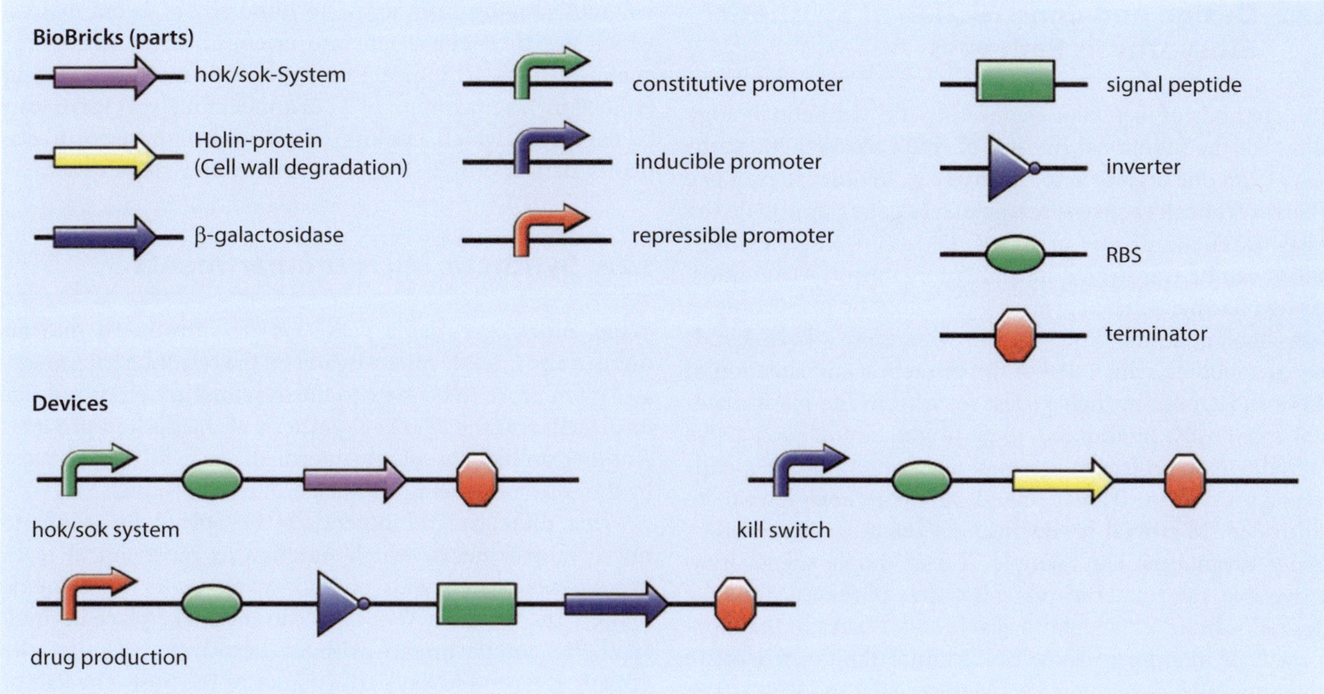

■ **Fig. 5.14** Exemplary symbols of some BioBricks that can be combined into certain functional assemblies. (VIT Vellore Team 2011)

similar way to Lego bricks. Roughly speaking, three different levels of complexity can be distinguished. Simple building blocks (*parts*) become units (*devices*), which, in turn, are used to assemble complex system units (*systems*). Building blocks are, for example, genes, promoters (DNA segments that control the expression of genes) or transcription regulators. One *device* consists of the fusion of individual building blocks, e.g., a combination of promoter, gene and terminator. A system unit performs a complex task, such as the synthesis of a biotechnological product at a specific growth phase. *BioBricks* represent the basis of complex, hierarchical systems upon which the concept of synthetic biology is based (▶ Excursus 5.4; ■ Fig. 5.14).

Analogous to electronic circuits in microprocessors, *BioBricks* can be combined to construct complex biological regulatory circuits and sensor circuits or artificial biosynthetic pathways with inverters, switches and amplifiers, which should fulfil their task independently of the respective cellular environment. The required functionality of the constructed *BioBricks*-based circuits in basically every biological system, however, is not always sufficiently guaranteed, due to the complexity of even the simplest microorganisms, and must always be re-evaluated. Also, the large repertoire of *BioBricks,* which currently comprises around 20,000 building blocks, has not yet been sufficiently characterised to guarantee their free combinability.

Excursus 5.4 The iGEM Competition

Based on the *BioBrick* principle, the *international Genetically Engineered Machine* (iGEM) competition for students takes place every year, giving committed students from the life sciences and engineering sciences the fascinating opportunity to creatively deal with synthetic biology. In small groups, one for each participating university, the students design and construct molecular machines in laboratories on the basis of the *BioBrick* principle for the purpose of tackling current problems in areas such as "food and energy," "environment" and "health and medicine." All participants meet at the Massachusetts Institute of Technology (MIT) in Cambridge (USA) to present their projects and select the winners in various categories. In recent years, this competition has produced many interesting biotechnological concepts and has inspired young scientists to embrace the challenges of modern biotechnology and the concepts of synthetic biology. The high financial expenditure of the participating groups is mainly borne by sponsors, and committed university institutes provide the necessary support voluntarily.

5

5.2.2 Design and Construction of Synthetic Biosynthesis Pathways

The methods of *metabolic engineering* and synthetic biology allow for the functional transfer of entire biosynthetic pathways from one organism to another, e.g., in order to produce pharmaceutically active substances in larger quantities on the basis of renewable raw materials. For example, genes from plants can be transferred into microorganisms. Many interesting pharmaceutical agents are produced by plants only in very small quantities and together with many other secondary metabolites, which makes the extraction and isolation of these substances in their purest form from the plant material very costly. In addition, plant production of such active ingredients is subject to seasonal fluctuations and, depending on product yield and demand, very large areas have to be cultivated. Microbial production, on the other hand, offers many advantages. For example, it uses simple sugars from renewable raw materials as substrates, produces only the desired substance, is highly scalable and, thanks to the rapid growth of microorganisms, can produce the desired active plant ingredients very quickly (Marienhagen and Bott 2013).

In the course of such work, several genes were identified for the synthesis of a precursor of the plant natural product artemisinin from the annual mugwort (*Artemisia annua*) in baker's yeast (*Saccharomyces cerevisiae*) (◘ Fig. 5.15). Artemisinin is an important active ingredient in the fight against multi-resistant *Plasmodium falciparum* strains, the pathogen that elicits malaria, which killed more than 400,000 people worldwide in 2015 (WHO 2015). About 90% of these deaths occur in Africa, where expensive therapies are not available. In order to be able to microbially produce substances such as artemisinin, the respective biosynthetic pathway must first be reconstructed on the computer using the enzymes involved from the respective plants (or other organisms) (◘ Fig. 5.16). The genes required for the biosynthesis of the molecule are then optimised for expression in microorganisms, chemically synthesised and assembled on plasmids in combination with necessary control elements such as promoters and ribosome-binding sites. Even when the enzymes involved have been optimized through *protein engineering* for their task in the chosen microbial cell factory, their formation in functional form in this "host" alone is not sufficient for the economic production of the desired substance. For high product yields, the natural metabolism of the host must also be modified so that the implanted metabolic pathway is sufficiently supplied with the required basic building blocks and co-factors.

The clever design of a synthetic biosynthesis pathway for artemisinic acid from the annual mugwort and extensive modifications to the metabolism of baker's yeast for optimal integration of the new enzymes led to the construction of a strain of baker's yeast that can produce up to 25 g of artemisinic acid per litre from sugar. By scaling up the process, large quantities of artemisinic acid can be produced efficiently and cost-effectively in industrial bioreactors, with volumes ranging from a few to hundreds of cubic metres, which are then converted into artemisinin in subsequent chemical steps. In comparison, on a cultivation area of 1 ha (10,000 m^2), a maximum of 2 t of annual mugwort leaves can be harvested, which, engineered even for high-performance plants, deliver only 2–3 kg of an artemisinin-rich extract.

5.2.3 Synthetic Microcompartments

When new biosynthetic pathways are introduced into an industrially suitable microorganism, the resulting intermediates often prove to be toxic to the host and inhibit or prevent its growth and the effective synthesis of the desired product. Another problem is volatile intermediates, which can be lost by the cell before being converted into the product.

This difficulty can potentially be solved by synthetic microcompartments, which function as biochemical reaction spaces coated with protein envelopes or lipid membranes. The toxic reaction steps can then take place in these synthetic compartments without negatively affecting the growth and production performance of the host. The loss of volatile intermediates can be minimized by concentrating the enzymes that form and consume them in the microcompartments, making product formation more efficient. Synthetic biologists have copied this trick from the microorganisms themselves. For example, some cyanobacteria, which carry out photosynthesis, fix carbon dioxide (CO_2) in carboxysomes. These are microcompartments that contain the two enzymes ribulose bisphosphate carboxylase (Rubisco) and carboanhydrase. Rubisco catalyses the actual CO_2 fixation, while carboanhydrase accelerates the adjustment of the equilibrium between bicarbonate (HCO_3^-) and CO_2, and thereby improves the availability of CO_2 required as substrate by Rubisco. It is postulated that the carboxysomes concentrate CO_2, and thus minimize the side reaction of Rubisco with molecular oxygen.

So far, however, synthetic microcompartments are still in the conception and testing phase and are currently not used in industrial production processes. This is due to the fact that the assembly of the compartments and their loading with the target enzymes is not yet easily controllable.

5.2.4 Synthetic Genomes

The design of synthetic genomes is a central issue of synthetic biology. Such genomes contain only those genes that a microorganism needs under certain conditions for its growth and for the formation of a desired product. All other genes, which the wild-type microbe only needs occasionally, but which are not essential for its growth and product formation, are missing. In this field of research, two basic approaches can be distinguished, namely, the *de novo* synthesis of genomes from chemically synthesized DNA (*bottom-up*-approach) and the minimization of genomes of existing organisms (*top-down*-

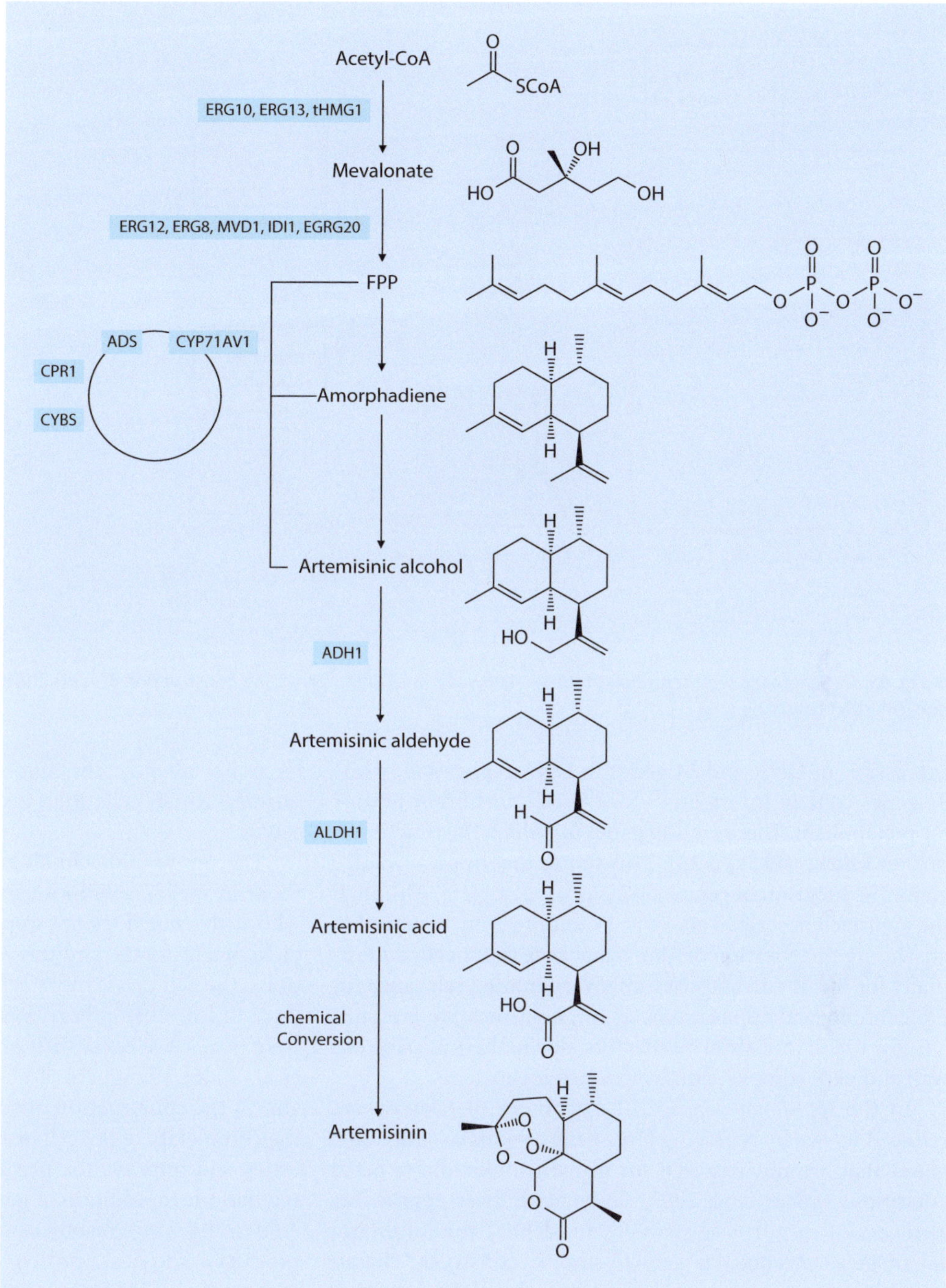

Fig. 5.15 Optimized synthetic biosynthetic pathway for artemisinic acid in baker's yeast. In order to be able to produce artemisinic acid with baker's yeast, the cell's own metabolism must first be modified so as to increase the supply of precursors. For this purpose, eight endogenous genes must be overexpressed. The functional expression of four genes from the annual mugwort then allows for the synthesis of artemisinic acid, which is then chemically converted into the actual active substance artemisinin in a final step. (Paddon and Keasling 2014)

approach), which were performed, for example, with *E. coli*, *Bacillus subtilis* and *Corynebacterium glutamicum*.

One leader in the development of the bottom-up approach is the *J. Craig Venter Institute* (JCVI), which, in recent years, has achieved significant breakthroughs with *Mycoplasma* bacteria as model organisms. *Mycoplasma* species have a very small genome and lack a cell wall, allowing for the transfer of very large DNA fragments into the cell. The researchers have succeeded in constructing a 1079 kb genome of *Mycoplasma mycoides* entirely based on chemically synthesized oligonucleotides, assembled in baker's yeast (Fig. 5.17), and then transferred into a recipient cell of *Mycoplasma capricolum*, whose own genome was subsequently removed by antibiotic selection. The transplanted cells showed the properties expected from the donor genome and were able to replicate themselves (Gibson et al. 2010). In this way, a viable organism with a foreign genome was created for the first time, named *Mycoplasma mycoides* JCVI-syn1.0.

In a further project of this institute, the genome of *M. mycoides* JCVI-syn1.0, which contained 910 genes, was further reduced in a *design build test* cycle by removing genes that were not essential for growth (Hutchison et al. 2016) (Fig. 5.18). This resulted in *M. mycoides* JCVI-syn3.0 with a genome of 521 kb and 473 genes. 195 genes are involved in the

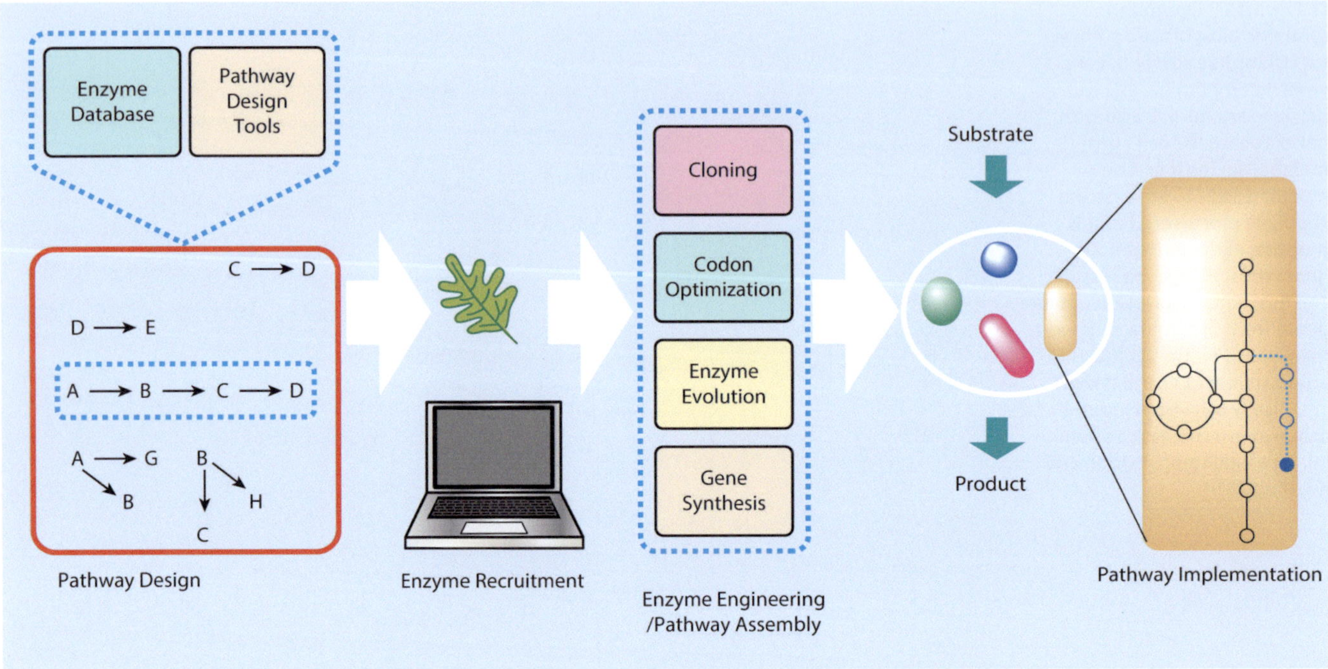

Fig. 5.16 Schematic representation of the workflows for the optimization and construction of synthetic biosynthetic pathways in microorganisms. (Martin et al. 2009)

expression of DNA and 34 genes in DNA replication, while 84 genes contain information for the cell membrane and 81 for metabolism. There are 149 genes for which the function is not yet known (Fig. 5.19). This shows that our knowledge about the function of genes is still very incomplete. Although the approach described above is groundbreaking in regard to answering the question of how many genes are needed minimally for life, it currently has only very limited relevance for biotechnological applications, as mycoplasmas are not suitable for use in industrial bioreactors, due to their missing cell wall and very complex nutrient requirements.

In the *top-down* concept, the genomes of bacteria are reduced in size by removing large gene regions or individual genes that are not required for growth under the selected conditions (Feher et al. 2007). In general, these approaches also serve two purposes, namely, to identify the minimum set of genes needed for growth and to construct "chassis strains" with improved properties for industrial production processes. Both for *Escherichia coli* and for *Bacillus subtilis*, strains with a genome reduced by more than 35% were described (Hirokawa et al. 2013; Tanaka et al. 2013). The number of genes identified as essential is currently 295 for *E. coli* and 253 for *B. subtilis*. Most of these genes are involved in protein biosynthesis, protein quality control, metabolism, cell wall biosynthesis, and cell division (Juhas et al. 2014). However, these figures apply to growth in the LB medium with glucose, which, due to the presence of yeast extract and tryptone, already contains many substances required for growth, which cells then no longer have to produce themselves.

The more biotechnologically relevant approaches are those in which only those genes or gene clusters are deleted whose absence does not impair the growth rate and the cell yield in a minimal medium with glucose. Under these conditions, the cell needs more genes, but is able to grow and produce in low-cost culture media. Corresponding approaches were, e.g., developed with *E. coli* and *C. glutamicum* (Posfai et al. 2006; Mizoguchi et al. 2008; Unthan et al. 2015). The aim in the construction of such "chassis strains" is to reduce the complexity of the cell in order to remove unwanted properties and improve the predictability of metabolism and its regulation. In addition, a genome-reduced strain should be able to use more resources for the formation of the desired products. The basic principle of these approaches is exemplarily shown in Fig. 5.20 for the industrially important *Corynebacterium glutamicum*, whose genome has been reduced by 13%. In fact, some of the genome-reduced strains of *E. coli* and *C. glutamicum* showed beneficial properties, e.g., an improved production of the amino acid L-threonine (Mizoguchi et al. 2008), reduced mutation rate and improved electroporation rate (Posfai et al. 2006), and improved protein production (Unthan et al. 2015). It is expected that genome-reduced strains will play an increasingly important role in the development of industrial production strains.

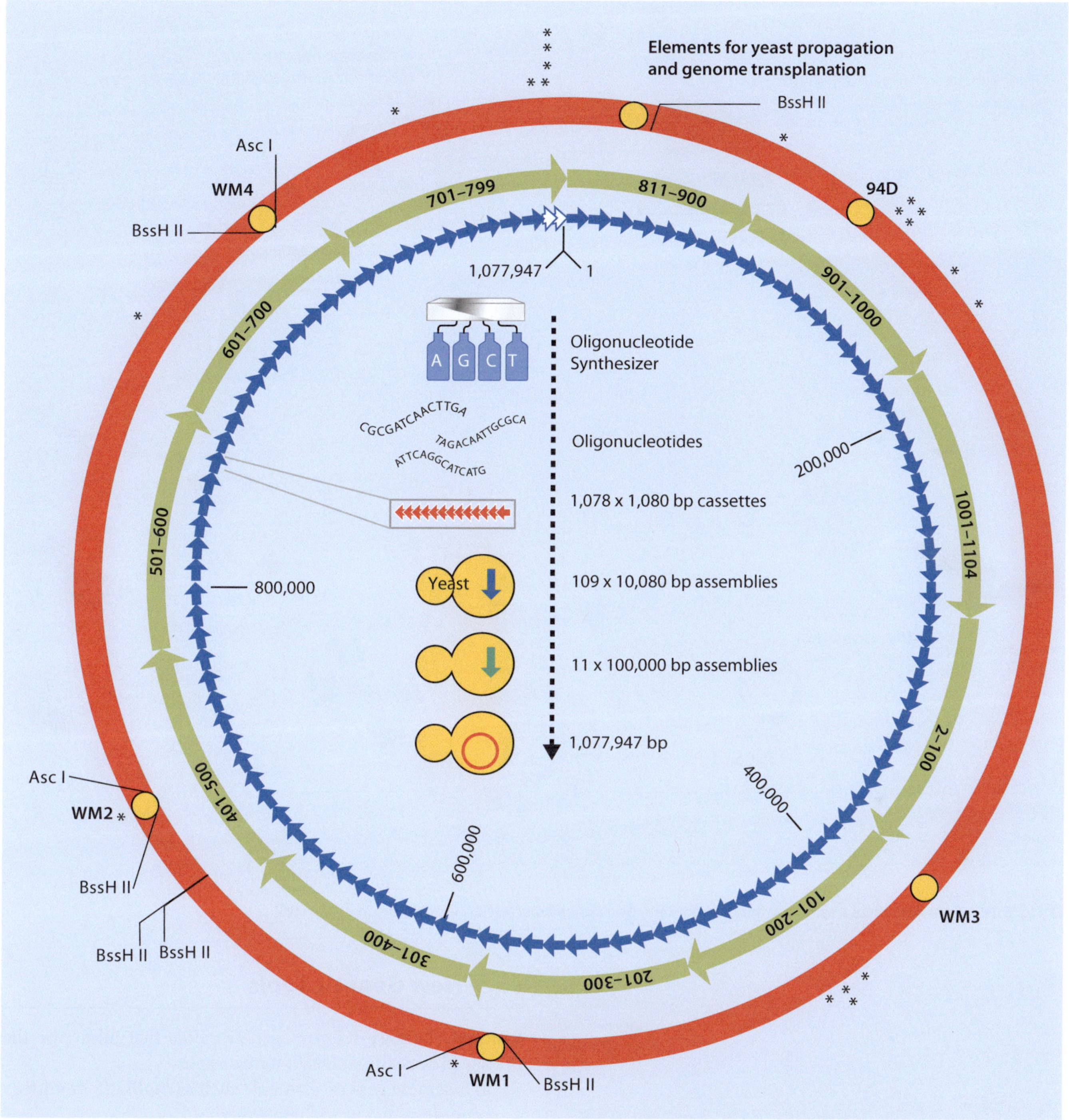

Fig. 5.17 Strategy for the assembly of a synthetic *Mycoplasma mycoides* genome from 1078 overlapping DNA fragments in baker's yeast. In the first step, ten 1080 bp DNA fragments each, which had been produced from overlapping synthetic oligonucleotides, were recombined to generate 11 approximately 100 kb DNA fragments (green arrows). In the final step, these 11 fragments were recombined to form the complete genome (red circle). Assembly was performed by homologous recombination in baker's yeast. The yellow circles indicate the deviations of the synthetic genome from the natural genome. Among other things, four regions were introduced, which were used as watermarks (*water mark*, WM) and the genetic elements needed for growth in baker's yeast and genome transplantation. The stars indicate point mutations in the synthetic genome. (Gibson et al. 2010)

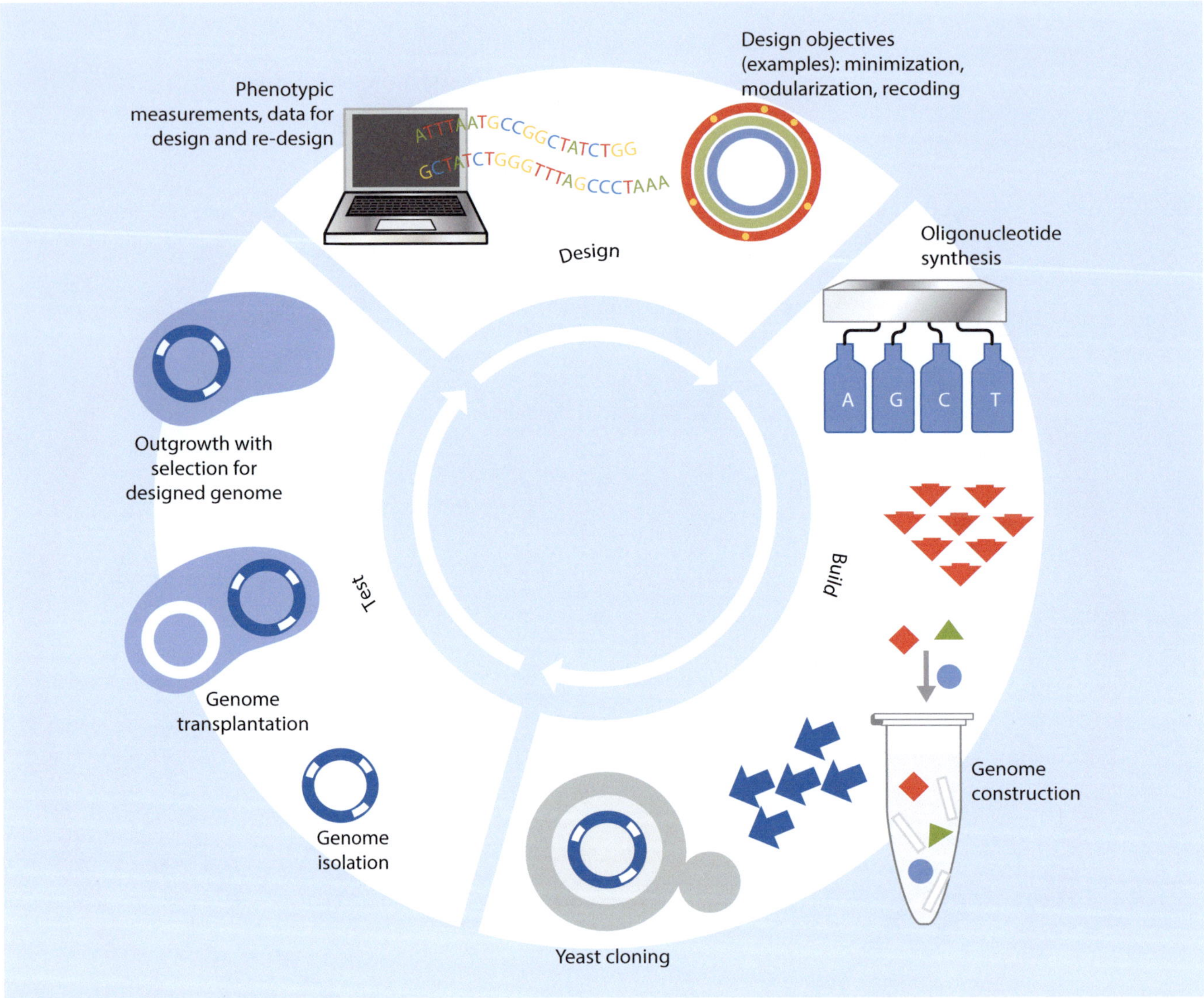

Fig. 5.18 Construction of a synthetic minimal genome by a *design build test*-cycle. (Hutchison et al. 2016)

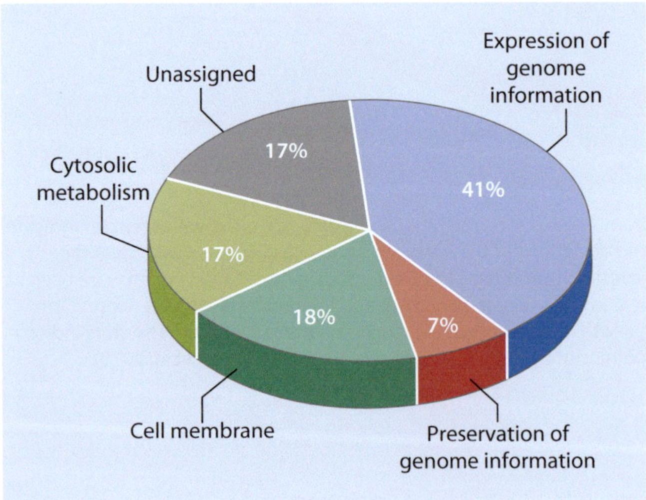

Fig. 5.19 Genetic information of the minimal genome of *Mycoplasma mycoides* JCVI-syn3. (Hutchison et al. 2016)

5.2.5 New Genetic Tools

Synthetic biology requires genetic tools that allow for the precise processing of DNA, for example:

- to assemble genetic elements such as *BioBricks* or entire genomes quickly and efficiently,
- to insert DNA specifically into genomes,
- to change single or small numbers of bases in genomes, and
- to delete DNA from genomes, as in the production of chassis strains.

In classic gene technology, restriction endonucleases and DNA ligases are used for the targeted assembly of DNA fragments. Restriction endonucleases are enzymes that recognise defined DNA sequences and introduce cuts within or next to these sequences. The matching interfaces of two DNA fragments are then reconnected by the DNA ligase. Particularly

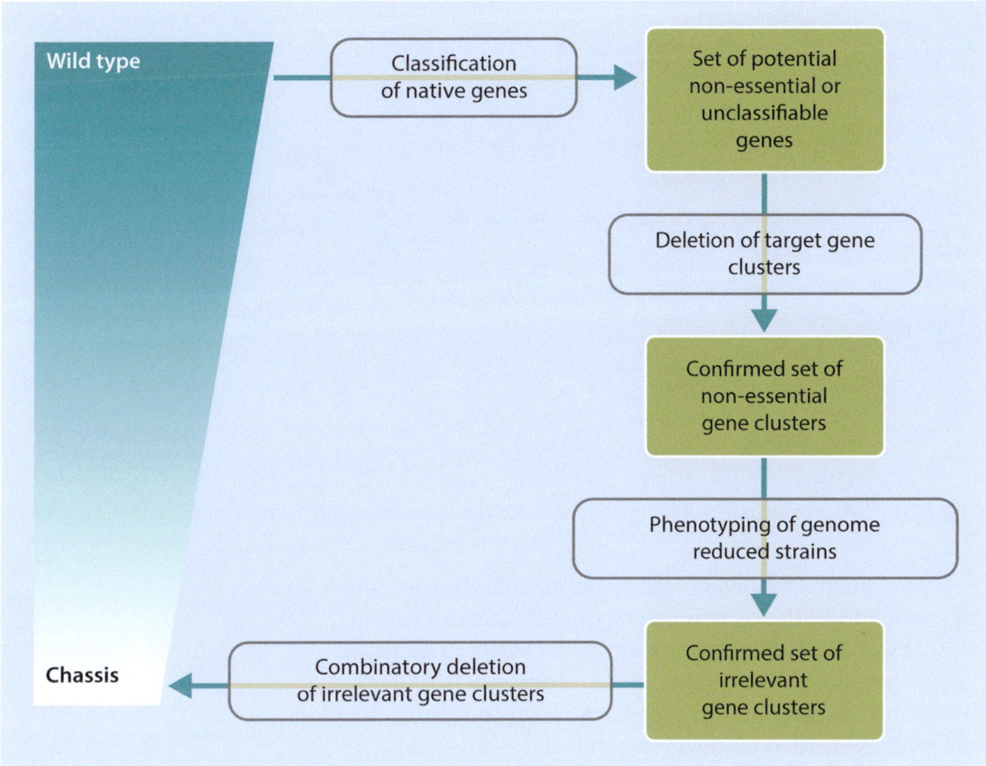

Fig. 5.20 The procedure of developing a "chassis strain" from a wild-type strain by systematically eliminating gene clusters. (Unthan et al. 2015)

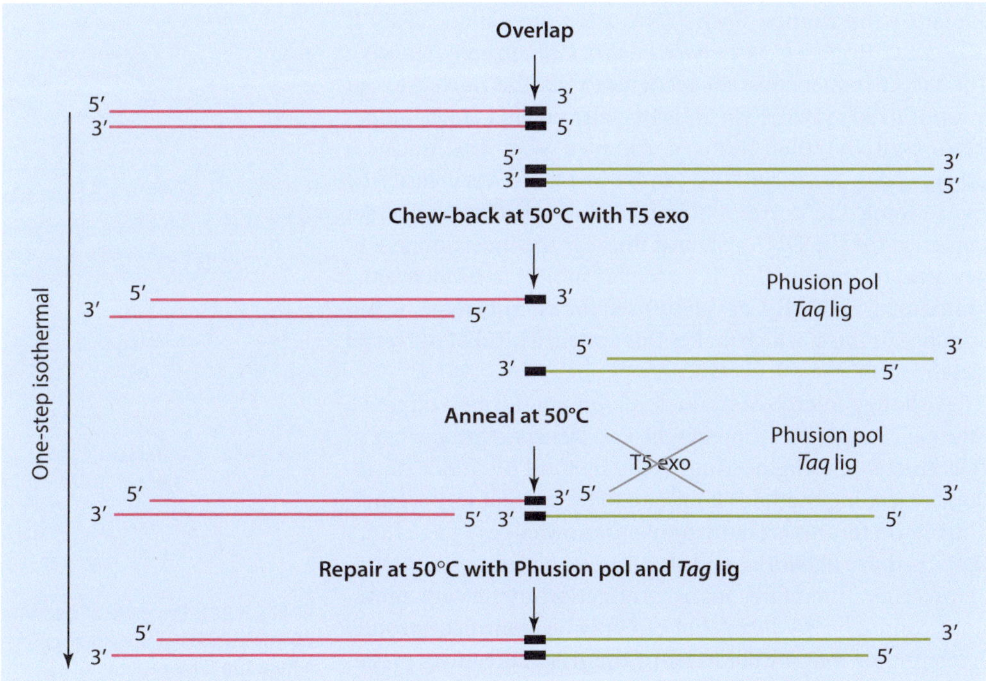

Fig. 5.21 Assembly of DNA fragments by means of *Gibson assembly*. (From Gibson 2011)

in the course of the development of synthetic genomes, methods have been developed that allow for the simultaneous assembly of several DNA fragments, for example, via ***Gibson assembly*** (Fig. 5.21) or homologous recombination in baker's yeast.

Recombineering is a tool for the rapid introduction of insertions, deletions, or point mutations in chromosomes or plasmids using PCR products or synthetic oligonucleotides (Court et al. 2002). It is based on recombination proteins of bacteriophages, such as the proteins Exo and Beta from the phage Lambda, which can efficiently recombine DNA sequences with identical regions of 35–50 bp via homologous recombination (Fig. 5.22). The method was developed for *E. coli*, but is now also used for other biotechnologically relevant bacteria, such as *C. glutamicum*, *B. subtilis*, *Pantoea ananatis* and *Lactococcus lactis*.

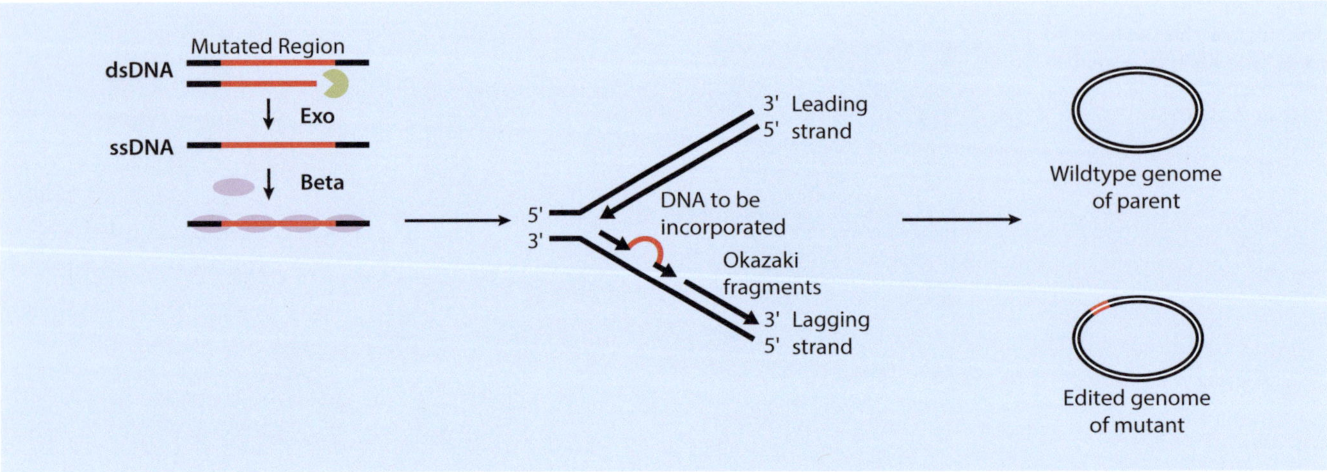

Fig. 5.22 Selective alteration of the genome through *recombineering* with the proteins Exo and Beta of the phage Lambda. (Bott and Eggeling 2016)

A tool that offers completely new possibilities for the targeted modification of genomes is **CRISPR-Cas** (Jinek et al. 2012). It was discovered during studies on a bacterial immune system that incorporates small pieces of foreign DNA from viruses into its genome and uses these to cut the viral DNA with nuclease Cas9 once the viruses infect the cell again. The foreign viral DNA fragments inserted into the genome are separated by short repeating DNA segments, some of which contain palindromes. Such DNA areas are called CRISPR (*Clustered Regularly Interspaced Short Palindromic Repeats*). The target sequence is not recognized via the nuclease, but in optimized systems via an artificially created *single guide*-RNA (sgRNA) that forms a complex with the nuclease (Fig. 5.23). Such sgRNAs can be produced very easily by transcribing the corresponding DNA, such that the target sequence for the DNA cut, and thus for the insertion site of new genetic information, is no longer subject to fundamental limitations. CRISPR-Cas can be used for microbial, plant and animal genomes, which makes this system a kind of universal genetic tool.

Although microbial strain development through targeted, rational changes of the metabolic and transport properties of cells (metabolic engineering) is the method of choice nowadays, screening methods continue to be of great importance. The reason for this is our incomplete knowledge of gene functions, cellular networks, and structure-function relationships of proteins. Therefore, many production-increasing mutations cannot be predicted, but can only be identified through screening of mutant libraries. In the past, screening procedures were usually very labor-intensive and time-consuming. Nowadays, high-throughput screening assays are developed, for example, through the use of **transcriptional regulators as biosensors.** In combination with a suitable target promoter and a reporter gene coding for a fluorescent protein, such regulators enable the transformation of the concentration of a desired metabolite within the cell into a fluorescence signal, which can be measured at a speed of more than 10,000 cells per second using fluorescence-activated cell sorting (FACS) (Fig. 5.24). Individual cells with increased metabolite concentration, and thus fluorescence, can therefore be separated in minutes to hours from libraries containing millions of cells generated by random mutagenesis. The mutations responsible for the increased metabolite formation are then detected by sequencing and further analysis. In this way, "productive

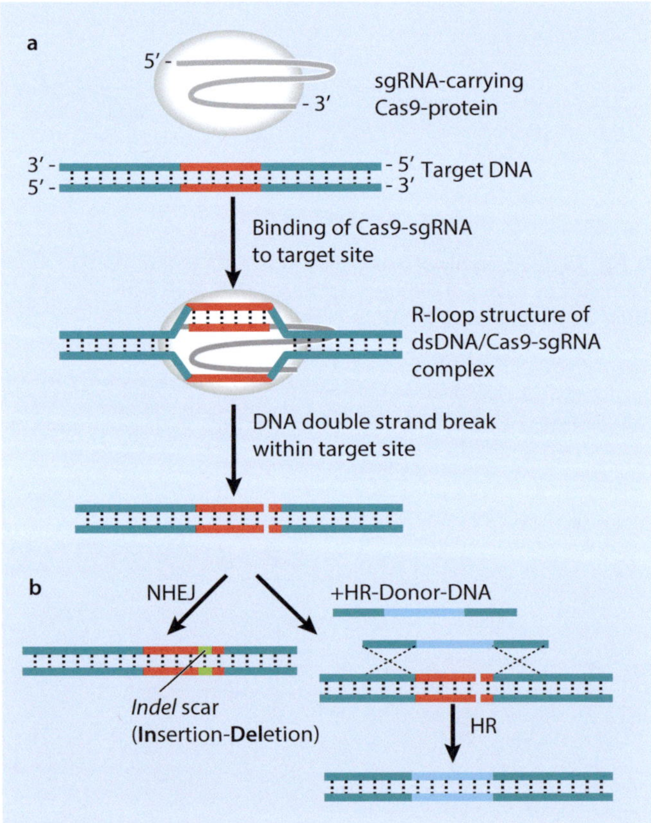

Fig. 5.23 Principle of genome editing with CRISPR-Cas9. **a** The first 20 nucleotides of the *single guide*-RNA (red) determine the target site where the Cas9 protein introduces a double strand break (DSB) after the formation of an R-loop structure. **b** DSB are either repaired via non-homologous end joining (NHEJ) or in the presence of donor DNA through homologous recombination (HR), which allows for precise editing of the gene sequence. (Pul et al. 2016)

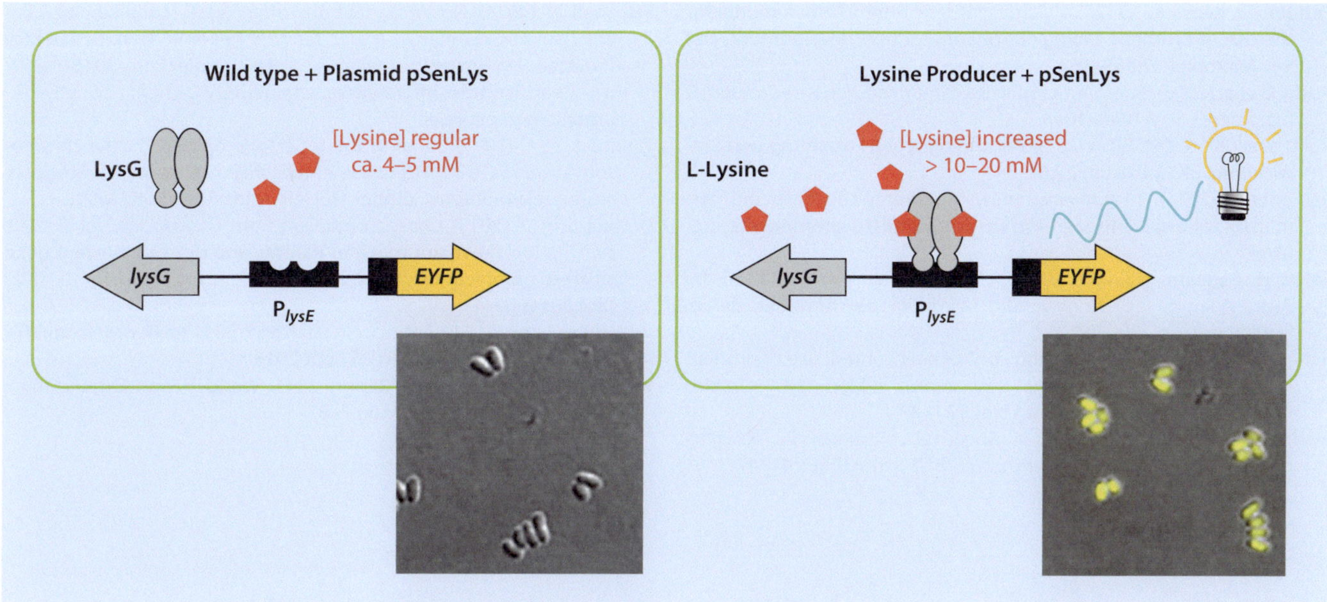

Fig. 5.24 Basic principle of transcription factor-based biosensors for the conversion of the intracellular concentration of a product, in this case, the amino acid L-lysine, into a fluorescence signal that can be detected at the single-cell level. (Bott and Eggeling 2016)

mutations" can be identified very quickly and then combined to obtain efficient production strains (Eggeling et al. 2015). In this way, for example, a point mutation in the gene *murE* of *C. glutamicum* was identified, which was sufficient to transform the wild-type strain into a lysine producer (Binder et al. 2012). This method has also proven to be suitable for the optimization of individual enzymes and for saturation mutagenesis of certain amino acid positions (Schendzielorz et al. 2014; Siedler et al. 2014; Binder et al. 2013).

References

Aquabounty. https://aquabounty.com. Accessed 14 Sept 2016

Bessler C et al (2003) Directed evolution of a bacterial α-amylase: towards enhanced pH-performance and higher specific activity. Protein Sci 10:2141–2149

Binder S, Schendzielorz G, Stäbler N, Krumbach K, Hoffmann K, Bott M, Eggeling L (2012) A high-throughput approach to identify genomic variants of bacterial metabolite producers at the single-cell level. Genome Biol 13:R40

Binder S, Siedler S, Marienhagen J, Bott M, Eggeling L (2013) Recombineering in *Corynebacterium glutamicum* combined with optical nanosensors: a general strategy for fast producer strain generation. Nucleic Acids Res 41:6360–6369

Bobo J (2015) First bioeconomy summit, Berlin. http://gbs2015.com/fileadmin/gbs2015/Downloads/Bioeconomy_World_Tour.pdf. Accessed 14 Sept 2016

Bott M, Eggeling L (2016) Novel technologies for optimal strain breeding. In: Yokota A, Ikeda M (eds) Amino acid fermentation (Chapter 13). Springer, Tokyo

Carlson ED et al (2012) Cell-free protein synthesis: applications come of age. Biotechnol Adv 30(5):1185–1194. https://www.ncbi.nlm.nih.gov/pmc/articles/PMC4038126/figure/F2/

Carlson DF, Lancto CA, Zang B, Kim ES, Walton M, Oldeschulte D, Seabury C, Sonstegard TS, Fahrenkrug SC (2016) Production of hornless dairy cattle from genome-edited cell lines. Nat Biotechnol 34(5):479–481. https://doi.org/10.1038/nbt.3560. Accessed 14 Sept 2016

Castellanos-Hernández OA et al (2011) Genetic transformation of forest trees. In: Alvarez M (ed) Genetic transformation. Chapter 10. Intech; ISBN 978-9533073644, Rijeka. https://doi.org/10.5772/868

Court DL, Sawitzke JA, Thomason LC (2002) Genetic engineering using homologous recombination. Annu Rev Genet 36:361–388. https://doi.org/10.1146/annurev.genet.36.061102.093104

Daniell J, Köpke M, Simpson SD (2012) Review: commercial biomass syngas fermentation. Energies 5:5372–5417. https://doi.org/10.3390/en5125372, Energies ISSN 1996-1073, www.mdpi.com/journal/energies

Eggeling L, Bott M (2015) A giant market and a powerful metabolism: L-lysine provided by *Corynebacterium glutamicum*. Appl Microbiol Biotechnol 99:3387–3394

Eggeling L, Bott M, Marienhagen J (2015) Novel screening methods-biosensors. Curr Opin Biotechnol 35:30–36

Feher T, Papp B, Pal C, Posfai G (2007) Systematic genome reductions: theoretical and experimental approaches. Chem Rev 107:3498–3513

Gibson DG (2011) Enzymatic assembly of overlapping DNA fragments. Methods Enzymol 498:349–361

Gibson DG et al (2010) Creation of a bacterial cell controlled by a chemically synthesized genome. Science 329:52–56

Hirokawa Y et al (2013) Genetic manipulations restored the growth fitness of reduced-genome *Escherichia coli*. J Biosci Bioeng 116:52–58

Hutchison CA 3rd et al (2016) Design and synthesis of a minimal bacterial genome. Science 351:aad6253

Jinek M et al (2012) A programmable dual-RNA-guided DNA endonuclease in adaptive bacterial immunity. Science 337:816–821

Jones S (2016) How to cut cost in fermentation and bio-catalysis. BIO World Congress on Industrial Biotechnology, San Diego (USA) 18.-20.4.2016

Juhas M et al (2014) *Bacillus subtilis* and *Escherichia coli* essential genes and minimal cell factories after one decade of genome engineering. Microbiology 160:2341–2351

Marienhagen J, Bott M (2013) Metabolic engineering of microorganisms for the synthesis of plant natural products. J Biotechnol 163(2):166–178. https://doi.org/10.1016/j.jbiotec.2012.06.001

Martin CH et al (2009) Synthetic metabolism: engineering biology at the protein and pathway scales. Chem Biol 16:277–286

Mizoguchi H et al (2008) Superpositioning of deletions promotes growth of *Escherichia coli* with a reduced genome. DNA Res 15:277–284

Paddon CJ, Keasling JD (2014) Semi-synthetic artemisinin: a model for the use of synthetic biology in pharmaceutical development. Nat Rev Microbiol 12:355–367

Posfai G et al (2006) Emergent properties of reduced-genome *Escherichia coli*. Science 312:1044–1046

Pul U et al (2016) CRISPR in der biotechnologischen Forschung und Entwicklung. BIOspektrum 22:62–64

Rec-energy (2016). http://www.renewable-energy-concepts.com/german/bioenergie/biogas-basiswissen/biogaszusammensetzung.html

Sahm H, Eggeling L (2009) Mikrobielle Aminosäurefermentation. In: Antranikian G (ed) Angewandte Mikrobiologie. Springer, Berlin, Heidelberg, pp 290–304

Schendzielorz G et al (2014) Taking control over control: use of product sensing in single cells to remove flux control at key enzymes in biosynthesis pathways. ACS Synth Biol 3:21–29

Siedler S et al (2014) SoxR as a single-cell biosensor for NADPH-consuming enzymes in *Escherichia coli*. ACS Synth Biol 3:41–47

Staropoli N (2016) No-till agriculture offers vast sustainability benefits. So why do many organic farmers reject it? Genetic Literacy Project. https://www.geneticliteracyproject.org/2016/06/02/no-till-agriculture-offers-vast-sustainability-benefits-so-why-do-organic-farmers-reject-it/

Tanaka K et al (2013) Building the repertoire of dispensable chromosome regions in *Bacillus subtilis* entails major refinement of cognate large-scale metabolic model. Nucleic Acids Res 41:687–699

Unthan S et al (2015) Chassis organism from *Corynebacterium glutamicum* – a top-down approach to identify and delete irrelevant gene clusters. Biotechnol J 10(2):290–301. https://doi.org/10.1002/biot.201400041

VIT Vellore Team (2011) Biobrick circuit. http://2011.igem.org/Team:VIT_Vellore/Project/. Accessed 26 Sept 2016

WHO (2015) World Malaria Report 2015. World Health Organization, Geneva. ISBN: 978 92 4 156515 8

The Bioeconomy from the Point of View of Innovation Economics

Andreas Pyka

"Add as many mail-coaches as you please, you will never get a railroad by so doing." – With these words, the economist Joseph Schumpeter symbolized the discontinuity of progress, a fitting characterization of the transition to the bioeconomy as well. (© spiritofamerica/Fotolia torsakarin/Fotolia)

J. Pietzsch (ed.), *Bioeconomy for Beginners*, https://doi.org/10.1007/978-3-662-60390-1_6

6.1 The Discontinuity of Progress

After more than 200 years of industrial production, large portions of humankind are wealthier than ever before. At the same time, industrial production is closely linked to the exploitation of natural resources. The influence of human activity has reached global dimensions as can be seen most clearly from the accumulation of climate-damaging gases in the atmosphere. This endangers human survival on planet Earth. Continuing "business as usual" is no longer an option. But how can the future be shaped, and humanity provided with a high or even increasing level of welfare, without continuing to risk the natural conditions of life? At the beginning of the twenty-first century, many economies worldwide are linking their answers to this question with the knowledge-based bioeconomy. Is this really a way out? This will be examined in the following from the perspective of innovation economics.

Among economists there is wide agreement that technological progress is the main driver of quantitative growth measured by the per capita income of economies. However, far less agreement exists on the qualitative characteristics of economic development: while the mainstream-oriented branch of economics, neoclassical economics (often referred to as the "economic sciences"), is limited to the purely quantitative view, and thus remains within its short term orientation, Neo-Schumpeterian economics assumes the qualitative perspective, and thus places change in fundamental economic structures over longer periods of time at the centre of its analysis.

Processes of change can be attributed both to incremental innovations and to structural changes, such as the emergence of new industries and the disappearance of old ones. To simplify matters, one can assume that incremental technological improvements in the sense of gradual improvement innovations build on already existing technological solutions, while structural changes are triggered by radical technological breakthroughs (major innovations) that question larger production contexts. They can, if necessary, lead to drastic changes, in the sense of "creative destruction" (Schumpeter 1943) of the world production system as a whole (▶ Excursus 6.1).

This chapter deals with a fundamental transformation of production systems: The overcoming of the *lock-in* of the current production system in fossil energy sources (Unruh 2000) and the simultaneous establishment of a knowledge-based bioeconomy (Pyka 2017; Pyka and Buchmann 2017). There is no doubt that this is a radical, qualitative and long-term transformation process that must be considered in the innovation-economic approach of Neo-Schumpeterian economics.

Excursus 6.1 The Great Cycles of Innovation

By 1939, in his *Business cycles*, Schumpeter had already revived Kondratieff's "Theory of long waves," explaining that this is a process of economic development that is quite normal in the long term. Most famous is his picture for clarifying the discontinuous character: "*Add as many mail-coaches as you please, you will never get a railroad by so doing*" (Schumpeter 1934). The first long wave began around 1800 with industrialization and was driven by the basic technology of steam engines and cotton processing. The wide availability of steel and railways then determined the second long wave, from around 1850 onwards, which, in turn, was replaced at the beginning of the twentieth century by electrical engineering and the chemical industry. With mass production and the automotive and petrochemical industries, the third long wave started rolling in the middle of the last century. As a result, a second fossil energy source, crude oil, moved to the centre of production activities, alongside coal. Since the 1980s, a fifth long wave is emerging which is reflected in solutions related to information and communication technology. At the beginning of the twenty-first century, another paradigmatic change of this kind is now on the horizon, albeit with a great difference from the previous upheavals. While the past cycles were driven by economic bottlenecks and the need to overcome them technologically, in the twenty-first century, humankind is faced with the crucial question of how it can restore the ecological sustainability of economic activity. A central role in this process of change, which is characterised by true uncertainty (Knight 1921), is to be played by the approach known as the knowledge-based bioeconomy (◻ Fig. 6.1).

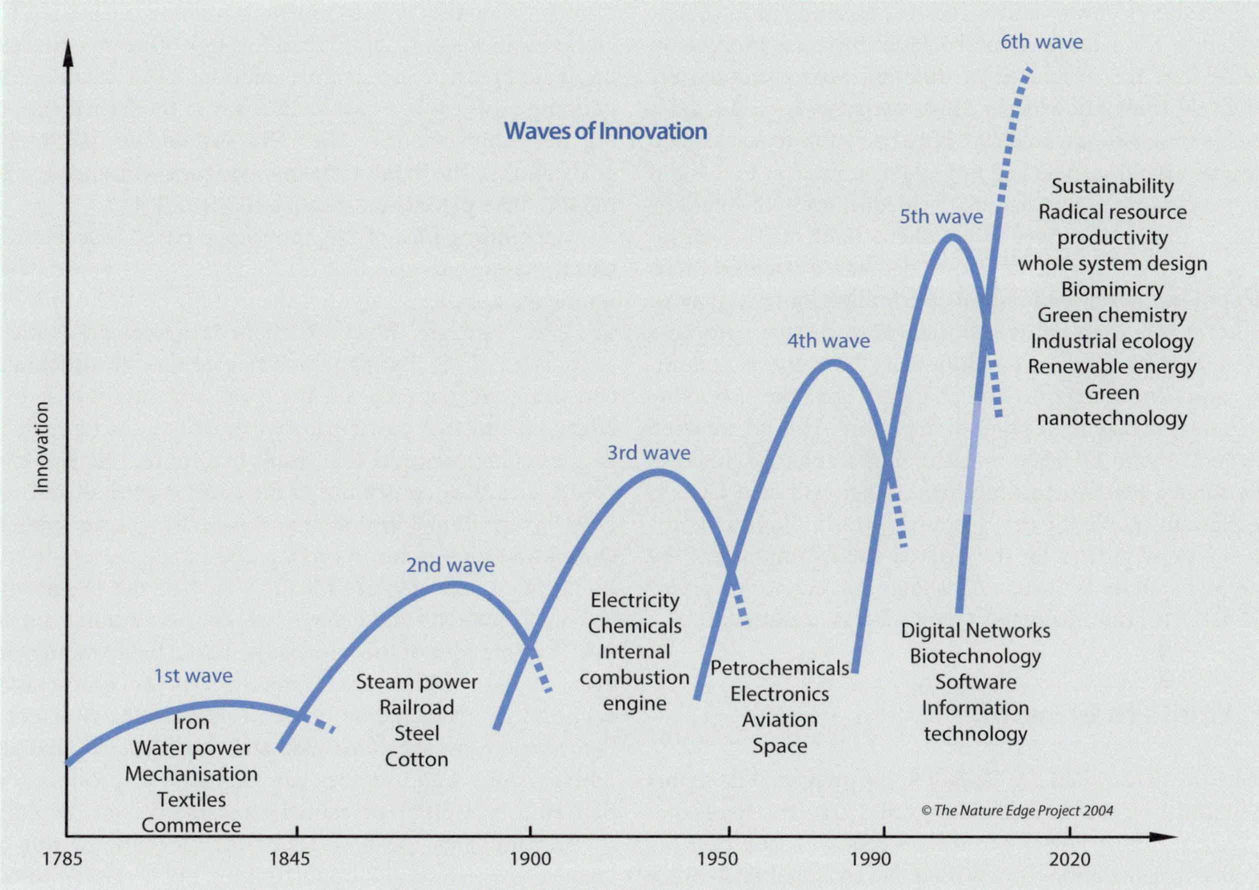

Fig. 6.1 According to Kondratieff's theory, the essential technologies of the sixth wave of innovation, which began at the turn of the twenty-first century, can be assigned to the necessity of ecological sustainability. (Own representation after von Weizsäcker 2010)

In the meantime, the literature discusses numerous alternative terms for structural changes and processes of change that affect the entire production system of the world economy. Freeman and Dosi call them "techno-economic paradigm changes" (Dosi 1982; Freeman 1991), Sahal uses cartographic analogies and refers to "technological guideposts" that are pointing to new "technological avenues" (Sahal 1985). In all of the studies, it is emphasized that economic systems over larger periods of several decades are confronted again and again with enormous upheavals, which question practically all established production approaches. Even in cases in which a single technology triggers these upheavals, this technology alone is not alone responsible for comprehensive changes to be observed. Rather, it forms the basis for several complementary developments. Let us look at the combustion engine, for example. It is part of a package of interdependent technologies, such as advances in petrochemicals and the introduction of assembly line production. The integration of these technologies, in turn, triggers numerous infrastructural developments, such as the establishment of a network of petrol stations and the expansion of motorways. This goes hand in hand with behavioural changes. People settle in suburbs and exurbs around the megacities. They commute to work and shop in shopping malls outside of the city. This results in institutional changes. The policy field of spatial planning is established, and commuting allowances are introduced for tax purposes. These are merely representative examples of the complex diversity of interdependent elements in fundamental processes of change. Only the interplay of all such elements allows a new paradigm to displace the old.

The Neo-Schumpeterian approach provides decisive clues as to how the forthcoming processes of change can take place. This will be clarified in ► Sect. 6.2. It briefly outlines the reflections of growth-pessimistic approaches, such as the *post-growth* or *degrowth* approaches, which enjoy great popularity. Then, it contrasts them with growth-optimistic approaches that uphold Schumpeter's intellectual heritage and rely on the creative forces of capitalist economies to overcome humankind's fundamental problems. Innovations are based on the discovery and successful dissemination of new knowledge. Knowledge-based societies organise innovation systems that are composed of different actors successfully combining their knowledge. This is what ▶ Sect. 6.3 deals with. No innovation would ever have been able to succeed on the market if consumers had not taken an interest in it and if their purchasing power had not helped innovative solutions to break through. ▶ Section 6.4 sheds light on the consequences of this insight. In knowledge-based societies, new concepts, in the sense of responsible innovation, will play an important role if an entire economy is to be steered onto a new, sustainable path of development. From these technology- and knowledge-driven changes, massive economic developments take their point of departure. This is discussed in ▶ Sect. 6.5. In addition to technological change, in a co-evolutionary process, institutional change will also have to take place to enable the new sustainable technologies to provide the prerequisites for the desired transformation of the economic system. ▶ Section 6.6 shows that economic policy must actively accompany this change if it is to succeed.

6.2 Limits to Growth?

As early as 1972, when the *Club of Rome* published its report "The Limits to growth" (Meadows et al. 1972), the status quo in Western industrialized economies calls into question the capitalist organization concerning its sustainability. Since then, the conservation of resources through growth abstinence on the one hand and the decoupling of growth and the exploitation of resources on the other have been discussed as two fundamentally different solution strategies for society. The first idea can be summarized by the keywords "abstinence" and "downscaling." Its proponents call for a move away from a lifestyle based on consumption and the increasing deployment of resources (Kallis et al. 2012; Blewitt and Cunningham 2014). This demand goes hand in hand with a mistrust of the adaptability of market-oriented economic systems, which are not expected to be able to change through endogenous market forces in the direction of greater sustainability. The most extreme versions ask for a return to small-scale regional agriculture or subsistence farming. Only in this way could a way of life and economy be made possible that is sustainable and that conserves resources. It is easy to see that this notion is in line with the neoclassical view, which refers to economic growth solely in regard to existing economies and their quantitative change, without taking into account the dynamics of change.

The second way, on the other hand, is characterised by the idea that innovation, market forces, structural change and urban lifestyles are part of the solution to the sustainability problem. It can thus be assigned to the Neo-Schumpeterian view. Especially in the late 20th and early 21st centuries, the capitalist-oriented economy has impressively demonstrated its global power for change: Through creative entrepreneurship in free markets, such as in China, for example, more people could be brought out of poverty in a short time (one of the 17 sustainable development goals of the United Nations until 2030) than through 50 years of development aid before. New creative solutions can reform our way of doing business in a sustainable way in the future, supporting the achievement of the UN's sustainable development goals and, at the same time, allowing growth and development to take place (Mazzucato and Perez 2015).

The guiding idea of the knowledge-based bioeconomy is based on the premise that abstinence, in the sense of economic dismantling, is neither the first goal nor the only solution. In principle, however, there is agreement with the supporters of the first approach that certain production and consumption patterns of the past urgently need to be changed, and that participatory elements must be included. In particular, concepts that result in a more intensive use of goods, and thus contribute to the conservation of resources (*sharing economy*), are of importance. The same applies to closed material cycles, recycling and intelligent waste treatment. Such concepts are ideally suited to the triggering of learning processes and behavioural changes among consumers. The core idea of the knowledge-based bioeconomy, however, is that, within the framework of a comprehensive economic transformation process (Geels 2002), new technological solutions are demanded and provided, i.e., that alternative goods and services are demanded, produced and delivered in a different, namely sustainable way. Exploiting the technological possibilities of the bioeconomy not only creates new investment opportunities, but is also a prerequisite starting point for socio-economic and cultural change – which will only succeed if consumers accept bio-based products and ask for appropriate solutions from companies. As a result, innovation, functioning markets and changing consumer attitudes become complementary conditions for creating a sustainable production system.

Representatives of the Neo-Schumpeterian school (Dosi et al. 1988; Lundvall 1992, 1998; Nelson 1993) point to the systemic character of innovation processes in knowledge-intensive economic sectors. So-called innovation systems consist of different actors (including companies, research institutions, political actors, and consumers) and the links between these actors (e.g., flows of goods, research and development cooperations, knowledge transfer relations, consumer-producer relations). Such connections are the prerequisite for mutual learning and joint knowledge development for the purpose of solving complex innovation tasks. Such systems are dynamic and co-evolutionary. This makes them enormously complex, because, over time, both the actors and their knowledge and the links and interactions between them are exposed to changes.

According to this systemic understanding, technological paradigms are defined as "…a set of procedures, a definition of the 'relevant' problems and of the specific knowledge related to their solution" (Dosi 1982). Applied to the knowledge-based bioeconomy, the problem is the substitution or saving of carbon-based materials and energy with bio-based materials and energy, for which very heterogeneous technological processes across the entire depth and breadth of the value chains are used. It is also about the development of economic complementarities, in the sense of the cross-fertilization of different fields of knowledge. The expansion of value chains through the possibilities of digitisation will play an important role because it will increase value creation in new sustainable areas of CO_2-neutral production, e.g., in autonomous electromobility or the expansion of intelligent power grids. However, the concept of technological paradigms implies that a paradigm shift is not always possible. A *window of opportunity* for the paradigm shift will only open up if several interconnected technologies are developed and the demand-side and institutional conditions are in place that are conducive to it. Only when these prerequisites for the emergence of a new bioeconomic innovation system are in place can the transformation process succeed and gain momentum.

6.3 Innovation Systems and Knowledge

A first indication of the development of innovation systems can be found in the theory of industrial life cycles, which emphasizes the pronounced dynamics in the emergence and maturation process of industries (Audretsch and Feldman 1996). Industrial development is therefore typically divided into four phases (Fig. 6.2):

1. Development phase (new knowledge creates the condition for innovation)
2. Entrepreneurship and growth phase (many entries of smaller innovative companies into the new industry)
3. Saturation and consolidation phase (development of industry standards, mergers and acquisitions, as well as market exits)
4. Downturn phase (oligopolistic competition in only moderate innovative industries)

In order to understand the transformation into a knowledge-based bioeconomy, the findings of the industrial life cycle theory are of great importance, especially for the first phases of its emergence and growth, although the bioeconomy is, of course, not a self-contained branch of industry. Rather, the bioeconomy is characterized by its cross-sectoral character. On the one hand, new sectors will emerge, such as bioplastics, waste management and biorefineries. On the other hand, the technological possibilities of the bioeconomy will trigger new dynamics to already existing sectors such as agricultural vehicle construction, battery technology and pharmaceutical production among others. It can therefore be assumed that the establishment of bioeconomic technologies will lead to the emergence of new industries and, in parallel, to new impulses for the development dynamics of existing industries. In the sense of co-evolution, these processes will be accompanied by the adaptation of old and the development of new institutions (e.g., the Renewable Energy Act or the Greenhouse Gas Emissions Trading Act), the adaptation of consumer habits and the creation of new educational opportunities.

The development patterns of the bioeconomy and the way in which new companies are created are influenced primarily by the national institutional framework (Casper et al. 1999; Whitley 1999). Institutions can be defined as "a set of rules, formal or informal, that actors generally follow, whether for normative, cognitive, or material reasons" as well

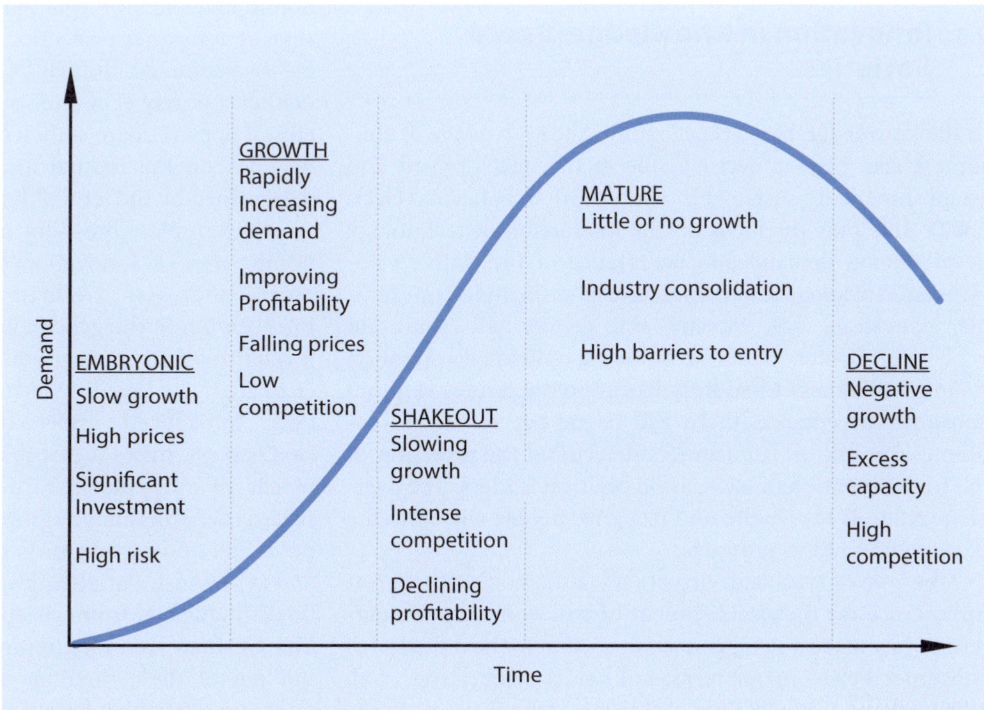

Fig. 6.2 Origin, maturation and decay of industries, measured as demand for their products over time, according to the theory of industrial life cycles. (Klepper 1997)

as "organizations as durable entities with formally recognized members, whose rules also contribute to the institutions of the political economy" (North 1990; Hall and Soskice 2001). One of the most important prerequisites for the transformation towards a bioeconomic production system is the knowledge base of an economy built up by the education and research system (Geels 2002). On the one hand, there is still a great deal of uncertainty with regard to the future competencies required for a bioeconomy; on the other hand, numerous individual fields of knowledge that play an important role in the transition have already been identified, such as synthetic chemistry, process engineering, genetic engineering, food technology and computer science. To generate an innovation system, it is necessary to understand the dynamics of these knowledge fields and the way in which they can be recombined with other knowledge fields and corresponding actors. The combination of different fields of knowledge (*cross-fertilization*) is often responsible for the emergence of major technological opportunities. For example, the fusion of information and database technology and molecular biology has led to the creation of the bioinformatics sector as a completely new branch of industry, which finally was the basis for the these days florishing big data service industries. At the same time, the combination of different areas of knowledge is confronted with great uncertainty, which makes public innovation policy an important factor. A supportive research and development policy should therefore identify development paths from analysis of the dynamics of knowledge and networking, which indicate the areas in which intensified research and development efforts must be undertaken in order to close existing gaps and build bridges between hitherto unconnected fields of knowledge (Burt 2004; Zaheer and Bell 2005).

6.4 Innovation in Knowledge-Based Societies

In the knowledge-based bioeconomy, the knowledge of consumers also plays a decisive role in the development and establishment of sustainable consumption patterns (Geels 2002). This puts the focus on the interaction of technology development, demand and acceptance of innovative solutions and sociological variables. The latter include, for example, education, age, income and gender, all important explanatory factors that determine the individual's attention to and willingness to address bioeconomic issues. Without consumer acceptance, there will be no successful bioeconomic innovations. Consumers determine the direction of the transformation process, as do political leaders. The overall question is how aware and receptive people will be to the bioeconomy and its products.

The role of (real and virtual) social networks is of great importance for the establishment of new consumption patterns. They make a significant contribution to the diffusion of consumer behaviour patterns and values (Robertson et al. 1996; Valente 1996; Nyblom et al. 2003; Deffuant et al. 2005). New studies show that attitudes are important for the formation of social relationships, and that social relationships, in turn, have a significant influence on behaviour and attitudes. In the field of renewable energies, for example, it was, in many cases, only the initiative of public utilities customers that led to a "green" orientation in regional electricity supply. In individual cases, such citizens' initiatives have even installed investor communities that are themselves involved in the energy industry.

But not everything that is technically possible is also socially desirable. Critical questions must therefore be dealt with in democratic processes. In the field of bioeconomy, among others these questions include the use of genetically modified organisms in agriculture. They promise efficiency benefits in terms of productivity and land and water consumption. Critics point out, however, that long-term health or ecological risks cannot be conclusively ruled out in their use. Accordingly, technology development takes place depending on consumer acceptance and attitudes, and is thus dependent on the level of education within an economy. This raises the question of a society's openness to innovations, which is fundamentally associated with uncertainty. The term ***Responsible Innovation*** summarises the responsible design of development, which is currently being discussed with high priority by European policymakers. A comprehensive working definition of *Responsible Innovation* has been developed by Von Schomberg (2011). He describes it as "a transparent, interactive process by which societal actors and innovators become mutually responsive to each other with a view to the (ethical) acceptability, sustainability and societal desirability of the innovation process and its marketable products (in order to allow a proper embedding of scientific and technological advances in our society)." It is the question of whether innovations are judged exclusively on their economic efficiency or whether other aspects also play a role such as consumer protection or ecological criteria. Based on the discussion on biofuels ("food vs. fuel"), it can be seen that neither a purely economic approach nor a one-dimensional ethical approach are sufficient. The quality of the discussion depends on the mutual understanding, which, in turn, is determined by the level of knowledge of the participants.

Modern plant breeding and seed production is a bioeconomic area of innovation in which justice concerns are openly addressed. German consumers are sceptical about intervention in the genetic material of food plants, although it often remains unclear what the individual points of criticism are. New breeding techniques, which, since 2012, have been introduced under the name of *genome editing* (▶ Chap. 5), make it possible to modify the DNA building blocks of crop plants in a targeted manner. Researchers regard these methods as groundbreaking because they enable potentially powerful plants to be cultivated in a short time and at low cost. Varieties developed in this way can no longer be distinguished from varieties from conventional breeding. The German Central Committee on Biological Safety does not regard these methods as genetic engineering in legal terms, in particular, because they do not involve the recombination of genetic material. Since these procedures are not explicitly mentioned in the law regulating genetic engineering, the legislature is now faced with the task of clarifying whether they should be regarded as genetic engineering at all. The result will influence the dissemination potential and the acceptance of *genome editing*. Here, too, there is a renewed need to include education and information policy in shaping the transformation towards a knowledge-based bioeconomy (▶ Chap. 8).

Within the concept of ***social innovation*** (Hanusch and Pyka 2013), active civic participation in the innovation process is even more evident. According to the understanding of the European Union, this term includes innovations "that are social both as to their ends and their means and in particular those which relate to the development and implementation of new ideas (concerning products, services and models), that simultaneously meet social needs and create new social relationships or collaborations, thereby benefiting society and boosting its capacity to act" (European Parliament 2013a). By strengthening cooperative behaviour, social innovations primarily make an important contribution to rural development and promote the economic resilience of these regions. Rural cooperatives, such as regional producer and marketing associations, winegrowers' cooperatives and tourism associations, can contribute to developing regional competitiveness while respecting ecological and social criteria. This can open up new opportunities in the bioeconomy for rural regions that are particularly affected by demographic changes and the associated depopulation.

6.5 The Economics of Change

The previous remarks made clear that the transformation into a bio-based economy is an extremely complex process of change for the current economic system. There are very many different actors involved in different roles who contribute different knowledge. At the same time, this process will involve not only innovative adaptations in existing industries, but also the emergence of new industries and the dropping out of mature industries. In addition to substitutive relationships between new bio-based industries and traditional oil-based industries, there will be numerous important complementary relationships that will provide dynamic impulses for the transformation process (▶ Chap. 7). First and foremost there are the possibilities and applications of digitisation, which can be used to replace numerous petroleum-based products and energy-intensive services with bits and bytes. One example of this is the paper industry, which, in a particularly resource-intensive manner, produces tons of paper for our daily newspapers that have to be transported, first to the printing plants and then to the customer. Changes in the behaviour and attitudes of customers who consume newspapers in digital form completely eliminate this resource requirement. Through the coordination of decentralised and small-scale bioeconomic technologies and processes, digitisation opens up new opportunities, for example, in energy production and through so-called intelligent networks (*smart grids*) in power transmission. Digital coordination will affect the overall composition of many economic sectors. The coexistence of large diversified companies and highly specialised small technology companies will often be a potential solution. Finally, digitization also enables the effective organization of consumer platforms in the sense of *sharing economy* approaches. The successful emergence and diffusion of bioeconomy-relevant knowledge depends on dynamic innovation networks (Pyka 2002), in which different actors share existing knowledge and jointly create new knowledge. In the innovation networks, the demand side, represented, for example, by consumer associations and politicians, will also play a prominent role and help to establish innovation networks in the early phases of technology development.

Also in the knowledge-based bioeconomy investment and economic growth will be a crucial prerequisite for employment, international competitiveness and income generation. The bioeconomy can make an important contribution to increasing investment by providing new investment opportunities through fundamental innovations, and thus bringing the large amount of liquidity currently available to productive use, which, in turn, accelerates the technological paradigm shift. The emergence of new major investment opportunities represents a typical pattern for the early phases of a new techno-economic paradigm: Carlota Perez (2010, 2014), for example, identifies three waves of industrialization, the first being the *Great British Leap*, the second being the *Victorian Boom*, and the third being the combined post-war accomplishments of the *Belle Èpoque* in Europe, the *Progressive Era* in the USA, and the German economic miracle (*Wirtschaftswunder*), as phases of enormous economic growth triggered by a fundamental transformation of the economic system.

The time path of the transformation process represents another critical component that has so far gained little attention. On the one hand, there is a hurry to reduce carbon-based production methods; on the other hand, frictions will occur in the transformation process that are caused, for example, by a shortage of skilled workers. In this context, the so-called *sailingship effect* (Howells 2002), which can often be observed in eras of revolutionary innovations, could be advantageous. When, in the middle of the nineteenth century, new steamboats threatened the existence of the established sailing ship technology, sailboat builders suddenly undertook innovation efforts that they had not considered for many decades, if not centuries. Due to the threat posed by innovative technologies, their predecessor technologies are therefore subject to adaptation reactions designed to prevent them from being forced out of the market quickly. Fuel-efficient internal combustion engines and hybrid drive technologies, for example, represent such adaptation reactions to the emergence of electric vehicles. In terms of environmental policy, however, both the old and the new technologies pursue the same objective, namely, a reduction in noise and exhaust emissions. This is an advantage, because it allows the new technology to take more time for development. The transformation process into a bioeconomy will also be characterized by a co-existence of traditional and bio-based industries over a long period of time. During this time, it will also be important to further advance relevant innovation processes in traditional technologies. This co-existence increases the degree of complexity of change. At the same time, however, it creates time for the development of the bioeconomy and prevents the early introduction of immature technologies, which could cause failure of promising approaches.

The distributional effects of the transformation process continue to be important for social acceptance. A bio-based economy on an industrial scale will, to a large extent, be a knowledge-based economy. It will generate additional demand for highly qualified workers, while the opportunities for the low-skilled will continue to deteriorate. In addition, jobs for low-skilled workers in traditional industrial production will disappear. On the other hand, there will be demand for other goods and services whose value-added and labour-market-relevant compensation potential is still unclear. The question of the extent to which companies are well prepared for the bioeconomy must also be asked. The transformation process will ensure that competencies responsible for past success are devalued through innovation. Incumbent companies will be confronted by the question of how they will deal with the *not-invented-here-syndrome* to overcome their "business myopia" and how to actively shape the transformation process in order to maintain value creation on established sites.

Thus, the distribution effect has an important regional component: Does the bioeconomy strengthen the divergence processes between the regions or does it lead to stronger convergence? Promising, but rarely realized approaches, are networks that are based on the principle of *smart specialisation* (Foray et al. 2009) that combine regional strengths along value chains in the best possible way. In this way, polarization tendencies can be avoided that, in addition to the concentration of economic power, also lead to political and cultural concentrations and the formation of distinct center-periphery structures. So far, however, it is unclear how stable and functional politically-induced networks are vis-à-vis self-organised networks and to what extent politics can influence them. Initial findings, however, suggest that the withdrawal of state coordination bodies from networks may lead to a tendency towards disintegration (Green et al. 2013).

From the transformation towards a knowledge-based bioeconomic production system, it is expected that the negative consequences of economic growth in terms of environmental pollution, resource consumption, climate change and energy consumption will be resolved in a sustainable way. Which contribution can be expected from individual areas, how complex feedback loops will influence competitiveness and whether rebound effects may counteract the positive effects of the transformation process are all questions that are closely linked to the fundamental uncertainty of the innovation process. Answers cannot be anticipated. Institutional rules would be one way of reducing such uncertainties, at least in part. For example, it would make sense for oil-producing countries to commit themselves to reducing their production volumes in line with the declining demand for oil caused by the bioeconomy. Ultimately, all actors involved in the transformation into a knowledge-based bioeconomy – from companies to private households to politicians – must learn to abandon optimization approaches and profit maximization principles. The complexity and uncertainty of this process calls for a willingness to experiment (*trial and error*) afforded by all actors.

6.6 Transformation as a Political Priority

Since the Industrial Revolution, socio-economic systems have been exposed to permanent transformation processes. While these development processes have so far been driven by open-ended innovation processes, the bioeconomic transformation process is characterized by the fact that its socially and politically desired direction is clearly defined. In the past, major technological upheavals have largely overcome bottlenecks based on scientific or economic constraints, thereby shifting the socio-economic system along new trajectories without giving direct instructions to the direction of the development process. However, with the massive accumulation of carbon dioxide in the atmosphere since the Industrial Revolution and the threat to current ecosystem services at the beginning of the twenty-first century, it is clear that global thresholds have almost been surpassed. This restricts the level of freedom of future developments if one does not want to irreversibly damage natural conditions for human life and biological diversity on earth. It is yet unclear whether this transformation process will succeed in a targeted manner and how it can be controlled by political influence in order to achieve the socially existential goals.

New technological developments alone are not enough to transform the socio-economic system, but will initially only create the necessary potential for radical changes affecting the economy as a whole. Only a broad societal commitment to a specific use of these technologies will lead to converging trajectories and synergies that can ultimately initiate the paradigm shift (Pérez 2014) – i.e., the commitment to try out all developmental directions that are linked to corresponding investments, innovations and the ability to cope with fundamental insecurity through politics. The "green growth paradigm" based on bio-based technologies can be such a direction, bringing together the potential of different technological developments and making them flourish. This requires political decisions supporting a reorientation of macroeconomic research and innovation activities, the exploration of new energy sources, improvements in the productivity of natural resources and new sustainable ways of living and production (Pérez 2014). In addition, such a transformation process creates opportunities for economic development in catching-up economies without overexploiting global natural resources and the environment. It will be decisive for the success of the bioeconomic transformation process that it is given a direction by politics and society (Mazzucato and Perez 2015).

This includes, for example, the development of new products within emerging bioeconomic innovation systems. In this perspective, innovations require the interaction of the actors along value chains that might lead to the development of new industries. In the past, for example, the provision of cheap electricity led to the spread of refrigerators and freezers in private households, which, in turn, led to innovations in frozen food and packaging. Similarly, in a bioeconomy, the establishment of a *sharing economy* may lead to new digital coordination platforms and the establishment of sustainable designs among product manufacturers. This would eliminate the resource-wasting phenomenon of planned obsolescence that shortens product life cycles and create new sectors such as repair and maintenance services. Networking and cluster formation, which lead to a reduction of uncertainty and to self-reinforcing effects, are particularly important for long-term development. In addition, social changes and changing lifestyles are both an expression and a driver of this transformation process (Mazzucato and Perez 2015).

Therefore, the role of governments goes beyond simply correcting market failures. Rather, government action prepares the ground from which new markets can emerge and thrive in the first place by creating investment security and reducing risks and uncertainty (Mowery et al. 2010). The transition from the invention phase to the innovation phase, i.e., to the expansion of bioeconomic activities in the markets, is a high-priority task of innovation and business start-up policy. To realise a growth path on the basis of the bioeconomy requires more than just the replacement of crude oil with renewable raw materials or renewable energies. What is needed is an innovation system that creates synergy effects, knowledge transfer and networks between manufacturers, suppliers and consumers. There is a need for a comprehensive transformation that encompasses the entire economy and renews the patterns of production and consumption that were established as a result of the previous transformation process.

The technological potential of the bioeconomy is therefore a necessary, but by no means sufficient, condition for the transformation process. A political decision is needed as to how this technological potential is to be used and which trajectories are to be developed and merged. The market in which innovations are profitable does not emerge by itself, but rather requires feedback loops between political decisions, corporate strategies and consumer preferences.

References

Audretsch D, Feldman MP (1996) Innovative clusters and the industry life cycle. Rev Ind Organ 11(2):253–273

Blewitt J, Cunningham R (eds) (2014) The Post-Growth Project: how the end of economic growth could bring a fairer and happier society. Green House, e-prints, Aston University, London

Burt RS (2004) Structural holes and good ideas. Am J Sociol 110(2): 349–399

Casper S, Lehrer M, Soskice D (1999) Can high-technology industries prosper in Germany? Institutional frameworks and the evolution of the German software and biotechnology industries. Ind Innov 6(1):114–139

Deffuant G, Huet S, Amblard F (2005) An individual-based model of innovation diffusion mixing social value and individual benefit. Am J Sociol 110(4):1041–1069

Dosi G (1982) Technological paradigms and technological trajectories: a suggested interpretation of the determinants and directions of technical change. Res Policy 11(3):147–162

Dosi G, Freeman C, Nelson R, Silverberg G, Soete L (1988) Technical change and economic theory. Pinter, London

European Parliament (2013a) Regulation (EU) No 1296/2013 of the European Parliament and of the Council of 11 December 2013 on a European Union Programme for Employment and Social Innovation ("EaSI"), Brussels. https://eur-lex.europa.eu/legal-content/EN/TXT/HTML/?uri=CELEX:32013R1296&from=DE

Foray D, David PA, Hall B (2009) Smart specialization – the concept. Knowledge Economists Policy Brief 9. http://ec.europa.eu/invest-in-research/monitoring/knowledge_en.htm

Freeman C (1991) Networks of innovators: a synthesis of research issues. Res Policy 20(5):499–514

Geels FW (2002) Technological transitions as evolutionary reconfiguration processes: a multi-level perspective and a case-study. Res Policy 31(8):1257–1274

Green L, Pyka A, Schön B (2013) A life-cycle based taxonomy of innovation networks – with focus on public-private collaboration. In: Gallouj F, Rubalcaba L, Windrum P (eds) Public-private innovation networks in services. Edward Elgar, Cheltenham, pp 113–138

Hall PA, Soskice D (2001) An introduction to varieties of capitalism. In: Hall PA, Soskice D (eds) Varieties of capitalism: the institutional foundations of comparative advantage. Oxford University Press, Oxford, pp 1–68

Hanusch H, Pyka A (2013) Social innovations in the perspective of comprehensive Neo-Schumpeterian economics. In: Ruiz Viñals C, Parra Rodríguez C (eds) Social innovation – new forms of organization in knowledge-based societies. Routledge, London, pp 29–43

Howells J (2002) The response of old technology incumbents to technological competition–does the sailing ship effect exist? J Manag Stud 39(7):887–906

Kallis G, Kerschner C, Martinez-Alier J (2012) The economics of degrowth. Ecol Econ 84:172–180

Klepper S (1997) Industry life cycle. Ind Corp Chang 6(1):145–182

Knight FH (1921) Risk, uncertainty and profit. Hart, Schaffner& Marx, Boston

Lundvall B-A (1992) National innovation systems: towards a theory of innovation and interactive learning. Pinter Publishers, London

Lundvall B-A (1998) Why study national systems and national styles of innovation? Tech Anal Strat Manag 10(4):403–422

Mazzucato M, Perez C (2015) Innovation as growth policy. In: Fagerberg J, Laestadius S, Martin BR (eds) The triple challenge for Europe: economic development, climate change, and governance. Oxford University Press, Oxford, p 229

Meadows D, Randers J, Behrens WW (1972) The limits to growth. Universe Books, New York

Mowery DC, Nelson RR, Martin BR (2010) Technology policy and global warming: why new policy models are needed (or why putting new wine in old bottles would not work). Res Policy 39:1011–1023

Nelson RR (1993) National innovation systems: a comparative analysis. Oxford University Press, Oxford

North DC (1990) Institutions, institutional change and economic performance. Cambridge University Press, Cambridge

Nyblom J, Borgatti S, Roslakka J, Salo MA (2003) Statistical analysis of network data – an application to diffusion of innovation. Soc Networks 25(2):175–195

Pérez C (2010) Technological revolutions and techno-economic paradigms. Camb J Econ 34(1):185–202

Pérez C (2014) Financial bubbles, crises and the role of government in unleashing golden ages. In: Pyka A, Burghof H-P (eds) Innovation and finance. Routledge, Milton Park, pp 11–25

Pyka A (2002) Innovation networks in economics – from the incentive-based to the knowledge-based approaches. Eur J Innov Manag 5(3):152–163

Pyka A (2017) The transformation towards a knowledge-based bioeconomy. In: Dabbert S et al (eds) Strategies for knowledge-driven developments in the bioeconomy – an international and interdisciplinary perspective. Springer, Heidelberg, New York

Pyka A, Buchmann T (2017) Die Transformation zur wissensbasierten Bioökonomie. In: Burr W, Stephan M (eds) Technologie, strategie und organisation. Springer, Berlin und Heidelberg

Robertson M, Swan J, Newell S (1996) The role of networks in the diffusion of technological innovation. J Manag Stud 33:333–360

Sahal D (1985) Technological guideposts and innovation avenues. Res Policy 14(2):61–82

Schumpeter JA (1934) The theory of economic development. President and Fellows of Harvard College, Harvard University Press, Cambridge

Schumpeter JA (1939) Business cycles. A theoretical, historical and statistical analysis of the capitalist process. McGraw-Hill Book Company, New York

Schumpeter JA (1943) Capitalism, socialism and democracy. Harper, New York

Unruh CG (2000) Understanding carbon lock-in. Energy Policy 28:817–830

Valente TW (1996) Social network thresholds in the diffusion of innovations. Soc Networks 18(1):69–89

Von Schomberg R (2011) Prospects for Technology Assessment in a framework of responsible research and innovation. In: Technikfolgen abschätzen lehren: Bildungspotenziale transdisziplinärer Methode. Springer VS, Wiesbaden, pp 39–61

Von Weizsäcker E (2010) Introduction, factor five: the global imperative. In: Von Weizsäcker E et al (eds) Factor five – transforming the global economy through 80% improvements in resource productivity. Earthscan, Routledge, Abingdon, pp 1–19

Whitley R (1999) Divergent capitalisms: the social structuring and change of business systems: the social structuring and change of business systems. Oxford University Press, Oxford

Zaheer A, Bell GG (2005) Benefiting from network position: firm capabilities, structural holes, and performance. Strateg Manag J 26(9):809–825

6

Bioeconomy as a Circular and Integrated System

Silvan Berg, Manfred Kircher, Nina Preschitschek, and Stefanie Bröring

Embedded in an agricultural region, this CropEnergies plant in the Belgian town of Wanze produces up to 300 million litres of bioethanol a year from around 800,000 tonnes of wheat and 400,000 tonnes of sugar beet. (© Martin Jehnichen, CropEnergies 2016)

J. Pietzsch (ed.), *Bioeconomy for Beginners*, https://doi.org/10.1007/978-3-662-60390-1_7

The transformation of our economic system, which is based on fossil resources, into a bioeconomy poses many challenges. The German Federal Ministry of Education and Research (BMBF) formulated the following in its concept paper "Bioeconomics as societal change", which addresses, in particular, the importance of social and economic research for the bioeconomy: "The term bioeconomy is associated with the vision of a 'natural economy', that is one based on the natural material cycle. The aim of the BMBF is to "develop a sustainable, bio-based economy with a strong focus on sustainability" and, on this basis, to ensure both global nutrition and the supply of energy sources and renewable raw materials for a wide range of industries and applications" (BMBF 2014). On the one hand, this addresses aspects of the circular economy. On the other hand, it is emphasised that the use of renewable raw materials should be made with the highest efficiency, and correspondingly be sustainable. Thus, it must be a fundamental goal of the bioeconomy to use the potential of renewable raw materials in a cascade-like manner best possible. This aim should therefore be to implement the bioeconomy as an integrated system with high resource efficiency, comparable to the established integrated system of the chemical industry (Biermann et al. 2011).

In ◘ Fig. 7.1, such cascade-like use of plant biomass is shown schematically. The first step of this use is the production of food and feed. It is and will remain the most important form of utilisation of biomass in the bioeconomy (c.f. ▶ Chap. 3). Global food security, especially with regard to the forecast of global population growth to more than 10 billion people in 2050, should always be given higher priority than any material or energy-related use (Swiaczny and Schulz 2009). The figure clearly shows that this primary objective can be meaningfully combined with the production of other high-quality products that are also relevant to value creation by using secondary and waste streams.

For example, at the second use stage, feed (e.g., rapeseed press cake as a by-product of rapeseed oil production) or ingredients (e.g., protein hydrolysate) can be produced from by-products of food production. In the third stage, platform chemicals can then be obtained that serve as the basis for bio-based plastics. Only in the fourth stage is the remaining biomass used to generate energy. All of these utilisation steps contribute to the total value-added potential of the biomass used. The aim of the bioeconomy is to exploit this potential as optimally as possible. In order to achieve this goal, new technologies and the successful introduction of innovations are necessary (▶ Chap. 8). On the other hand, there is a need for bioeconomy value chains, which meet the opportunities and requirements of cascade utilisation. This chapter describes the way of designing these value chains. However, it should be pointed out that bioeconomy value chains have long been established all over the world. The production of food and feed, construction timber, paper and textile fibres, for example, is traditionally bio-based. In the European Union, bioeconomy generated € 2.1 billion in 2013, employing 18.3 million people (◘ Fig. 7.2).

However, supported by the goal of transformation towards a bio-based economy, industries are increasingly focusing on agricultural commodities to which they had previously paid only scant attention. Their use represents the starting point for a wide range of applications (◘ Fig. 7.3). In addition, new technologies and bio-based processes and the development of new sustainable sources of raw materials (e.g., lignin-containing biomass or carbon dioxide) will lead to the emergence of completely new bioeconomic value chains (Kroner 2015). For the successful implementation of the bioeconomy, many different industrial sectors and the underlying value chains have to be newly connected (Boehlje and Bröring 2010). The related processes are explained below.

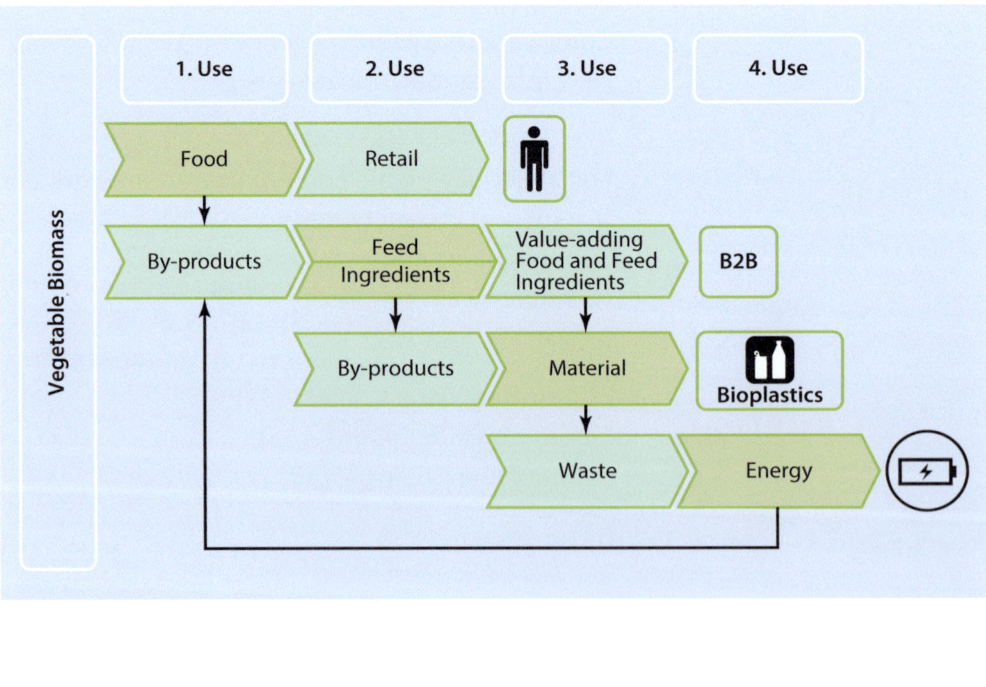

◘ **Fig. 7.1** Schematic representation of the cascade use of vegetable biomass. (Own representation)

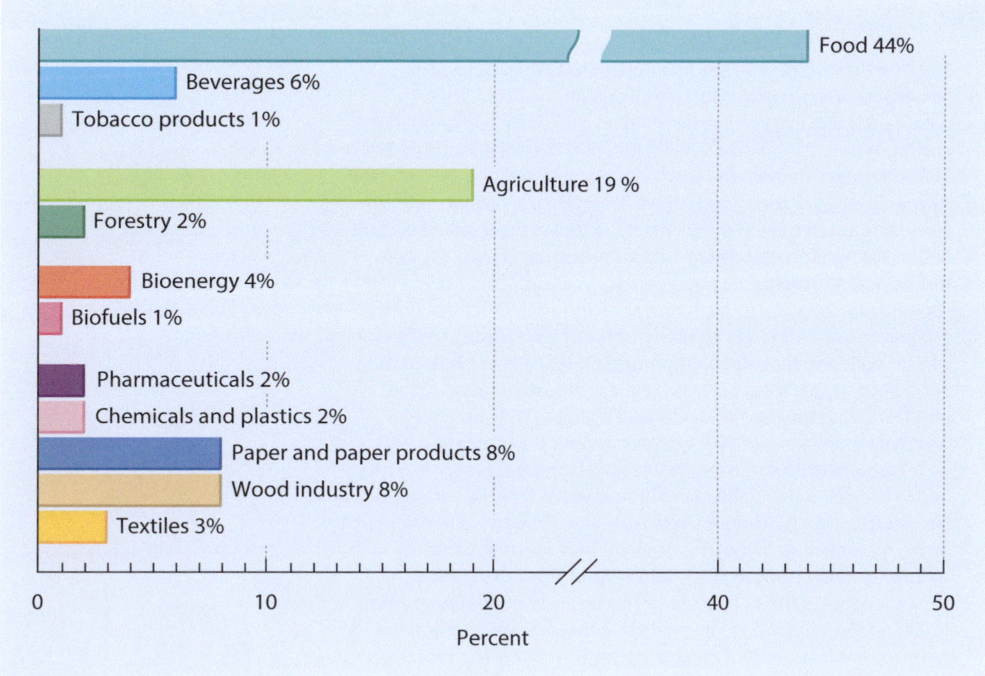

Fig. 7.2 Distribution of turnover of the bioeconomy in the EU (2013; € 2.1 billion). (Source: Piotrowski et al. 2016)

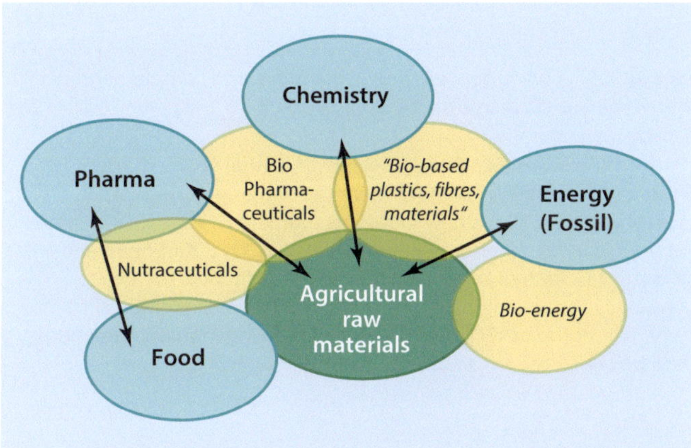

Fig. 7.3 Agricultural raw materials as a starting point for various industries. (Source: Boehlje and Bröring 2010)

7.1 The Emergence of New Value Chains

In order to manage the transformation process from a fossil-based to a bio-based economy, not only conventional and traditionally separate value chains have to be linked in a new way, but also completely new structures for value creation have to be established. At the industrial or organizational level, value chains describe the integrated production steps of a product or service from its original state to its sales state across the boundaries of companies and industries (Kaplinsky and Morris 2001). On the way to a bioeconomy, novel interdependencies of value chains often occur between two previously isolated industrial sectors. Such interdependencies or mergers are also referred to as convergence (► Excursus 7.1). They can take place at the technology, product or value chain levels. Convergence processes are driven in many ways and are determined by various influencing factors. In the bioeconomy, they are driven not only by the influence of cross-sector networks and new technology platforms, but also by changes in customer needs, standards and industry norms, as well as regulation (Berg et al. 2018, Carraresi et al. 2018). Frequently, several factors interact, as, for example, in the case of functional food (Bröring 2005). The converging processes between the food and pharmaceutical industries have been influenced by a strong customer demand for nutritional products that promise additional health benefits in addition to saturation, together with technological progress. Basically, three types of convergence can be distinguished in the emergence of new value chains: a) substitutive convergence, b) complementary convergence and c) a new networking of existing structures.

7.1.1 Substitutive Convergence

In substitutive convergence, the newly emerging value chain replaces the structures of the previously established value chains (1 + 1 = 1). However, this is a process that takes time, so that the old value chains do not immediately completely disappear, but rather the new value chain stands in competition with the established value chains (Bröring 2010a). Ultimately, however, it gradually replaces the old structures and becomes the only value chain (Song 2016). The principle of substitutive convergence at the value chain level is depicted in Fig. 7.5.

The best-known result of substitutive convergence is certainly the smartphone. It combines the original functions of a mobile phone, a portable music player, a handy digital camera and some basic data processing functions. Therefore, it is also called a hybrid product. A new value chain has emerged that replaces the original one. The extent to which this is really happening in full - distinct product categories such as digital

Excursus 7.1 What is Convergence?

The term "convergence" has its origin in the late Latin word *convergere*. As a general educational definition of convergence, dictionaries and encyclopaedias contain the terms "approximation" and "agreement of opinions, goals, etc." (Bibliographic Institute 2012). In the business context, the term convergence was first used in 1963 by the American economist Nathan Rosenberg, in dealing with the technological change in the US machine tool industry in the second half of the nineteenth century. In this context, he coined the term *technological convergence*. He defined it as a contrast among sequences of processes that happen next to each other but not in relation to each other. He pointed out that, in pre-industrial times, certain skills and manufacturing processes were of a specific nature and linked to specific processes, whereas industrialisation in the nineteenth century was characterised by the establishment of a relatively small number of very similar production processes. This development was particularly noticeable in the machinery industry and in the general metalworking industries. As a result, various industrial sectors, which previously had no relationship to each other, such as the manufacture of firearms, sewing machines and bicycles, ultimately had a common technology base (Rosenberg 1963).

Since the 1970s, convergence has become the defining term for developments in the information, consumer electronics and telecommunications (ICT) sectors (Nyström 2008). The practical relevance of this became clear in 1977, when the Japanese company Nippon Electric Company (NEC) forecast a complete convergence of the computer and communications sectors by 1990, and aligned its corporate strategy accordingly (Hacklin 2008; Bröring 2010a). Although convergence is now also being discussed as a relevant phenomenon for other industrial sectors (e.g., between the food and pharmaceutical industries or between the agricultural and chemical industries) and there is an increasing number of scientific publications on the subject, there is still lacking a clear and widely accepted definition of convergence. In our understanding, convergence takes place not only at the level of technologies and industries, but also in the area of scientific disciplines (e.g., NanoBiotech) and markets, products and entire value chains (Duysters and Hagedoorn 1998; Choi and Valikangas 2001; Bröring 2005; Curran 2013). In Fig. 7.4, the convergence process is illustrated schematically using the example of two converging industries. Either the new industrial segment resulting from convergence is substituted for the two original industrial sectors (1 + 1 = 1) or a new industrial segment (1 + 1 = 3) is created at their interface (Bröring 2005; Christensen 2011; Karvonen and Kässi 2011). In the case of substitutive convergence, the two originally separate industrial sectors A and B will be replaced, whereas, in the other case, the newly emerging industrial segment will be characterised as complementary to A and B.

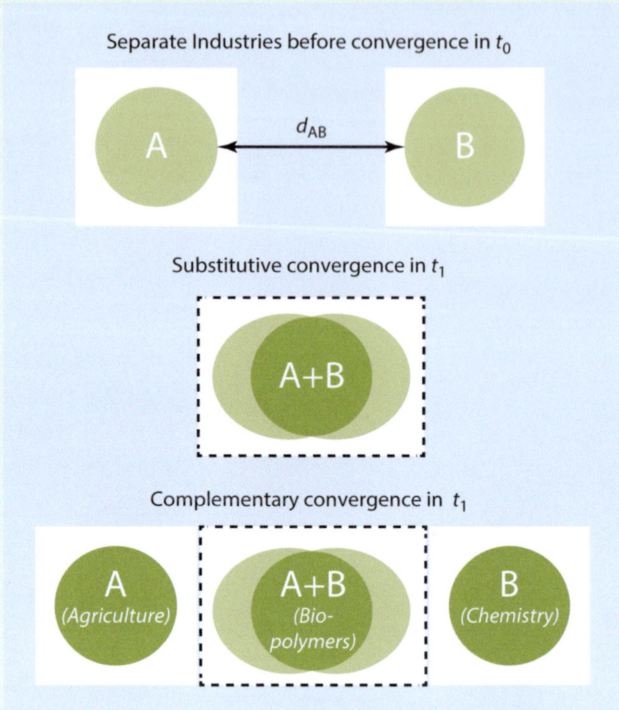

Fig. 7.4 Convergence types: substitutive and complementary convergence. (Source: Bröring et al. 2017)

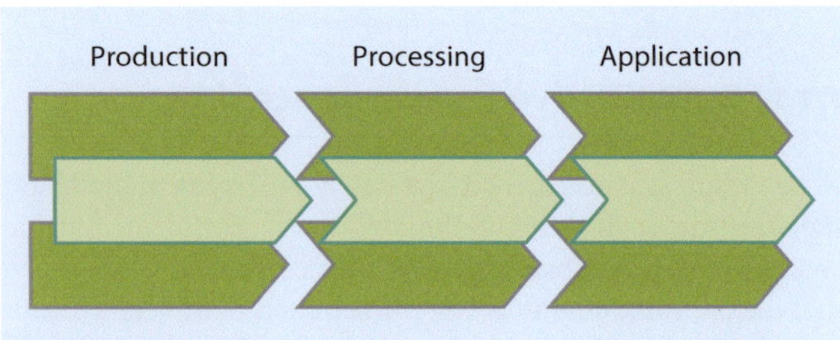

Fig. 7.5 Substitutive convergence in value chains. (Based on Bröring 2010a)

cameras and music players do still exist - can be discussed (Daurer et al. 2012). Nevertheless, it should be noted that a process of displacement has taken place here.

In the bioeconomy, ethylene, the most important platform chemical, with an annual production of more than 150 million tonnes, offers an example of substitutive convergence. Ethylene is produced by cracking crude oil and natural gas and in the gas fraction of crude oil distillation (Behr et al. 2010). It is the starting material for an entire family tree of organic chemistry products. In a technically simple process, ethylene can also be produced from bioethanol obtained from carbohydrates. In the production of bioethylene, which can substitute the conventionally produced version, the original value chains from oil and sugar refining converge. All further processing steps and the performance profile of the end products remain unchanged (▶ Chap. 4).

7.1.2 Complementary Convergence

Through complementary convergence, a new value chain is created that joins the existing structures and, from then on, exists and acts in conjunction with them (1 + 1 = 3). The new value chain realizes synergies with and combines functionalities from the adjacent value chains (Fig. 7.6) (Song 2016).

The development of the functional food value chain is one example of such a complementary convergence process (Bröring 2005). It does not replace the respective value chains

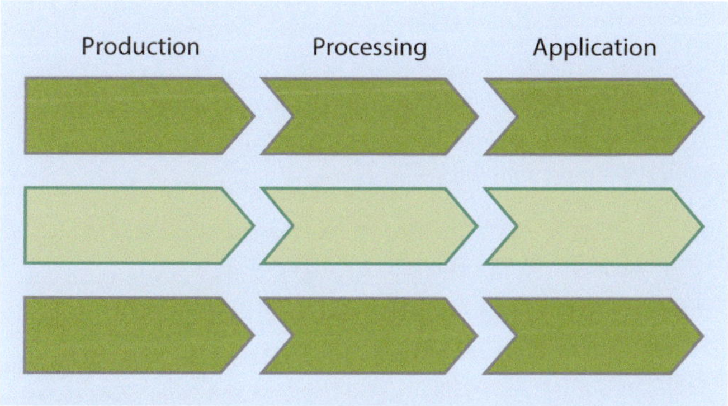

■ **Fig. 7.6** Complementary convergence in value chains. (Based on Bröring 2010a)

of the food and pharmaceutical industries, but rather functions as a complementary structure between them. This is best illustrated by a concrete product. A cholesterol-lowering margarine certainly does not completely replace other fatty food spreads, nor does it represent an equivalent alternative to medications for the prevention of coronary heart disease in cases of high cholesterol levels. On the contrary, this margarine meets the growing need among many consumers to easily integrate products with a positive additional health benefit into their everyday lives (Bornkessel et al. 2014).

In the bioeconomy, such a complementary process can be found in the production of alternative protein sources. However, it is still in the development stage (▶ Sect. 5.1). Protein from chicken eggs is an important raw material for the food industry, since it is particularly valuable for humans because of its amino acid profile and digestibility (Pichler 2013). As plants are also naturally capable of protein biosynthesis and storage, an obvious choice is to try to produce a protein in plants that is as identical to chicken egg white as possible. If successful, the existing value chains of egg production, from chicken breeding to food producers and plant production, from seeds to food trade, would be partially supplemented by the much shorter value chain between the production of vegetable "chicken egg white" and food production (Bobo 2015).

Excursus 7.2 The Example of Sugar Production from Sugar Beet

The production of bioethanol from sugar beets and its dehydration into bioethylene illustrate the great potential offered by connecting the respective value chains of the agricultural, energy and chemical industries. It illustrates, at the same time, both substitutive convergence and the cascade use in the sense of the use of waste streams.

In the bio-based production of ethanol and ethylene, after washing and cutting, the sugar beets are processed into a thin juice, and then a thick juice. The residual materials, such as beet pulp, are processed into animal feed. The thick juice obtained can then be used classically for the production of sugar by means of crystallization and centrifugation. It can also be fermented, and the resulting bioethanol is separated by means of a distillation process. The remaining vinasse is used as animal feed or organic fertilizer (■ Fig. 7.7) (Harms 2003).

First and foremost, bioethanol is currently used as a biofuel or biofuel additive to generate energy. However, its material use also plays an increasingly important role. Ethanol, for example, is used in the cosmetics industry as an additive in the manufacturing process. It is also tested in biorefineries as a starting material for the production of ethylene (▶ Chap. 4). At present (2016), however, the material use of bioethanol remains the exception. It is estimated that around 80% of the bioethanol produced is used for fueling and burning. In the interest of a sustainable bioeconomy, material use should be expanded and made more attractive in the future (▶ Chap. 8).

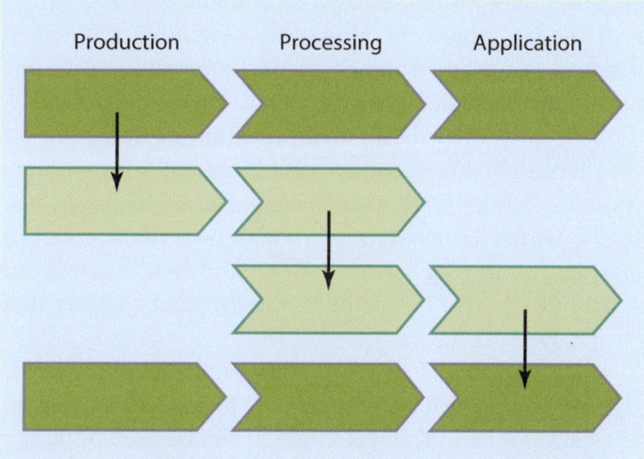

■ **Fig. 7.7** Extraction of ethanol and ethylene from sugar beet. (Own representation)

7.1.3 New Interconnectedness

In line with the fundamental idea of a circular economy, new bioeconomy value chains can also be created through the innovative networking of existing value chains. In this case, existing value chains are neither replaced nor supplemented directly at their interface. Rather, there are new links between individual stages of existing value chains, from which new independent value chains continue to emerge (■ Fig. 7.8). This networking can also be classified as a special case of complementary convergence (Fahrni 2008). Such new networks take place, for example, when ancillary and waste flows are added to value in a cascade of uses (▶ Excursus 7.2).

However, this is only possible if the by-products of a value chain are recognised as relevant inputs for other value chains. A concrete example of this can be found in the field of biogas production: Material flows in multi-stage process chains ultimately achieve a quality that, due to low concentrations or a complex, non-standardisable composition, seems to make them suitable only for sewage treatment plants (Guenther-Lübbers and Theuvsen 2015). In fact, in many places, aqueous wastewater streams are disposed of in this way, despite

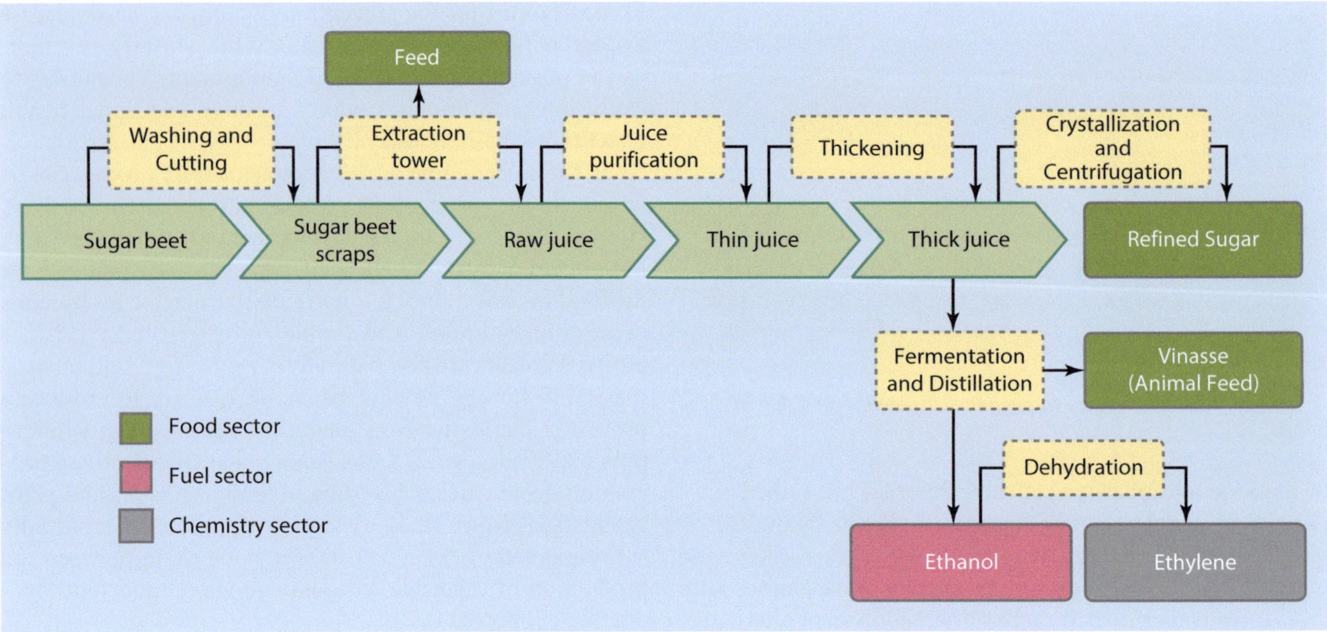

Fig. 7.8 New networking in value chains. (Own representation)

7

the fact that they would basically be suitable for the production of biogas. However, biogas fermentation requires a minimum volume and a certain media composition in terms of carbon, nitrogen and other nutrients. Therefore, the material flow of only one process chain is often not sufficient - a hurdle that can be overcome by combining flows from different sources. The operation of the Frankfurt-Höchst industrial park, which has a biogas plant with material flows from the production of biopharmaceuticals and food waste from the surrounding area in one of its biogas plants, shows that this works (▶ Sect. 7.3.3).

Synthesis gas (CO, CO_2, H_2) offers another possibility for the use of biomass, and also the organic portion of household waste (Hackelöer and Kneißel 2015). It can be used both as a source of electricity and as a carbon source in chemical production (▶ Chap. 4; ▶ Chap. 5.1). An even more important option in terms of volume is offered by the steel industry. There, synthesis gas is used as a reducing agent and emitted either directly or after conversion into electricity. In gas fermentation, synthesis gas from steel production can be used materially (Schöß et al. 2014). In fact, ethanol is already being fermented on this basis. Further products are in preparation. In this way, the value chains of steel production and the fuel and chemical industries are linked in a completely new way. Comparable options exist for the use of carbon dioxide from the energy and cement industries (▶ Sect. 7.3.3).

7.2 Conditions for the Creation of Bioeconomy Value Chains

The formation of new value chains generally poses a number of challenges. Stable bioeconomy value chains should exist no later than the point at which the oil price has reached a threshold that makes bio-based raw materials competitive. It is therefore already necessary today to work on technological developments that build the basis for a bio-based economy in such a way that they are in harmony with the development of supply and demand (market). Such a coordination between technology and market development is the central idea of road mapping (Phaal et al. 2004). Development paths of products, services and technologies into the future are analysed, forecast and visualised (Möhrle and Isenmann 2008). However, in order to illustrate the emergence of bioeconomy value chains in the sense of road mapping, it is not sufficient to consider, analogously to the classical approach, only the levels of technology, products and the market. A reliable supply chain must also exist, i.e., biomass and its by-products must be available in an adequate business model with appropriate logistics. This extended road mapping approach is illustrated in ◘ Fig. 7.9 (Bröring 2016). Social requirements (A), market needs (B), bioeconomic product developments (C), supply chain concepts (D) and the required technology base (E) must be aligned along the value chain and projected and controlled into the future. This is a very complex task that requires a systemic approach. After all, new value chains will only develop successfully if the five categories mentioned can be interlinked efficiently and effectively. In economic literature, in this context, it is called *systemic innovation*. (Teece 2000; Bröring 2008). Thereby it defines innovations that do not take place singularly at one value-added stage (e.g., a new manufacturing process), but rather comprise various stages (e.g., biomass that is specially adapted to bioprocess technology, which, in turn, is flexible enough to be used in decentralised logistics concepts).

In the following, a brief description will be given of the individual areas that need to be coordinated.

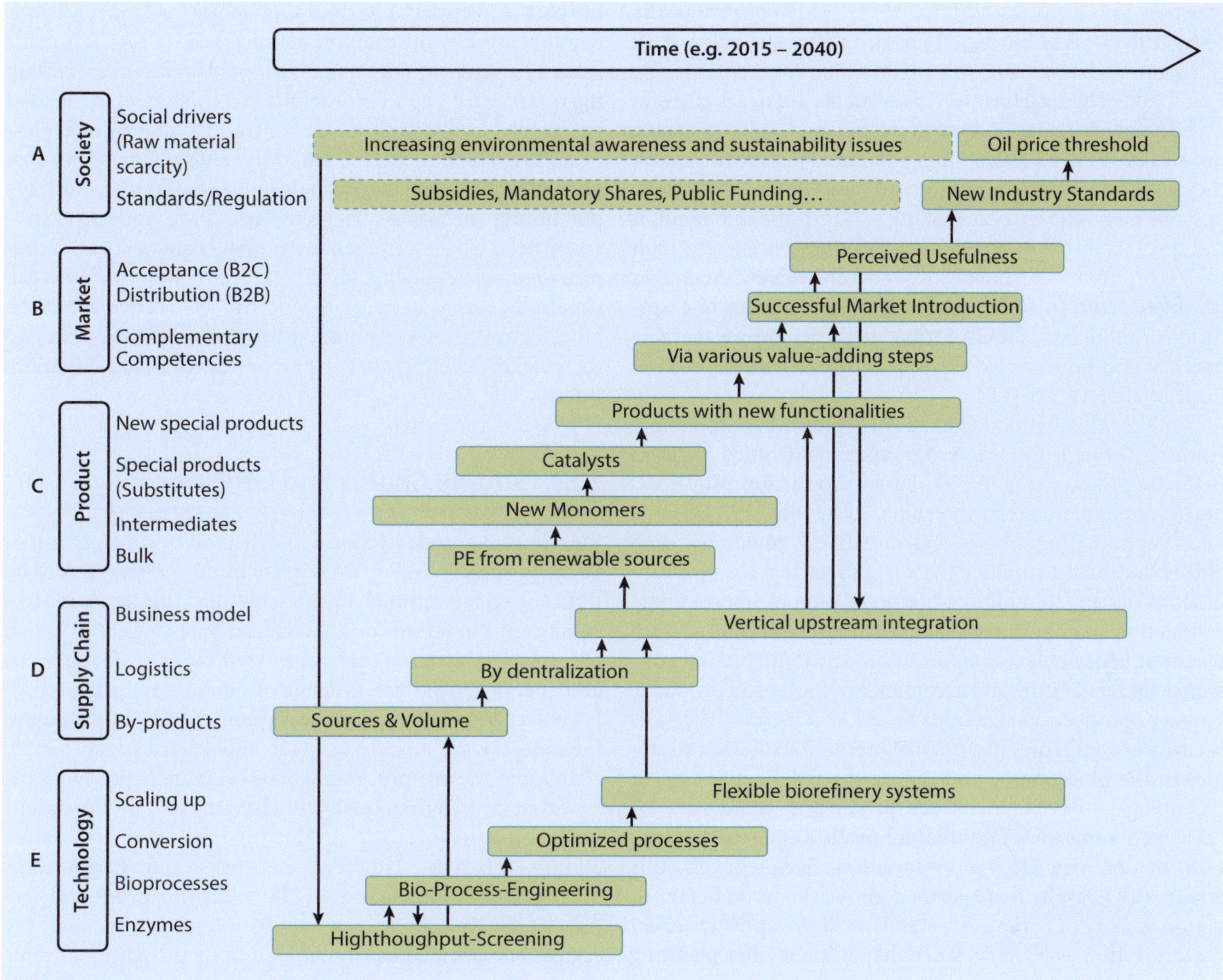

Fig. 7.9 Roadmap for the development of the bioeconomy in Germany. (Based on Bröring 2016)

7.2.1 Social Requirements and Political Framework Conditions

A 2008 study by the European Commission concludes that European citizens consider environmental protection as highly important. Although 56% of the German participants in the study (n = 1519) consider the topic of environmental protection to be very important and 40% quite important, Germany ends up in the lower third in this regard compared to its European neighbours (top positions: Cyprus, 94% very and 5% quite important; Sweden, 89% very and 10% quite important) (European Commission 2008). A strong environmental awareness has been observed since the end of the twentieth century (Stern 2000). The growing interest in environmental and sustainability issues is explained by post-materialist values. These values include cultural, social and intellectual needs, and are only developed by a society when existential security is given. However, it should be noted (► Chap. 8) that, despite increasing environmental awareness, the willingness to pay for sustainability in society is relatively limited, so that it can be seen as a gap between attitudes and behaviour (Kollmus and Agyeman 2002).

New bioeconomy value chains will therefore only establish themselves if they become economically competitive with conventional petroleum-based production systems. The question arises: What needs to be done to make bio-based production systems competitive? In addition to the efficiency of production systems, the price of raw materials is decisive for their competitiveness. Thus, there is a link between the price development of crude oil and that of raw agricultural materials. In the first decade of the twenty-first century, crude oil initially became drastically more expensive. With the onset of the global financial crisis in autumn 2008, the price fell only temporarily, returning to a price level of over USD 100 per barrel in 2014 as part of the economic recovery. As a result, crude oil-based products became much more expensive and bio-based products appeared to be an attractive alternative from an economic point of view. In the meantime, however, the price of crude oil has again fallen sharply and has remained at a relatively constant low level (as of June 2016). This constrains the competitiveness of bio-based products. The low price of oil is currently hampering the further development of bio-based value chains. However, investments are necessary and investment strategies of great importance in order to enable innovative, cost-effective bioeconomic value creation, and thus a successful transformation process. This is where politics can intervene in a controlling way. On the one hand, it can specifically promote bio-based production, e.g., through tax exemptions and concessions, and create new incentives for investment. In addition to setting such framework conditions, politicians should formulate clear and measurable goals, as was the case, for example, at the climate summit in Paris in 2015 (▶ Chap. 8).

On the other hand, politicians can intensify bioeconomic research through funding programmes and funds, be it in thematic projects (e.g., on bio-based production processes) or by strengthening partnerships, networks and clusters. Such support programmes and initiatives provide learning effects and help to reduce the manufacturing costs of bio-based products. In addition, harmonisation of international production standards and norms for bio-based products are useful. It is not only an important means to increasing consumer understanding and acceptance of bio-based products. Uniform production standards would also increase the sales security of companies that pursue internationalisation strategies in the bioeconomy and reduce the risk of investing in technological development. Yet, production standards also set a fixed framework for future innovation and could potentially hamper new ideas and innovations that go beyond this framework. Despite these partly negative effects on the innovation activity of companies, standards have a positive effect on innovation as a whole, since they initially offer planning security, and thus investment incentives for research and development (cf. Allen and Sriram 2000).

7.2.2 Market and Products

In addition to the economic competitiveness of bio-based products, their acceptance by customers - first in the business-to-business (B2B) sector, then in the business-to-consumer (B2C) sector - plays a key role in the successful transition to a bioeconomy (▶ Chap. 8). However, consumers, in particular, are often unaware of the properties of bio-based products and their importance for sustainable consumption. Even when this ignorance has been overcome, enlightened end consumers show a discrepancy between their desire to consume bio-based products and their willingness to pay a premium price for them (Carus et al. 2014). This discrepancy is exacerbated when product communication on the part of manufacturers and suppliers is inadequate. This is demonstrated by the example of polylactide (PLA). It is made from bio-based monomers of lactic acid and used as packaging material for cups, bottles, foils and containers in the food industry or as textile fibres. In contrast to conventional plastics, PLA is biodegradable, and thus fulfils a criterion of sustainability. This is an increasingly influential factor in society's purchasing decisions. Nevertheless, PLA products have rarely been labelled in such a way that customers can recognize their sustainability advantage. This communication gap should be closed in order to improve knowledge about the bioeconomy in society and, with growing acceptance, to increase the willingness to pay so-called Green Premium prices.

7.2.3 Supply Chains and Logistics

Bioeconomic production is dependent on raw materials that grow seasonally and - in contrast to oil - often cannot be transported far without considerable ado. This fact leads to a significant limitation for a globalized bioeconomy in which biogenic raw materials are considered the basis for material and energy-related use. Outside of the food industry, which is subject to different framework conditions, supply routes for industry are already of great importance today. Supply chains and the pre-processing of bio-economic products are, however, strongly regional, and characterised by the decentralised nature of their production processes and value-adding activities. However, networks and partnerships among the various actors in the value chain are still frequently lacking. Due to a lack of interconnectedness, the regional actors are firmly anchored in their traditional value chains. They are bound to certain customers and suppliers in the long term and overlook the potential of their raw materials. It would often require an intermediary or broker to act as a link between small- and medium-sized raw material producing companies and large processing companies from previously unknown value chains. In addition to its mediation function, such a broker could also coordinate the decentralised processing of biomass, and thus play a central role in the formation of new value chains. This opens up opportunities for innovative business models.

This must be accompanied by the conversion of the existing infrastructure. The infrastructure for an industrial value-added structure based on raw fossil materials has been optimised holistically over decades. The associated use of production facilities and material flows in the sense of joint production accordingly exhibits very high economic efficiency. Although technological progress always goes hand in hand with the reconstruction of the existing infrastructure, the investments required for the bioeconomic transformation process do not offer an attractive return on capital, at least in the short term. This particularly puts small- and medium-sized enterprises, which rely on bioeconomic innovation, in a dilemma, because they can only rarely finance the restructuring of the infrastructure from available equity capital. They are dependent on the commitment of larger companies or the willingness of external partners to invest.

7.2.4 Technological Complexity

The implementation of bio-based production processes also depends on internal company decisions. Accordingly, the benefits of bio-based production and the associated recycling of residues must become apparent at the level of the individual company. However, because a company's management is often guided by its previous research and production decisions, it finds it difficult to identify the potential of bio-based production processes or the development of new technologies in this area. Industry-specific path dependencies limit the absorptive capacity of new knowledge in this way (Cohen and Levinthal 1990). For the purpose of increasing the capacity of a company to absorb new knowledge, cross-industry cooperations and *Open Innovation* processes are excellent means. The opening of their internal innovation process enables companies to close competence gaps, and thus generate added value. However, they must first be willing to cooperate with other companies. They also should have sufficient network competencies to be able to successfully embark on *Open Innovation* (▶ Excursus 7.3) (Chesbrough 2003). They must coordinate the exchange of information and knowledge and decide on the organisational form of cooperation, which may take the form of vertical integration, mergers and acquisitions, or joint ventures (Borés et al. 2003; Bröring 2010b). Bio-based production processes often require the integration of knowledge from outside of the industry into existing production processes. At the company level, special expertise is needed to integrate new technologies into existing production processes. However, there is still a lack of special training opportunities for specialists in bioeconomics, as a search for bachelor's and Master's programmes shows. Of the 67,000 courses offered by 1400 European universities and colleges, only two are specifically aimed at training in the field of bioeconomy: MSc Pharmaceutical and Industrial Biotechnology at the Martin Luther University Halle-Wittenberg and MSc Bioeconomy at the University of Hohenheim (xStudy 2016).

Even if it is possible to recruit the appropriate specialists, the integration of new knowledge is associated with high efficiency losses, at least at the beginning, and is not rewarded with short-term profits. Therefore, in times of strong competitive pressure in the market as currently exists, it is often unattractive for companies to invest in bio-based technologies and production processes. A lack of attractiveness for investment also particularly slows down the development of biotechnological processes, which are usually associated with long development times and high financial expenditures and risks.

For an effective technology transfer in the bioeconomy, too, there is a need for close links and cooperation between industry, universities and public research institutions in order to develop new technologies and production processes and successfully introduce them into the market. For market success, it is essential to realize economies of scale, i.e., to enable lower product prices through increasing product volumes.

Excursus 7.3 Open Innovation

It caused a sensation when Henry Chesbrough advised companies at an OECD-conference in 2001 to open their innovation processes vis-à-vis external players. But the success of the postulated opening should concede a point to the Managing Director of the *Center for Open Innovation* at the University of California in Berkeley. His approach of an open innovation process is based on the insight that the integration of external knowledge does not only save in-house resources, but also boosts the creativity potential (Chesbrough 2003), because the creativity spectrum of the research and development departments of companies is normally narrowed by the experiences they have cumulated over decades. Thus, through cooperation with external researchers and developers, the company's internal horizon can be broadened, favouring the emergence of radical new innovation approaches. These are particularly important in the case of bioeconomy, in which the interaction of a multitude of specialists with different viewpoints is required. ◘ Figure 7.10 shows the interplay of external and internal knowledge in the process of open innovation.

In the past ten years, many large enterprises have established platforms on which they have tested the process of *Open Innovation*. For example, the Danish brewery group Carlsberg inaugurated an open innovation process in 2009 to lower high CO_2-output in the production of its packagings. Together with a large number of external specialists (e.g., from ecoXpac and the Technical University of Denmark), they were looking for solutions. The result was the *fibre bottle,* a bottle that is biodegradable and enables emission reductions of up to 80%. The brewery then invited other companies in the beverage industry to join its open innovation platform in order to make the new technology marketable (Berman 2016).

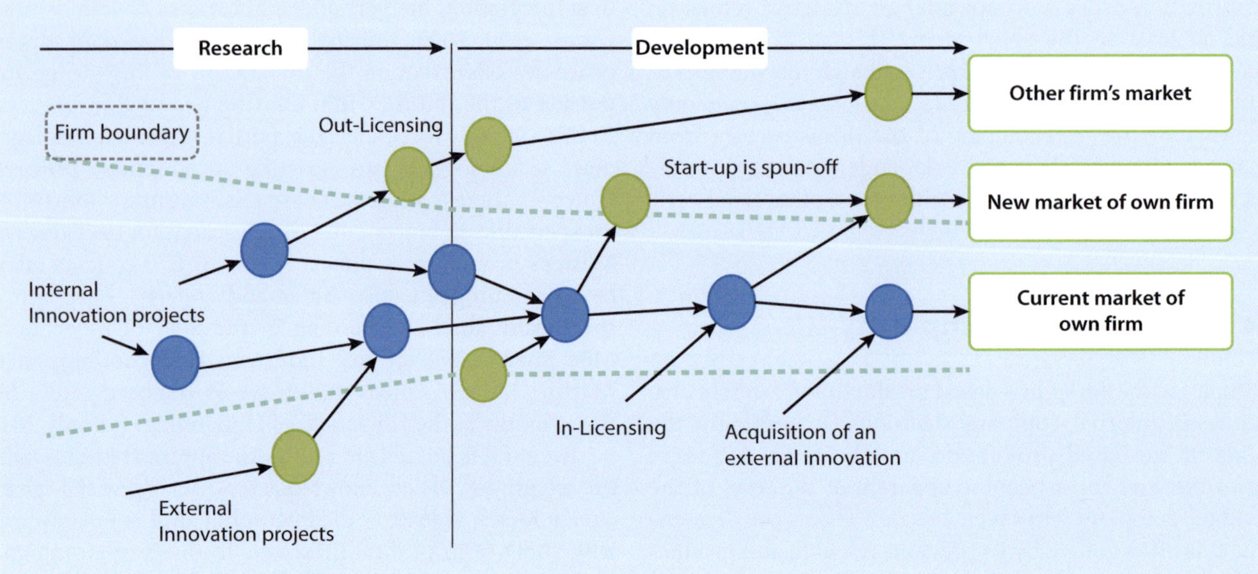

Fig. 7.10 The *Open Innovation*-lawsuit. (In the style of Chesbrough 2006)

7.3 Perspectives of Bioeconomic Value Chains

7.3.1 Established Value Chains

In recent years, new value chains, from agricultural cultivation to fuel production, have gained enormously in importance. In particular, the growth in production capacity for bioethanol is impressive (Fig. 7.11).

Which agricultural commodities are used depends on the regional conditions. In Brazil, sugar cane is mainly used for the production of bioethanol, in Europe, wheat starch, and in the USA, corn starch. In all cases, these are established agricultural commodities. Therefore these value chains are referred to as first generation value chains. A glance at the USA shows just how quickly these value chains have developed: in 2014, 40% of the corn harvest there went into the fuel sector; ten years earlier, the figure was only 0.7% (European Biofuels Technology Platform 2016). The prerequisite for this growth was the expansion of production capacities for bioethanol (Fig. 7.12).

It is notable that almost all production facilities were built in maize growing areas. This means that the fuel value chain is shifting spatially from fossil-based industrial centres to rural areas (▶ Chap. 8).

Ethanol can not only be used as a fuel, but also as a raw material for the production of ethylene, the most important basic chemical in terms of quantity (Behr et al. 2010). In fact, in Brazil, ethylene is produced from sugar, and it seems possible that certain stages of chemical value chains may shift from established industrial centres to biomass regions - with all of the long-term implications for the respective labour markets and regional levels of prosperity. Since, in the case of ethylene, this is a pure commodity switch, cost competition essentially determines whether the substitution will be successful. The situation is different if a bio-based product replaces the fossil-based alternative because it has a superior performance spectrum. An example of this is the bioplastics PLA (polylactide) and PEF (polyethylene furanoate). PLA is a biodegradable polymer, and PEF, as a component of beverage bottles, leads to higher gas tightness and material stability (Ißbrücker and von Pogrell 2013). This convinced Coca-Cola to use the PEF developed by the medium-sized company Avantium (Tullo 2012). In 2016, BASF announced the formation of a joint venture with Avantium to operate a production plant. Avantium was only founded in 2000 as a spin-off from Shell. This example can serve as a successful model for the emergence of a new value chain: An oil company looks for bio-based process alternatives and founds a spin-off; its successful product development is started by a user; his decision convinces a chemical company to invest in a sugar-based production plant.

Bio-based chemical products already have large production volumes today (Table 7.1). Products such as L-glutamic acid, L-lysine, citric acid and isomaltulose can only be produced bio-based and economically using biotechnological processes. Other products such as ethanol, polylactic acid and 1,3-propanediol have advantages over fossil-based alternatives because of their product properties and lower carbon footprint.

In the medium term, bio-based chemistry is expected to grow by 11% annually, with intermediates and polymers accounting for the largest share (72%) (Fig. 7.13). Today, the product spectrum is almost exclusively limited to mole-

Fig. 7.11 Development of production capacities for bioethanol and biodiesel. (Source: CropEnergies 2016)

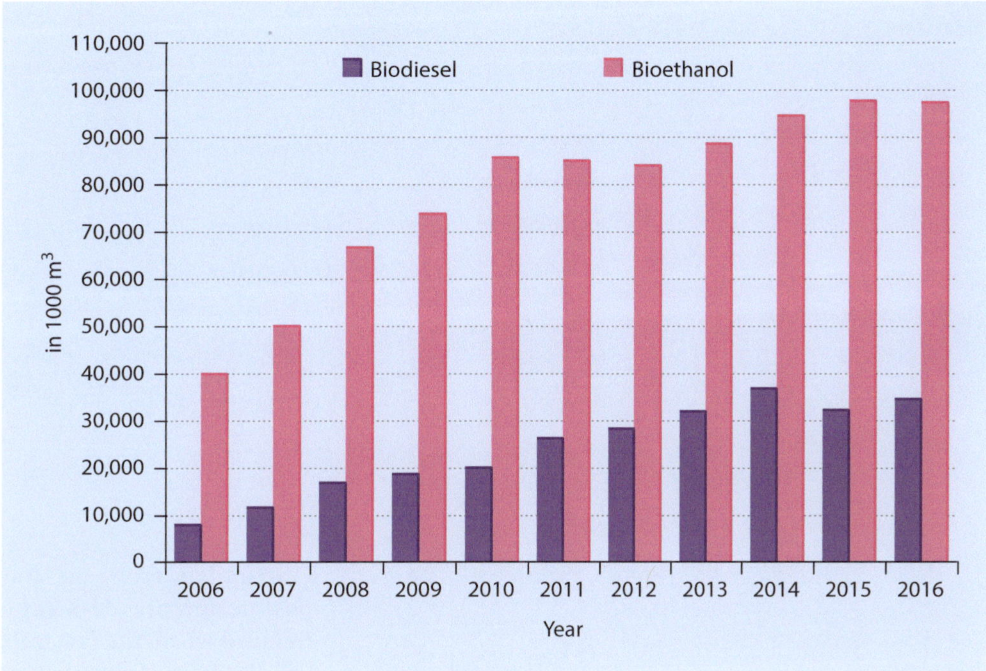

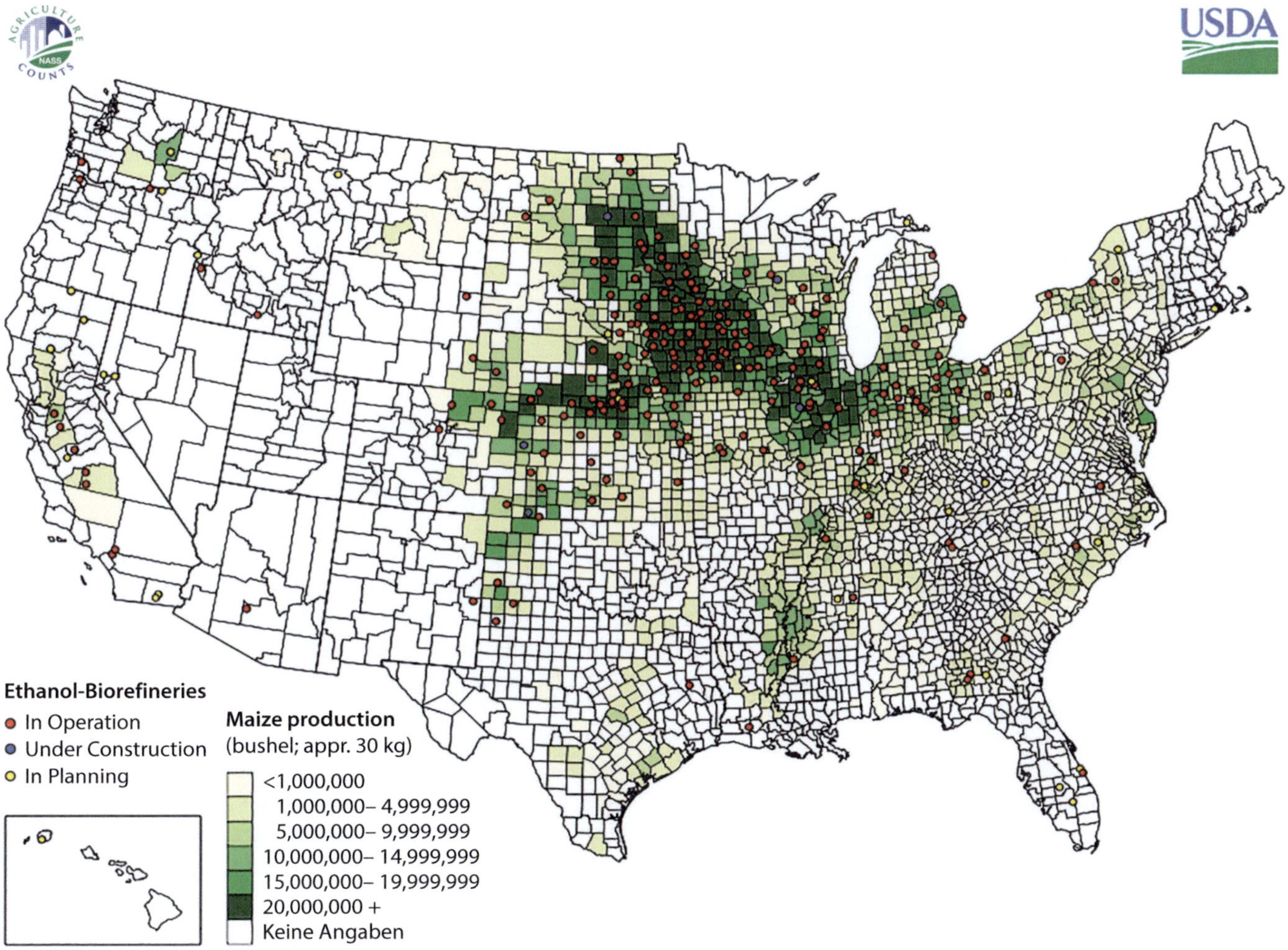

Fig. 7.12 Spatial distribution of ethanol biorefineries and maize acreage. (USDA 2013)

Table 7.1 Application and production capacity of bio-economic products. (Pelzer 2012)

Product	segment	Proceedings	Production volume (t/a)
Ethanol	biofuel	Fermentation	68634150
L-glutamate	food products	Fermentation	2160000
Citric acid	food products	Fermentation	1700000
L-lysine	animal feed	Fermentation	1480000
Polylactic acid (PLA)	polymers	Fermentation	140000
Isomaltulose	food products	Biotransformation	100000
Polyhydroxyalkanoate	polymers	Fermentation	50000
1,3-propanediol	polymers	Fermentation	60000

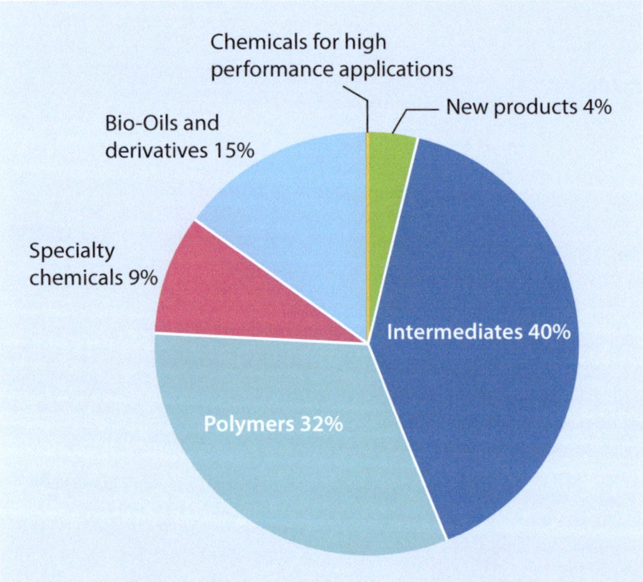

Fig. 7.13 Prognosis of bio-based chemistry according to product groups. (Allen 2015)

cules that are naturally provided by nature. Intensive research into synthetic biotechnology (▶ Chap. 5) will ensure that the bio-based product portfolio expands and becomes a real alternative to fossil-based chemistry.

7.3.2 Emerging Value Chains

In order to avoid competition with the food sector, increasing efforts are being made to use biomass that is not suitable for nutrition. Here, too, bioethanol is the pioneering product for the very large fuel market. In Brazil (Braskem) and the USA (a joint venture of Poet and DSM), plants for second-generation bioethanol based on non-edible sugarcane and corn biomass are already in operation, and in Europe, Clariant is working on straw-based processes for this purpose in Hungary (Al-Kaidy et al. 2014). These plants are also installed where the raw material is produced. New products will also follow from such second-generation plants. In 2017, Energochemica will commission a second-generation ethylene plant in Slovakia. Corbion Purac is working on the introduction of PLA based on lignocellulose.

In addition to agricultural waste biomass, wood can also be considered for these plants. In Leuna (Saxony-Anhalt), a pilot plant for the use of beech wood as a carbon source is operated (Amann 2014). New value chains are developing, from forestry and related sectors to the fuel and chemical sectors. The paper industry, for example, processes very large quantities of wood. It now sees the opportunity to process lignocellulosic sidestreams from this production as raw materials for second-generation chemical products. Stora-Enso, one of the world's largest paper manufacturers, is conducting intensive research in this field.

7.3.3 Future Value Chains

Agriculture does not only supply raw vegetable materials. It also has the potential to integrate processing steps, and thus shorten value chains. This is because plants are capable of forming and storing polymers (e.g., polyhydroxybutyric acid, PHB; polyisoprene). Compared to the fermentative production of PHB, the value-added stages of sugar plant cultivation, sugar refining and fermentation are directly integrated into the plant production system. PHB has not yet achieved industrial success, but a production system for polyisoprene based on dandelion developed by Fraunhofer IME (Aachen) in cooperation with Continental is well advanced.

If successful, a further plant-based process will significantly shorten an established value chain. Chicken egg white,

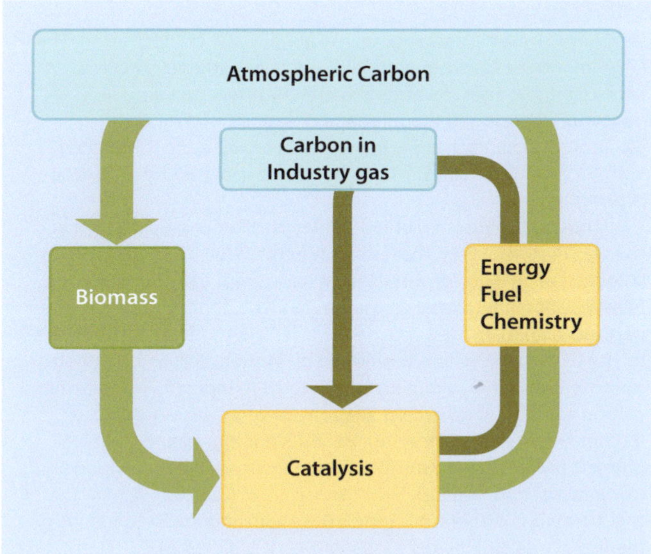

Fig. 7.14 Carbon cycle via biomass or the use of industrial material flows. (Source: Kircher 2015)

an important raw material in the food industry, is obtained from chicken eggs, and is therefore part of the animal breeding value chain. The start-up company Clara (USA) has set itself the goal of producing this animal protein agriculturally in plants. If this succeeds, the previous value-added stages of poultry breeding, including the necessary feed production, can be completely eliminated for certain applications of chicken egg white.

Traditional value chains range from raw materials and their transformation to consumer products. After use, this product is disposed of and classified as CO_2-emissions that are used to generate energy. The carbon cycle is ultimately closed by photosynthetic carbon fixation into vegetable biomass. New bioeconomy value chains will shorten this natural material flow, and thus increase raw material efficiency by reducing carbon emissions into the atmosphere (Fig. 7.14). For example, the carbon cycle can be partially or largely closed through technical recycling in industrial processes (▶ Excursus 7.4).

Excursus 7.4 Circular Economy Through the Use of Waste as a Resource

Fossil-based process chains are linear from raw material to consumer product. CO_2 is emitted and accumulated in the atmosphere because the capacity of the natural carbon cycle is not sufficient to convert the additional carbon introduced into biomass (Fig. 7.15). The imbalance in the carbon budget negatively affects the climate.

Bioeconomy processes based on biomass are also often linear. After use, bio-based consumer products, which are ideally biodegradable, should be used for cascading energy production or decomposted. Today, the CO_2 is also emitted into the atmosphere and introduced into the natural carbon cycle (Fig. 7.16). If the volume of carbon emitted were equal to that bound in new biomass, the carbon budget would be in balance, and thus climate-neutral. A circular bioeconomy is therefore of special importance for climate protection.

The 2015 UN Climate Change Conference in Paris agreed on a timetable to achieve a balanced carbon budget by the end of the century. In the second half of this century, a "balance between anthropogenic emissions of greenhouse gases from sources and their degradation by sinks" has to be established (European Commission 2016).

However, given the enormous global demand for bio-based raw materials, it is doubtful that the capacity of the natural carbon cycle alone will be sufficient to provide the biomass needed to achieve this balance. However, there are additional technical options available for closing the carbon cycle and establishing an industrial recycling economy for carbon, regardless of its origin. Already 2014, the European Union published the strategy "Towards a circular economy: A zero waste program for Europe" (EU Commission 2014). The aim is to recycle industrial material flows consistently and to avoid the release of waste into the environment. This also applies to carbon, regardless of whether it comes from biogenic or fossil sources (Fig. 7.17).

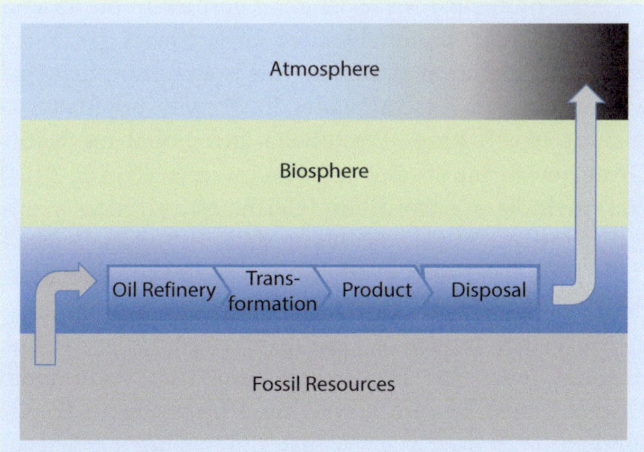

Fig. 7.15 The fossil-based processing chain is linear

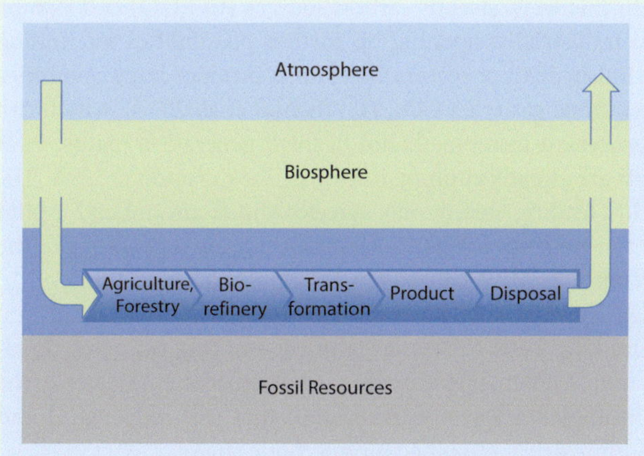

Fig. 7.16 The bio-based processing chain is linear

Since 2015, for example, fuels based on recycled CO_2 of fossil origin are treated as being equal to biofuels under certain conditions with regard to funding (Dammer and Carus 2015). Hence, priority is given to the environmental balance rather than to the origin of the raw material.

The economy also increasingly sees industrial carbon emissions as a raw material and is targeting corresponding production processes. Gas fermentation is well advanced (► Chap. 5), and is developing into a real wave of innovation. The large number of arising material flows is remarkable: Facilities processing biomass, municipal waste and steel mill emissions have already reached the pilot and demonstration scale (◘ Table 7.2). In each case, the natural carbon cycle is replaced with technical recycling: The carbon cycle is closed by industrial processes.

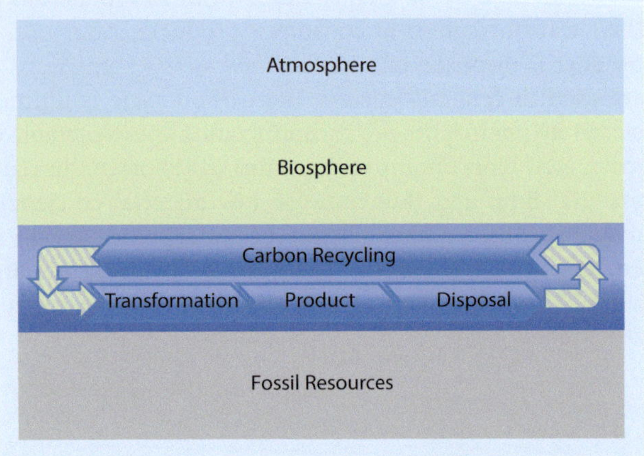

◘ **Fig. 7.17** The processing chain of the circular economy

Under the leadership of Thyssenkrupp, Siemens, Fraunhofer and the Max Planck Society, the Carbon2Chem (BMBF 2016) project was launched in 2016 to convert exhaust gases from blast furnaces into fuels, plastics or fertilizers on a large scale by chemical catalysis. The partners plan to invest € 100 million by 2025. Another contribution to closed carbon cycles is developed by Siemens AG, with synthetic photosynthesis. CO_2 is directly converted into industrially interesting carbon compounds (Siemens 2016). In Bavaria, a consortium led by IBB Netzwerk GmbH is developing biotechnological and chemical-physical processes for atmospheric and emission CO_2 as an industrial raw material (IBB Network 2016), and in NRW, CLIB2021 works together with partners in Flanders (Belgium) and the Netherlands on the use of industrial gases containing CO (CLIB2021 2015).

The industry's commitment proves that technically and economically attractive solutions for creating the sinks for anthropogenic carbon emissions demanded by the Paris Climate Summit are elaborated. The recycling economy thus emerging will complement and relieve the bioeconomy in a sustainable way.

Manfred Kircher

◘ **Table 7.2** Examples of carbon recycling by gas fermentation from different waste streams. (ArcelorMittal 2015; Lanzatech 2014; Abengoa 2015)

Carbon	Resource	Product	Company	Country
Fossil	steel mill emission	ethanol	Arcelor Central	Belgium
Organic	wood	ethanol	Lanzatech	USA
Mixed	municipal waste	ethanol	Abengoa	Spain

In this way, gaseous carbon sources (CO_2CO) can also be used industrially, opening up further possibilities for linking previously distinct sectors. Steel mills produce large quantities of synthesis gas (CO, CO_2, H_2) (Schöß et al. 2014), which can be processed into bioethanol in third-generation plants. Pilot plants are already running in China, Taiwan and the USA. The world's leading steel group, ArcelorMittal, announced a pilot plant in Belgium for 2017. All plants use the technology of the start-up company LanzaTech (USA), which was founded in 2005. The fact that the start-ups Syngip (Netherlands) and White Dog Labs (USA) are already working on basic third-generation chemical products can be seen as an indicator of the beginning of a wave of innovation that will reduce CO and CO_2-industries into bioeconomic value chains.

Also, municipal waste, which is gasified in the absence of oxygen, is a raw material source for synthesis gas, and thus for third-generation products. This is an option that enables waste management companies in densely populated areas such as North Rhine-Westphalia and global megacities to create new forms of value creation (Styczynski et al. 2014). In fact, in India, ConcordBlue (Germany) produces synthesis gas based on municipal waste and, in Japan, LanzaTech (USA) has used such synthesis gas in a pilot plant as a raw material for the fermentative production of bioethanol.

A further model of closing material cycles by linking material flows has been implemented in the Frankfurt-Höchst Industrial Park. Glycerine, a by-product of biodiesel production, is used as a raw material in a biopharmaceutical process. The waste water from this and another production plant is fed into a biogas plant, together with organic waste from the surrounding area. Some of this biogas is used to generate electricity, while another portion of it is fed into the public natural gas grid after purification via a joint venture between site operator Infraserv Höchst and energy supplier Mainova (■ Fig. 7.18).

In future, part of the biogas could also be oxidized into methanol, the most versatile platform chemical ever produced (▶ Chap. 4), and thus be materially recycled. Waste disposal and chemical, pharmaceutical and energy production will be linked in a completely new way. The integration of waste management value chains into those of a chemical site is the entry into a consequent raw material cycle for carbon. In the long term, carbon cycles and other cycles may be able to be closed completely on the basis of natural models (▶ Excursus 7.5).

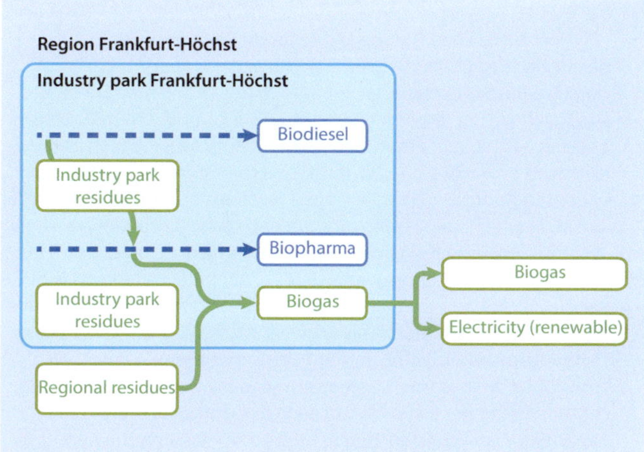

■ **Fig. 7.18** Linking and recycling the material flows of different value chains. (Source: Infraserv Höchst 2016)

Excursus 7.5 Cycles of Matter in Nature and Sustainability

Sustainable processes are those that do not consume more resources than those which are replaced within a reasonable period of time. The use of raw fossil materials could therefore, in principle, be sustainable, if the consumption of coal, gas and crude oil were in equilibrium with their regeneration. Under these conditions, the available quantity of raw fossil materials would remain constant over time. However, we are far away from fulfilling these conditions: today, there are about 22,000 Gt of coal resources worldwide (BGR 2015). It took about 300 million years to build up these coal reserves - beginning in the Carboniferous Era and continued until today (■ Fig. 7.19). On the other hand, there was a global use of 8 Gt in 2013, which corresponds to a stockpile that was built up, on average, over a period of 109,000 years; coal formation does not even come close to occurring continuously at such a rate. With current technology, however, a reserve of about 968 Gt of these resources can still be mined, which makes the unbalanced relation between availability and consumption even more explicit.

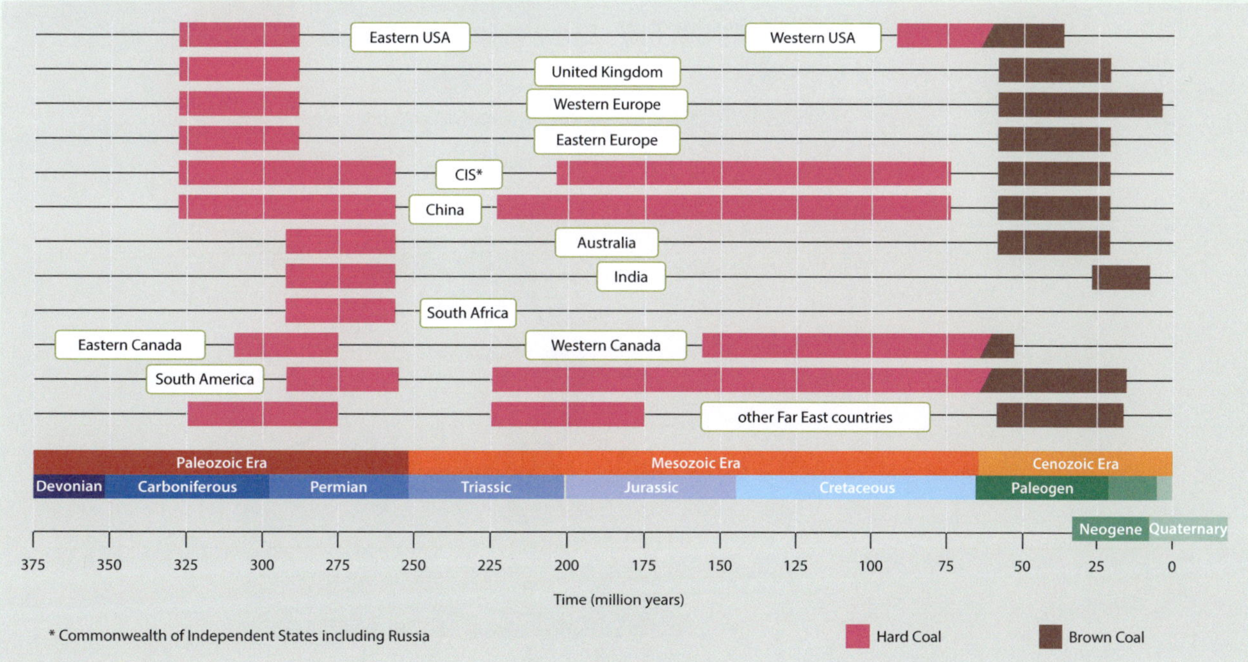

■ **Fig. 7.19** Periods of origin of the world's most important hard coal and lignite deposits. (Heinrich Böll Foundation 2015)

In order to keep the use and regeneration of resources in balance and to generate as little waste and residual materials as possible, it is particularly appropriate to use closed-loop processes such as those implemented in nature. For example, the metabolism of living cells is built around essential cyclic processes such as the citric acid cycle (◘ Fig. 7.20). It serves as a "hub" for the provision of various metabolites. Nine individual reactions are coupled in a closed circuit, in which the starting product and the end product are identical. The required target substances, which are used either for energy production or biosynthesis, are removed from the cycle at various points, and the starting substance is restored by incorporating precursor substances. This achieves high efficiency in the use of materials and energy. The prerequisite is that the substances are "controlled" in the cycle, i.e., they are spatially transferred from one reaction to the next, and that, in each reaction step, there is the capacity for complete and timely transformation into the next step.

In the economy, on the other hand, many current production processes are organized as linear utilization chains, with a raw material at the beginning and a product at the end. Raw materials are removed from the environment and residual materials are released into the environment in between or at the end. Such production processes use the capacity of the environment to deal with residual materials and convert them back into raw materials - albeit partly over geological timeframes. For example, released CO_2 is absorbed by plants and then becomes available again as biomass. However, the capacity of photosynthesis and uptake into the ocean, the two most important global processes, do not match anthropogenic CO_2 release. The resulting increase in CO_2 concentration and of other greenhouse gases has considerable consequences: The substances accumulate, with negative consequences such as global warming or eutrophication in the affected environmental compartments. The substances are usually diluted in the air or water or transported to other locations, and are therefore hardly accessible for technological recovery processes. A particular disadvantage is that the periods in which such residues can be converted are usually very long.

A basic concept in avoiding the disadvantages of linear production and using materials sustainably therefore lies in the implementation of cyclic processes similar to those established in nature. The aim is to release as few substances as possible into the environment and to achieve the highest possible proportion of internal substance conversion. Such an objective could - especially on a decentralised scale - be achieved, for example, with biogas facilities whose main product is methane. However, 40% of the gas produced is carbon dioxide. Instead of blowing this gas off into the atmosphere, it could be directed into greenhouses, algae photobioreactors or Clostridia cultures, and thus directly fed back at high concentrations for plants or microorganisms that produce highly efficient biomass from it. Another cycle can be realised if residual materials from biorefineries (e.g., nutrients or organic carbon) are processed in such a way that they can be reintroduced into agricultural production as quickly as possible. Such circulation systems (◘ Fig. 7.21) must not, however, be optimized to maximize individual substance transformations and products. Rather, they must be designed and optimized as a holistic system from systemic design and their individual steps must be coordinated appropriately.

Ulrich Schurr

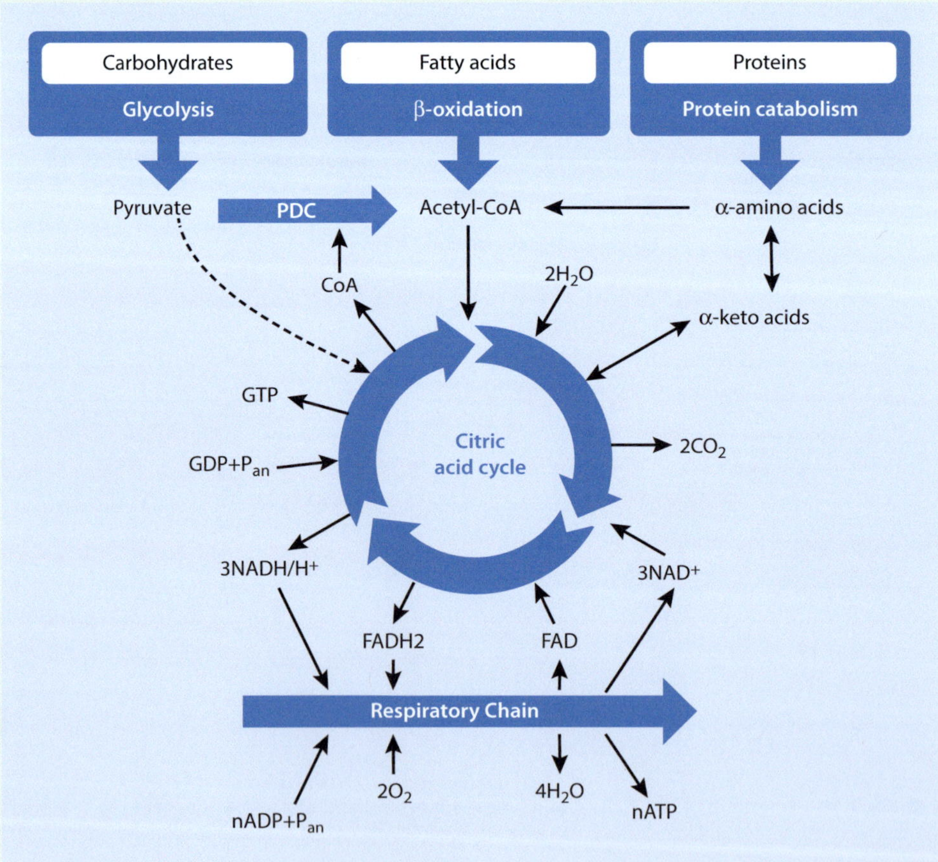

◘ **Fig. 7.20** The citric acid cycle as an example of a coupled cycle with input and output points

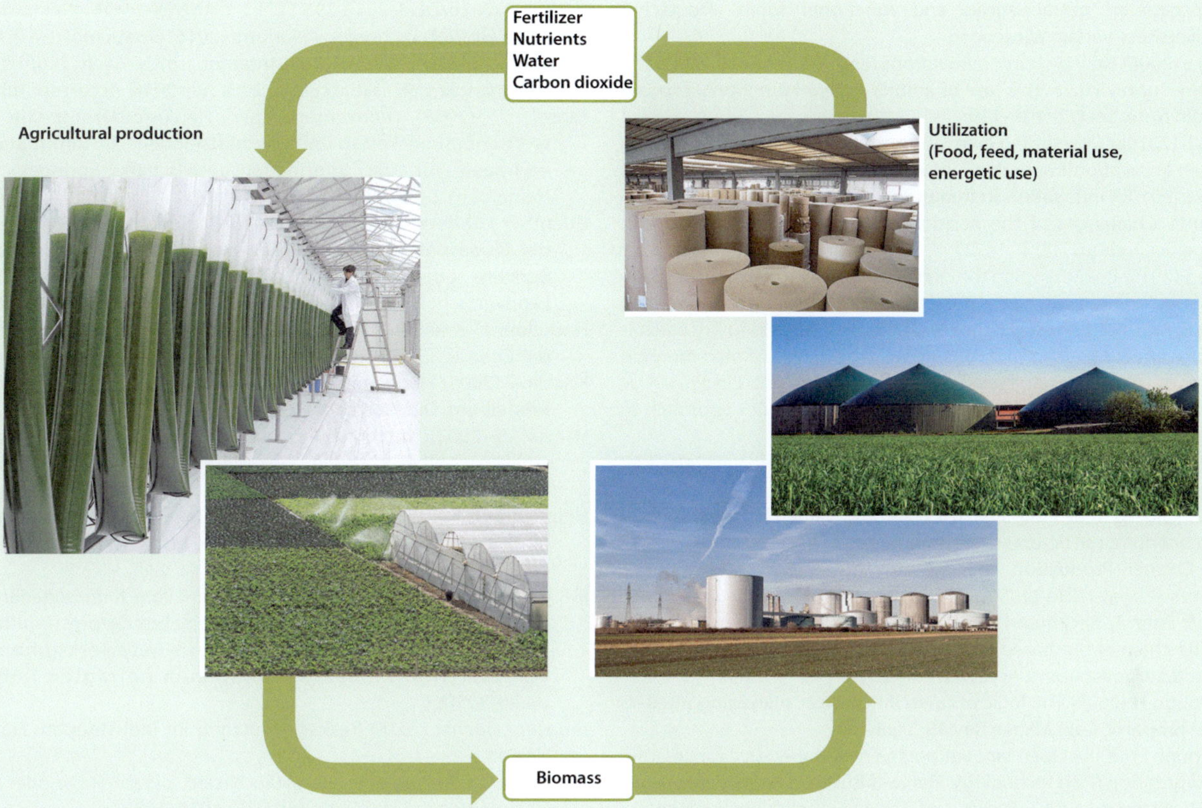

Fig. 7.21 Integrated biorefinery concepts combine the production of raw materials and conversion systems. (According to Forschungszentrum Jülich)

References

Abengoa (2015) Abengoa will build the first biorefinery producing biofuels from MSW in the USA. http://www.abengoa.com/web/en/noticias_y_publicaciones/noticias/historico/2015/05_mayo/abg_20150505.html

Allen J (2015) Global biobased economy trends: renewable chemicals & biofuels (Vortrag im Rahmen des Bio World congress on industrial biotechnology am 20.7.2015 in Montral, Canada)

Allen R, Sriram R (2000) The role of Standards on Innovation. Technol Forecast Soc Chang 64:171–181

Al-Kaidy H, Duwe A, Huster M, Muffler K, Schlegel C, Sieker T, Ulber R (2014) Biotechnologie und Bioverfahrenstechnik-Vom ersten Ullmanns Artikel bis hin zu aktuellen Forschungsthemen. Chemie Ingenieur Technik 86(12):2215–2225

Amann M (2014) Laccase und ligninhaltige Prozessströme. Chemie Ingenieur Technik 86(9):1517–1518

ArcelorMittal (2015) ArcelorMittal, LanzaTech and Primetals Technologies announce partnership to construct breakthrough €87m biofuel production facility. http://corporate.arcelormittal.com/news-and-media/news/2015/july/13-07-2015

Behr A, Kleyensteiber A, Hartge U (2010) Alternative Synthesewege zum Ethylen. Chemie Ingenieur Technik 82(3):201–213

Berg S, Cloutier LM, Bröring S (2018) Collective stakeholder representations and perceptions of drivers of novel biomass-based value chains, Journal of Cleaner Production 200: 231–241. https://www.sciencedirect.com/science/article/pii/S0959652618322935

Berman A (2016) Carlsberg dreams of paper bottles. online im Internet: http://www.mnn.com/money/sustainable-business-practices/stories/carlsberg-dreams-of-paper-bottles. Accessed July 2016

Bibliographisches Institut (2012) Duden, das große Wörterbuch der deutschen Sprache, Online-Ausgabe, online im Internet: www.duden.de. Accessed July 2016

Biermann U, Bornscheuer U, Meier MA, Metzger JO, Schäfer HJ (2011) Fette und Öle als nachwachsende Rohstoffe in der Chemie. Angew Chem 123(17):3938–3956

BMBF (2014) Bioökonomie als gesellschaftlicher Wandel - Konzept zur Förderung sozial- und wirtschaftswissenschaftlicher Forschung für die Bioökonomie. BMBF, Bonn

BMBF (2016) Mit Abgas das Klima retten, https://www.bmbf.de/de/mit-abgas-das-klima-retten-3044.html

Boehje M, Bröring S (2010) The increasing multifunctionality of agricultural raw materials: three dilemmas for innovation and adoption. Int Food Agribus Manag Rev 14:1–16

Bobo J (2015) From burgers to biomes: food agriculture and the bioeconomy. In: First Bioeconomy Summit Berlin 2015. online im Internet: http://gbs2015.com/fileadmin/gbs2015/Downloads/Bioeconomy_World_Tour.pdf. Accessed: July 2016

Borés C, Saurina C, Torres R (2003) Technological convergence: a strategic perspective. Technovation 23(1):1–13

Bornkessel S, Bröring S, Omta SWF (2014) Analysing indicators of industry convergence in four probiotics innovation value chains. J Chain Netw Sci 14(3):213–229

7

Bröring S (2005) The front end of innovation in converging industries: the case of nutraceuticals and functional foods. Deutscher Universitäts-Verlag, Wiesbaden

Bröring S (2008) How systemic innovations require alterations along the entire supply chain: the case of animal-derived functional foods. J Chain Netw Sci 8(2):107–119

Bröring S (2010a) Developing innovation strategies for convergence - Is 'open innovation' imperative? Int J Technol Manag 49(1–3):272–294

Bröring S (2010b) Innovation strategies for functional foods and supplements. Challenges of the positioning between foods and drugs. Food Sci Technol 7(8):111–123

Bröring S (2016) Roadmapping the circular economy to align technological development with social needs. Working Paper, Uni Bonn

Bröring S, Andreae F, Preschitschek N (2017) How does industry convergence affect suppliers and consumers? An analysis of the emerging sectors of functional foods and biopolymers. In: Lindgreen A, Hingley MK, Angell RJ, Memery J (eds) A stakeholder approach to managing food. Routledge, London

BGR (2015) Energiestudie 2015. Reserven, Ressourcen und Verfügbarkeit von Energierohstoffen (19). – 172p

Carraresi L, Berg S, Bröring S (2018) Emerging value chains within the bio-economy: structural changes in the case of phosphate, Journal of Cleaner Production 183: 87–101. https://www.sciencedirect.com/science/article/pii/S0959652618304475

Carus M, Eder A, Beckmann J (2014) GreenPremium prices along the value chain of bio-based products. Nova-Institut, Hürth: www.bio-based.eu/novapapers. Accessed July 2016

Chesbrough H (2003) The logic of open innovation: managing intellectual property. Calif Manag Rev 45(3):33–58

Chesbrough H (2006) Open innovation: the new imperative for creating and profiting from technology. Harvard Business Press, Boston, MA

Choi D, Valikangas L (2001) Patterns of strategy innovation. Eur Manag J 19(4):424–429

Christensen JF (2011) Industrial evolution through complementary convergence: The case of IT security. Ind Corp Chang 20(1):57–89

CLIB2021 (2015) Internationale Vernetzung von NRW, Flandern und Niederlande wird durch CLIB2021 weiter ausgebaut. http://www.clib2021.de/news/news-clib2021?id=275

Cohen WM, Levinthal DA (1990) Absorptive capacity: a new perspective on learning and innovation. Adm Sci Q 35(3):128–152

CropEnergies (2016) Bioethanol: Dynamisches Wachstum. http://www.cropenergies.com/de/Bioethanol/Markt/Dynamisches_Wachstum Accessed: July 2016

Curran C-S (2013) The anticipation of converging industries - a concept applied to Nutraceuticals and Functional Foods. Springer Verlag, London [u. a.]

Dammer L, Carus M (2015) RED reform: European Parliament agrees to cap the use of traditional biofuels. What are the impacts on the bio-based material sector? http://news.bio-based.eu/media/2015/05/15-05-11_PR_RED_Reform_Material_Sector_nova.pdf

Daurer S, Molitor D, und Spann M (2012) Digitalisierung und Konvergenz von Online-und Offline-Welt. Z Betriebswirt 82(4):3–23

Duysters G, Hagedoorn J (1998) Technological convergence in the IT industry: the role of strategic technology alliances and technological competencies. Int J Econ Bus 5(3):355–368

Europäische Kommission (2008) Eurobarometer Spezial 295: Einstellungen der Europäischen Bürger zur Umwelt. http://ec.europa.eu/public_opinion/archives/ebs/ebs_295_de.pdf. Accessed July 2016

EU-Kommission (2014) Towards a circular economy: a zero waste program for Europe; COM 398: 1–14

Europäische Kommission (2016) ANHANG 1 des Vorschlags für einen Beschluss des Rates über den Abschluss im Namen der Europäischen Union des auf der 21. Konferenz der Vertragsparteien des Rahmenübereinkommens der Vereinten Nationen über Klimaänderungen vom 30. November bis zum 21. Dezember 2015 in Paris angenommenen Übereinkommens. Brüssel, 10.6.2016 COM(2016) 395 final. https://ec.europa.eu/transparency/regdoc/rep/1/2016/DE/1-2016-395-DE-F1-1-ANNEX-1.PDF

European Biofuels Technology Platform (2016) Bioethanol use in Europe and globally. Online im Internet: http://www.biofuelstp.eu/bioethanol.html. Last accessed 20th July 2016. Accessed: July 2016

Fahrni F (2008) Konvergenz als Herausforderung für neue Wertschöpfungsketten und Geschäftsmodelle. In: Informationelles Vertrauen für die Informationsgesellschaft. Springer, Berlin/Heidelberg, pp 347–356

Guenther-Lübbers W, Theuvsen L (2015) Regionalwirtschaftliche Effekte der Biogasproduktion: Eine Analyse am Beispiel Niedersachsens. Berichte über Landwirtschaft-Zeitschrift für Agrarpolitik und Landwirtschaft 93(2). https://doi.org/10.12767/buel.v93i2.74.g197

Hackelöer K, Kneißel B (2015) Kohle stofflich nutzen. Nachrichten aus der Chemie 63(10):1006–1008

Hacklin F (2008) Management of convergence in innovation, 1. Auflage, Heidelberg, Deutschland: Physica-Verlag

Harms AE (2003) Untersuchungen zum Futterwert von expandierten Trockenschnitzeln sowie von Vinasse beim Rind. Inaugural-Dissertation, Tierärztliche Hochschule Hannover

Heinrich-Böll-Stiftung (2015) Kohleatlas: Daten und Fakten über einen globalen Brennstoff. 1. Auflage, Berlin. https://www.boell.de/sites/default/files/kohleatlas2015.pdf

IBB Netzwerk GmbH (2016) CO2 vernetzt – vom Klimaschädling zum gefragten Rohstoff. http://www.pressebox.de/pressemitteilung/industrielle-biotechnologie-bayern-netzwerk-gmbh/CO2-vernetzt-vom-Klimaschaedling-zum-gefragten-Rohstoff/boxid/808099

Infraserv Höchst (2016) Kaskadennutzung im Industriepark Frankfurt-Höchst

Ißbrücker C, von Pogrell H (2013) Bio-basiert, bioabbaubar oder beides. Nachrichten aus der Chemie 61(10):1037–1038

Kaplinsky R, Morris M (2001) A handbook for value chain research, vol 113. IDRC, Ottawa

Karvonen M, Kässi T (2011) Patent analysis for analysing technological convergence. Foresight 13(5):34–50

Kircher M (2015) Sustainability of biofuels and renewable chemicals production from biomass. Curr Opin Chem Biol 29:26–31

Kollmuss A, Agyeman J (2002) Mind the gap: why do people act environmentally and what are the barriers to pro-environmental behavior? Environ Educ Res 8(3):239–260

Kroner WG (2015) Deutschland an der Schwelle zur Bioökonomie. Nachrichten aus der Chemie 63(9):917–918

Lanzatech (2014) Piloting for Woody Biomass Syngas to Fuel & Chemicals Plant. http://www.lanzatech.com/facilities/

Möhrle M, Isenmann R (2008) Technologie-Roadmapping: Zukunftsstrategien für Technologieunternehmen. Springer Fachmedien Wiesbaden

Nyström A (2008) Understanding change processes in business networks: a study of convergence in Finnish telecommunications 1985–2005. Åbo Akademi University Press, Turku

Pelzer S (2012) Maßgeschneiderte Mikroorganismen. Online im Internet: https://www.brain-biotech.de/blickwinkel/technologie. Accessed: Juli 2016

Phaal R, Farrukh C, Probert D (2004) Technology Roadmapping – a planning framework for evolution and revolution. Technol Forecast Soc Chang 71:5–26

Pichler S (2013) Alles Pro für Proteine? Eiweiß in der Humanernährung. Journal für Ernährungsmedizin 15(1):20–23

Piotrowski S, Carus M, und Carrez D (2016) European bioeconomy in figures. Ind Biotechnol 12(2):78–82

Rosenberg N (1963) Technological change in the machine tool industry, 1840-1910. J Econ Hist 23(4):414–443

Schöß MA, Redenius A, Turek T, Güttel R (2014) Chemische Speicherung regenerativer elektrischer Energie durch Methanisierung von Prozessgasen aus der Stahlindustrie. Chemie Ingenieur Technik 86(5):734–739

Siemens (2016) Künstliche Fotosynthese – aus Kohlendioxid Rohstoffe gewinnen http://www.siemens.com/innovation/de/home/pictures-of-the-future/forschung-und-management/materialforschung-und-rohstoffe-co2tovalue.html

Song CH (2016) Früherkennung von konvergierenden Technologien. Springer Fachmedien Wiesbaden

Stern P (2000) New environmental theories: toward a coherent theory of environmentally significant behavior. J Soc Issues 56(3):407–424

Styczynski ZA, Stötzer M, Lombardi PA (2014) Primärenergien. In: Dubbel. Springer, Berlin/Heidelberg, pp 783–795

Swiaczny F, Schulz R (2009) Wachstum der Weltbevölkerung und nachhaltige Tragfähigkeit. J Verbr Lebensm 4(2):136–144

Teece DJ (2000) Managing intellectual capital: organizational, strategic and policy dimensions. (Clarendon Lectures in Management Studies). Oxford University Press, Oxford, NY

Tullo AH (2012) Coke plays spin the bottle. Chem Eng News 90(4):19–20

USDA (2013) Corn for Grain 2012 production by county and location of ethanol plants as of 2013. Online im Internet: https://www.nass.usda.gov/Charts_and_Maps/Ethanol_Plants/U._S._Ethanol_Plants/EthanolPlantsandCornProdUS. Accessed Jul 2016

xStudy (2016) Study in Europe: find bachelor and master programmes. Online im Internet: http://xstudy.eu/. Accessed: July 2016

Criteria for the Success of the Bioeconomy

Stefanie Bröring, Chad M. Baum, Olivier K. Butkowski, and Manfred Kircher

Partial view of a distillation column for bioethanol in Brazil (© Usina Nova Gália Ltda.)

J. Pietzsch (ed.), *Bioeconomy for Beginners*, https://doi.org/10.1007/978-3-662 60390-1_8

The establishment of bioeconomic value chains has implications for companies, regions, jobs and consumers. The most important factors on which the successful transformation into a sustainable bioeconomy will depend are raw material supply, technological progress, production costs, ecological sustainability, and social acceptance. The transition from the fossil-based economy to the bioeconomy will take decades, especially since, at this point in time, most bio-based value chains remain in competition with their fossil-based counterparts. It must be borne in mind, however, that not all areas currently dominated by the fossil-based economy will be replaced by bio-based processes. In the energy industry, for example, non-bio-based processes must also pursue a greater degree of sustainability, for example, through the use of wind, water, and solar energy. The transition to more bio-based forms of economy must therefore be oriented towards three specific dimensions of sustainability, which together take into account the potential for increasing competition of numerous economic sectors for scarce biomass resources. This results in fundamental conflicts of objectives that must inevitably be resolved if the transition is to succeed. The two most important general prerequisites for a successful transition to future bio-economies are therefore contingent on solutions of this nature being attained. After all, new innovations, whether products or process technologies, must be competitive if they are to attain a foothold in the market. This means that innovations in both the business-to-business area (B2B) and the business-to-customer area (B2C) also require the active involvement of customers.

8.1 Resolving Conflicts Among Sustainability Goals and the Relevance of Eco-innovations

Attainment of a more sustainable form of life must include economic, social and ecological objectives. The transition to a bio-based economy is accordingly oriented in a similar fashion.

The bioeconomy can make important contributions to **economic sustainability**. It has the potential to maintain the international competitiveness of those sectors concerned while also facilitating improvements in productivity, innovation and resource efficiency. Nevertheless, for such goals to be feasible, innovative, bio-based, and sustainable products and production methods are needed, not least to protect European economies from being "vulnerable and insecure to dwindling supplies and volatile markets" during the envisioned transition away from the broad reliance on fossil fuels (EC 2012).

With regard to **social sustainability**, the bioeconomy further promises to create jobs at local and regional levels. This helps, on the one hand, to maintain social standards and reduce poverty. What is more, there is a further opportunity for fairer distribution of economic benefits, particularly through the greater economic development of rural and coastal areas. For instance, the productive use of by-products such as plant residuals and wheat straw as crucial raw materials for bio-based functional ingredients or packaging can lay the foundation for more regionally integrated networks (EBP 2014; EC 2014).

The bioeconomy also offers the opportunity of no longer having necessarily to rely on fossil resources to meet the growing global demand for food, bioenergy, textiles and other end products (OECD 2009). It can thus (partly) mitigate the consequences of climate change while also improving resource efficiency. The third and final essential dimension of the bioeconomy is therefore ecological in nature, with specific regard to the closing of material cycles between, for instance, suppliers and buyers in a more sustainable fashion (EC 2015; ▶ Chap. 7). Accordingly, a bioeconomy oriented towards improving sustainability can be expected to make crucial contributions within the dimension of **ecological sustainability** as well.

On the other hand, as a result, all of those resources allocated to bio-based sectors are not available for other uses of value creation. Accordingly, when it comes to resource and biomass usage or the means of production more generally, companies in the bio-based sectors are not only in competition with their fossil-based counterparts, but also with one another – whether within a particular sector or, indeed, across all of the different sectors that comprise the bioeconomy. The need to apportion existing biomass resources among many different application areas therefore results in another kind of goal conflict, though this time in relation to the specific needs and challenges of societies to which priority is assigned. One relevant example here is the potential conflict between the objective of food security and that of the energy-related use of biomass. Generally speaking, goal conflicts can therefore be said to arise whenever the achievement of one goal results in a more limited achievement of another.

In order to solve goal conflicts of this nature, principles for the bioeconomy must be established for the purpose of guiding the transition to a bio-based economy. These principles, reflecting the particular needs and challenges of a society, specifically define the priorities – or, perhaps, rules – according to which firms, stakeholders, and other actors then make their decisions. A good example of how such priorities might look is the overarching principles set out by the *Standing Committee of Agricultural Research* (SCAR). As an important advisory body for research and innovation policy in the European Union, the SCAR's proposals have assumed a pioneering role in the transition to a bio-based economy. Within this framework, the fact that food security and sustainable yields are given higher priority than energy-related uses of biomass provide one potential solution that would allow the bioeconomy to resolve its conflicts among sustainability goals, not only in the near future, but in the longer term as well.

SCAR (2015): Five Principles for the Bioeconomy

1. Food first: assign priority to food security above other goals
2. Sustainable yields: ensure that the amount harvested does not exceed or impair the potential for regrowth

3. Cascading approach: biomass should first be assigned to the highest-value use
4. Circularity: take specific care to reduce, reuse, and recycle production waste
5. Diversity: pursue diversification in the output, scale, processes and technology of production

As such, these principles highlight one way in which the bioeconomy can be successfully implemented that is compatabile with the broader functioning of a free market economy. More crucially for the purposes of the present chapter, their application also demonstrates why **eco-innovations** could represent a solution for overcoming the possible goal conflicts between different objectives (► Excursus 8.1). The concept of eco-innovation is defined differently in the literature. According to Kemp and Pearson (2007), an "*eco-innovation is the production, assimilation or exploitation of a product, production process, service or management or business method that is novel to the organization (developing or adopting it) and which results, throughout its life cycle, in a reduction of environmental risk, pollution and other negative impacts of resources use (including energy use) compared to relevant alternatives.*" By comparison, Ekins (2010) refers to an eco-innovation as "*a change in economic activities that improves both the economic and the environmental performance of society.*" This definition shows the relevance of eco-innovation as a potential means to resolve goal conflicts (■ Fig. 8.1) among the various dimensions of sustainability (Hasler et al. 2016).

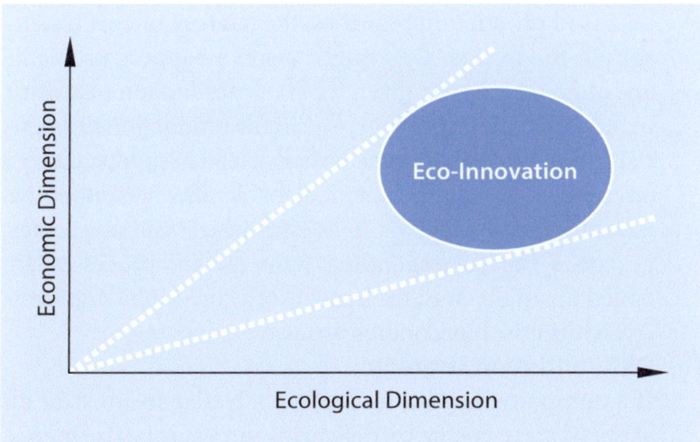

■ **Fig. 8.1** Eco-innovation as a means to resolve goal conflicts among dimensions of sustainability

8.2 Competitiveness

No matter the particular product application that one envisions, the establishment of the bioeconomy depends, to a large extent, on one criterion: whether or not it is competitive with the established product offerings currently on the market. For this reason, the starting point for such discussions is the economic status quo, that is, the predominantly fossil-based products and technologies that are widely entrenched. Idealistic motives alone can therefore contribute relatively little to the ultimate success of bioeconomic products and an effective transition to a more bio-based economy – unless competitiveness is broadly attained. What this could entail is described below.

8.2.1 A Theoretical Perspective on the Competitive Advantages of the Bioeconomy

Like companies in all sectors of the economy, companies in the bioeconomy also strive to gain a competitive advantage over their competitors. Generally speaking, this involves the following: "*A firm is said to have a sustained competitive advantage when it is implementing a value creating strategy not simultaneously being implemented by any current or potential competitors and when these other firms are unable to duplicate the benefits of this strategy*" (Barney 1991). The ability to implement a value-adding strategy that cannot be simultaneously implemented by competing companies, nor imitated by them in the short tem, is therefore a prerequisite for a sustainable competitive advantage to be attained. From the perspective of a firm, the existence of and emphasis on an inherent strategic orientation that guides all entrepreneurial decisions and activities is thus at the core of the pursuit and maintenance of competitive advantage in a general sense.

From a resource-oriented perspective (Penrose 1959; Wernerfelt 1984), the competitive advantages of firms are linked to their ability to adapt dynamically to changing market conditions (Prahalad and Hamel 1990; Teece et al. 1997, Teece 2016). Those much-quoted *dynamic capabilities* are therefore also indispensable within the dynamic environment of the bioeconomy (see Teece et al. 1997, 2016). With the help and application of such skills, bio-based companies have the potential to generate (and maintain) above-average

Excursus 8.1 Phytase as Eco-innovation

The typical goal conflict between the ecological and economic dimensions of sustainability can be resolved, in principle, through eco-innovations. One example of an eco-innovation that can both increase the competitiveness of bio-based products and benefit the environment is the enzyme phytase. This enzyme is widely used in animal feed to increase the bioavailability of phosphorus, which is important, given that phosphorus is an essential nutrient for building bones. A crucial component of the naturally occurring phosphorus in plants is stored as phytate or phytic acid. However, if phosphorus is in the form of phytate, it cannot be digested by monogastric animals such as poultry and pigs, and, as a result, much of this phytate is ultimately excreted and ends up as an environmental pollutant. However, if phytase itself were to be used for feeding purposes, it would be possible not only to improve feed conversion rates (e.g., efficiency of feed) (► Sect. 2.1), but also increase profitability while decreasing resource use – in sum, with the overall impact on the environment thereby positively influenced as decreasing levels of phytate are excreted.

profits. With specific regard to their competition with fossil-based companies, moreover, this would entail that they no longer need to abide strictly by their rules, but rather, by seeking to implement their new strategies, could disruptively engage in the creation of new rules for competition. As a result, such a *modus operandi* would enable bio-based firms to challenge established technological platforms and call into question the effectiveness and desirability of existing market practicies (Christensen 1997; Nameroff et al. 2004).

And yet, it is broadly true that the standards that currently dominate within a particular market necessarily determine the rules of competition. Given that many fossil-based products and technologies have assumed the status of **dominant designs** across the economy as a whole (see Utterback and Abernathy 1975), it is necessary to consider how this constrains the potential opportunities and strategies available to proponents of the bioeconomy. In the first place, this implies that bio-based products perhaps ought not initially to attempt to compete against the dominant design, not least because of the advantages of incumbency that this status affords. Establishment of a product as the dominant design, as detailed in the previous paragraph, bestows advantages, not just in relation to economies of scale, established customers, supplier networks, and the like, but, moreover, in the specific terms of competition that are set – and even what it means for a firm to be "competitive" – in a sector. Once established, the fact that this particular design is taken up as the industry standard thus institutes additional, technology-specific barriers to market entry, both in the form of technical specifications that need to be met or market expectations for clients, consumers, and users by which new products are inevitably measured. Irrespective of the particular advantages of bio-based offerings, whether actual or prospective, there is the potential that such products will be evaluated on terms that are not wholly favorable to them: namely, those that were instituted and emphasized by their fossil-based counterparts. Every attempted introduction of innovative products and technologies must therefore factor in the necessity of so-called **switching costs**, whereby, at least for a time, such products have to beat existing products "at their own game." With respect to the bioeconomy, there must therefore be significant incentives or clear reasons that would lead potential customers to entertain the switch to bio-based products or processes.

Against this background, the question arises as to what strategic options are available to a company seeking to survive and thrive in a particular competitive environment? Generally speaking, all of these, of course, involve the positioning of a company vis-à-vis its competitors in the relevant market. Nonetheless, if one has a look at real-world examples, it is readily apparent that companies pursue a range of different strategies when aiming toward competitive advantage. Accordingly, Porter (1980) proposed a classification scheme of "generic strategies" from which companies can choose (◘ Fig. 8.2). These three strategies differ along two dimensions: (1) in terms of the underlying basis of the competitive advantage (uniqueness for customers vs. cost advantage) and

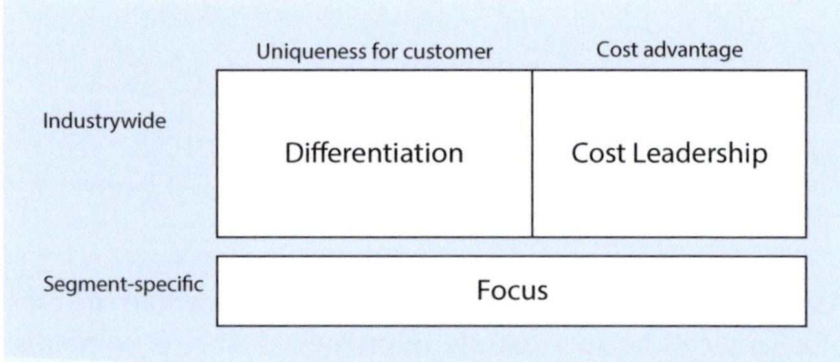

◘ **Fig. 8.2** Generic strategies of competitive advantage. (Following Porter 1980)

(2) the size and scope of the targeted market, that is, industrywide vs. segment-specific:

(a) **Cost-leadership strategy:**
According to this strategy, the key to achieving competitive advantage is linked with the pursuit and attainment of a (relatively) superior cost structure. On the basis of such an advantage, the product or service of the company can be offered to the whole market at a lower price than that of the competition. This strategy is therefore ideal within those market contexts in which products are sufficiently standardized, thereby allowing for greater production volume and higher overall efficiency. In this regard, it is also relevant to keep in mind the empirical demonstrations of the learning curve, whereby per-unit production costs are found to decrease by 30% with each doubling of the production volume (Henderson and Gälweiler 1984). Due to the advantageous cost structures of the established crude oil-based production branches, the strategy of cost leadership in the bioeconomy might therefore appear problematic at first. However, given that new production platforms in industrial biotechnology, e.g., in the production of essential amino acids for animal feeds, can lead to significant cost advantages, there is the potential for significant competitive advantages to emerge vis-à-vis established synthesis routes, i.e., once the corresponding fermentation processes are scaled up. In this way, the benefits of a cost-leadership strategy within the bioeconomy are also demonstrated.

(b) **Differentiation strategy:**
If a company decides to become a leader in most or all areas of an industry by delivering superior performance in relation to a particular attribute (e.g., higher product quality or improved service offerings), it might instead pursue a strategy of product differentiation. In doing so, the firm strives to distinguish itself from its competition by pursuing quality-related improvements wherever feasible and desirable – or perhaps focusing mostly on those domains most likely to be meaningful and impactful in the minds of its consumers and clients. At the same time, it must also seek to minimize costs in areas that are not relevant for its ability to differentiate itself. Here, food packaging made of polylactide (PLA) is an example from the context of bioeconomics. With regard to its performance as a packaging material (impermeability, weight, etc.), PLA hardly differs from the fossil-based alterna-

tives, even if the overall carbon footprint is significantly lower. Consequently, if its customers acknowledge it as a crucial feature of overall product performance, the company can utilize this innovation as a way to set itself apart from its competitors. If viewed as a peripheral aspect, however, then PLA-based packaging might simply come to be seen as a kind of "window dressing" – leaving aside the question of how such an innovation may affect the final product costs. This example thus demonstrates how a firm wishing to pursue a differentiation strategy must have a good knowledge of its customer base and, as a key aspect of its business operations, engage with and communicate the relevance of such differences for the activities and offerings of their clients. If this is achieved, further differentiation could also take place in relation to the raw materials used for PLA: e.g., there are producers of PLA presently striving for production on the basis of lignocellulose, which, as a raw material, has the further advantage of not being suitable for food consumption.

(c) **Focus strategy:**
If a company aims to be competitive in only one or a few areas of its particular industry, it can adopt a focus strategy limited to maximizing its advantages in a select number of customer segments. Within this chosen niche, it can either offer its products at a lower price than the competition, i.e., by pursuing a cost focus, or, instead, distinguish itself by means of superior quality performance, i.e., by adopting a differentiation focus. Such a strategy could be especially suitable for those situations marked by a significant level of competition and/or those in which it proves quite difficult to grab a large share of the market. Take the example of the Brazilian firm Braskem, which decided to invest in a bioethylene plant able to produce basic bio-based chemicals from ethanol, even in spite of the enormous difficulty of competing with oil refineries on the basis of a cost-leadership strategy. Similarly, the firm Energochemica also announced that trial operations would begin at its new cellulosic ethanol plant in Slovakia by the end of 2018. In both cases, the focus of such investments on a relatively niche market enables the firms to gain a foothold and, despite the higher costs involved with producing bio-based ethylene, to begin working their way up the "learning curve." In taking the opportunity to gain experience with the new raw materials and discover and implement improvements in the transformation processes, the firms become better able to optimize their costs and, perhaps, to eventually scale up production and pivot towards relevance in the broader market(s).

8.2.2 The Status Quo: The Established Fossil-Based Economy

The fossil-based economy is essentially based on oil, gas and coal. These carbon sources almost exclusively contain carbon (Table 8.1); their annual consumption is equivalent to a total of 11 billion tonnes of carbon per year.

Table 8.1 Composition (%) and calorific value (MJ/kg) of coal, oil, gas

	C	H	N	O	MJ/kg
Natural gas	75–85	9–24	Traces	Traces	32–45
Mineral oil	83–87	10–14	0,1–2	0,5–6	43
Hard coal	60–75	6	Traces	17–34	25–33
Lignite	58–73	4,5–8,5	Traces	21–36	22

Source: Kircher (2016)
C carbon, *H* hydrogen, *N* nitrogen, *O* oxygen

More than 95% of this is used to generate various forms of energy: electricity, heat, fuel and the operation of energy-intensive industries. Worldwide, 3.9 billion tonnes of oil, the most important raw material in the chemical industry, are produced every year. With a carbon content of 85%, this corresponds to 3.3 billion tonnes of carbon. The lion's share of this, 92%, is devoted to energy generation. Only 8%, or 300 million tonnes, of carbon is utilized for chemical products. Oil production is controlled by very large companies: Number one is the state-controlled Saudi-Aramco, which accounts for 11% of world production, with over 400 million tonnes of oil per year. The largest private company is Exxon Mobil (USA), which ranks fifth.

Petroleum is produced in only a few regions of the world. 52% of all recoverable resources are located in the Middle East and North Africa, 20% in Latin America and the Caribbean and 13% in North America (World Energy Council 2011). Accordingly, the main oil-exporting nations are Saudi Arabia, Russia and Iran. The largest oil consumers, on the other hand, are the USA, China and Japan. A very efficient logistical infrastructure has developed between producers and consumers, because oil is liquid, and therefore easy to handle, has a high carbon and energy density, can be stored, and has a relatively homogeneous composition regardless of its origin. 60% of crude oil is transported by ship between seaports. 40% is pumped overland through pipelines. In Germany, for example, 2400 km of pipelines have been laid, connected to the ports of Rotterdam (Netherlands), Genoa and Trieste (Italy) and to the oil wells in Adamovo (Russia; 3000 km away). The leading clusters of the chemical industry have emerged where the logistical connection to the oil supply is guaranteed: The ARRR region (Antwerp-Rotterdam-Rhine-Ruhr) in the Netherlands, Flanders in Belgium, North Rhine-Westphalia (Germany), Houston (USA), Shanghai (China), Jurong (Singapore) and Jubail (Saudi Arabia).

In the refineries, crude oil is refined into liquid gas (propane, butane), fuel (petrol, diesel), naptha, heavy oil and bitumen. This means that the value chain for gases and fuels is very short: The raw material oil is extracted and then transported to the refinery, and, from there, directly to distribution to the end consumers:

Production → Refinery → Distribution → Consumption

The long-chain hydrocarbons bitumen and, especially, naphta are the raw materials for the chemical industry. They are split in the cracker into shorter alkanes, cycloalkanes and aromatics. The manageable number of around 300 refinery products is the starting material for the enormous variety of organic basic, fine and specialty chemicals. The most important basic chemicals in terms of volume are ethylene (>150 million tonnes) and propylene (85 million tonnes), which are obtained from crude oil or natural gas through refining, with a raw material yield of around 95%. Basic chemicals are chemically processed into chemical products, further into components, and finally into the end products that we deal with in everyday life. These are, for example, lubricants, detergents and reflective road markings. Such products have a specific function, and thus an added value. Chemical products therefore achieve, on average, a sevenfold higher added value than energy sources. With their much longer material value chain compared to energy recovery, their job potential is also much greater:

$$\text{Production} \rightarrow \text{Refinery} \rightarrow \text{Cracker} \rightarrow \text{Ethylene} \rightarrow \text{Ethylene derivative} \rightarrow \text{Component} \rightarrow \text{End product} \rightarrow \text{Distribution} \rightarrow \text{Consumption}$$

The oil industry is thus characterised by very efficient logistics, large refineries and chemical clusters, which, in the sense of Porter's competitive strategies, provides it with the optimal basis for a strategy of cost leadership. It thus has considerable competitive advantages and is characterised by high market entry barriers.

8.2.3 Challenges and Requirements for a Competitive Bio-economy

8.2.3.1 Logistics

Just like the oil economy, the bioeconomy can supply energy, fuel and chemicals. Its raw material, biomass, however, has a much lower carbon and energy density than oil. Without drying, it contains water, and is therefore biologically unstable. In addition, it is chemically very complex and its composition varies depending on its plant origin (see ◘ Table 8.2).

◘ Table 8.2 Composition (%) and calorific value (MJ/kg) of biomass and components

	C	H	N	O	MJ/kg
Biomass	45	6	2	42	6,8
Wood	50	6	3	41	14,4–15,8
Glucose	40	7	0	53	15,6
Sucrose	43	6	0	51	16,5
Vigour	44	6	0	50	17,5
Lignocellulose	44	6	0	50	10–25

Source: Kircher (2016)
C carbon, *H* hydrogen, *N* nitrogen, *O* oxygen

Also, with regard to biomass, certain regions of the world are particularly productive. For sugar production, for example, we have the state of São Paulo in Brazil. In the USA, the state of Iowa harvests 20% of worldwide corn. More than 20% of the world's wheat is grown in Russia, Ukraine and Kazakhstan. Malaysia is a leading producer of palm oil. Even if there are particularly productive raw material regions in the bioeconomy, agricultural production, unlike oil production, is seasonally limited and spread over very large areas cultivated by numerous independent farmers. The crop is transported from the arable land by tractor and truck, i.e., in batches of around 25 t, which limits the transport radius for further processing to around 50 km. The production plants must therefore not be built at too great a distance from the sites of biomass production (► Chap. 7). There are 350 sugar refineries in Brazil and 70 palm oil refineries in Malaysia. The bioeconomy will therefore develop a strong regional character wherever primary biomass is produced and processed. However, the capacity of biorefineries remains relatively low as a result, usually only reaching about 1% of the capacity of an oil refinery, which entails a considerable cost disadvantage. Only the extracts from these early processing stages exhibit a carbon and energy density that makes transport to more distant locations economically feasible. As in the fossil-based economy, producing and consuming regions are therefore not identical and must be linked by logistics, which are, however, much more cost-intensive. The challenges associated with the logistics around bio-based value chains are one of the major differences between the bio-based and fossil-based economies.

8.2.3.2 Availability of Biomass

Plant biomass consists of 40–55% cellulose, 10–35% hemicellulose and 18–41% lignin. From this, an average carbon content of around 45% can be estimated. The entire photosynthesis of nature fixes about 105 billion tonnes of carbon annually, which corresponds to about 210 billion tonnes of biomass per year. 14 billion tonnes of this biomass are produced agriculturally (equivalent to 7 billion tonnes of carbon) and used for food, fibre, chemicals and fuel products. The comparison of the amount of agriculturally produced carbon with today's carbon consumption from oil, coal and gas of 11 billion tonnes per year illustrates the dimension of the challenge that already exists for the provision of biomass as a raw material, in particular, taking into account the growing world population with increasing food demand (► Chap. 3).

8.2.3.3 Production Costs

Bio-based raw materials have a difficult position in direct competition with crude oil. This is illustrated by the example of ethylene explained above. For bioethylene, the route from sugar to bioethanol, which is catalytically hydrogenated into ethylene, is established. Technically speaking, the conversion from fossil-based to bio-based ethylene is not a problem. However, the extended value chain is leading to higher production costs:

Biomass production → Biomass harvesting / storage / transport → Sugar refining → Fermentation → Ethanol → Ethylene → Ethylene derivative → Component → End product → Distribution → Consumption

In addition, the costs of the already costly logistics from agricultural land are increased by the fact that considerably more raw material has to be transported than oil. The reason for this is the generally lower raw material yield of bio-based processes. For example, the theoretical yield of ethanol from glucose is 51%, which, in turn, can be dehydrated into ethylene with a theoretical yield of 61%. In relation to the sugar originally used, the theoretical yield is therefore only 31% (losses resulting from the process are not taken into account):

- 100 kg of glucose produce 51 kg of ethanol and 49 kg of CO_2.
- 51 kg of ethanol yield 31 kg of ethylene and 20 kg of H_20.

If lignocellulose is used as a raw material instead of sugar, the value-added chain is further extended, as its digestion requires additional process steps and costs.

Biomass production → Harvesting / storage / transport of biomass residues → Biomass preparation → Biomass saccharification → Fermentation → Ethanol → Ethylene → Ethylene downstream product → Component → End product → Distribution → Consumption

An overview of the costs of lignocellulose-based ethanol can be found in ▪ Table 8.3. In order to be competitive with sugar-based processes, the costs of enzymatic hydrolysis must be offset by the cost advantages of the raw lignocellulosic material.

▪ **Table 8.3** Cost breakdown of the combined production of ethanol based on cane sugar and bagasse (cane sugar-lignocellulose)

Cost factor	Costs (USD/1000 l)	Cost share (%)
Resource	115	40
Enzymes	70	25
Maintenance	10	4
Staff	5	2
Cost of capital	5	2
Other costs	80	27
Sum	**285**	**100**

Source: Junqueira et al. (2016)

It should also be borne in mind that only cellulose and hemicellulose, which account for 70–80% of lignocellulosis, can be saccharified. 20 to 30% is lignin and, accordingly, the theoretical raw material yield decreases from 31% to about 23% compared to pure sugar.

8.2.3.4 Product Quality Requirements

Numerous bio-based products are successful on the market both because they are bio-based and because there are no fossil-based alternatives. These include enzymes, L-amino acids and enantiomerically pure active pharmaceutical ingredients. Yet bio-based products, which have only incrementally improved properties compared to their fossil-based competitors, can also successfully compete in the market if a strategy of product differentiation is pursued (► Sect. 8.2.1).

An example of this is Polyethylene furanoate (PEF) , a biopolymer that has proven itself within the larger beverage bottle market in competition with the established and fossil-based polymer polyethylene terephthalate (PET), due to its higher gas tightness and material stability (► Chap. 7).

Property	PEF compared to PET
Oxygen bareer	10-times superior
Carbon dioxide bareer	4-times superior
Water bareer	2-times superior

PLA (compare ► Sect. 8.2.1.; ► Chaps. 5 and 7) is another bio-based polymer that can compete with fossil-based polymers such as polyethylene (PE) and polycarbonate (PC). Its durability makes it suitable for car interiors and the housings of household appliances. The breathability of PLA fabrics makes them attractive for sportswear. The transparency and compostability of PLA is in demand for packaging materials. The argument in favor of PLA is particularly convincing when it comes to its ecological footprint (▪ Table 8.4).

▪ **Table 8.4** Ecological footprint of different polymers

Polymer	Footprint (kg CO_2 eq/kg polymer)	
	Petroleum-based	Bio-based
Polycarbonate (PC)	5,0	
Polystyrene (PS)	2,2	
Polyethylene terephthalate (PET)	2,0	
Polypropylene (PP)	1,7	
Polyethylene (PE)	1,7	
Polylactid (PLA)		0,5

Source: Corbion (2016)

8.2.3.5 Prioritisation of Bio-based Products

Against the background of the limited resources and complex logistics described above, the question arises as to the conditions under which the bioeconomy could displace the oil economy. A simple switch from raw fossil materials to raw agricultural materials will not be possible. This would require more than doubling today's agricultural production in order to obtain a sufficient amount of carbon sources. But which economic sectors are absolutely dependent on carbon and where are the alternatives? There is no doubt that the area of nutrition has top priority. In order to meet the challenge of food safety alone, a considerable expansion of the production volume would already be necessary – in addition to this, further demand would arise from the consumption of biomass for material and energy-related use. But for the energy sector, the largest consumer of raw fossil materials, there are other important sources that are independent of carbon, and therefore also of biomass. Worldwide, the use of renewable energies from the sun and wind is being promoted above all, with hydropower and geothermal energy also playing a role; even nuclear power, which is not accepted in Germany, is an option for some countries. With regard to the implementation of EU Directive 2009/28/EC on the promotion of the use of energy from renewable sources, Germany has set itself the target of covering 18% of its primary energy needs with renewable energies by 2020. Today, the share of renewable energies is 12.6%; slightly more than half of this is bio-based (BMU and BMELV 2010) (▫ Fig. 8.3).

However, there are no alternatives to carbon for material conversion in organic chemistry. The current demand for carbon for organic chemical products, at around 300 million metric tonnes per year, appears manageable compared with the global agricultural production of 7 billion metric tonnes per year. In addition, the added value and job potential of material recycling are much greater than those for energy use. For these reasons, it makes sense to prioritise the use of the limited resource of biomass for the production of bio-based chemistry and utilize it only after all possible material applications for energy-related purposes have been exhausted, thus following a cascading usage (see Kircher 2015; ▶ Chap. 7).

However, the spectrum of raw materials used in biogas production shows how far, for example, Germany is from optimal cascade utilisation of bio-based raw materials. 50% of the nation's biogas is produced from corn from agricultural production, and thus on raw material and land use that could also serve primarily for food and secondarily for material use. By contrast, current use of industrial, municipal and agricultural residues as substrates in biogas fermentation stands at around 5% (▫ Fig. 8.4).

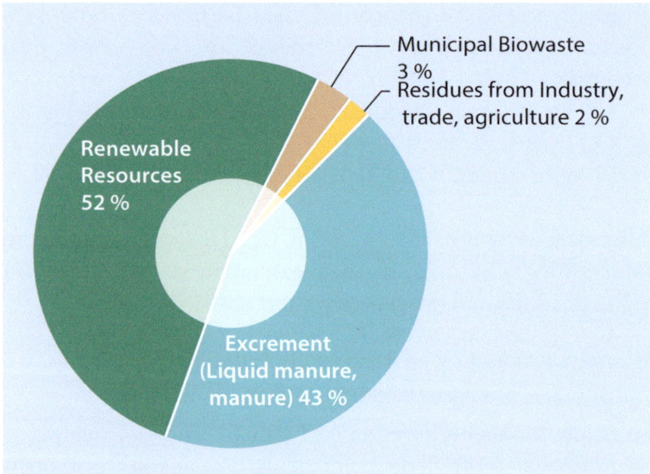

▫ **Fig. 8.4** Use of substrates in biogas plants in Germany in 2014 (mass in percent). (Source FNR 2015)

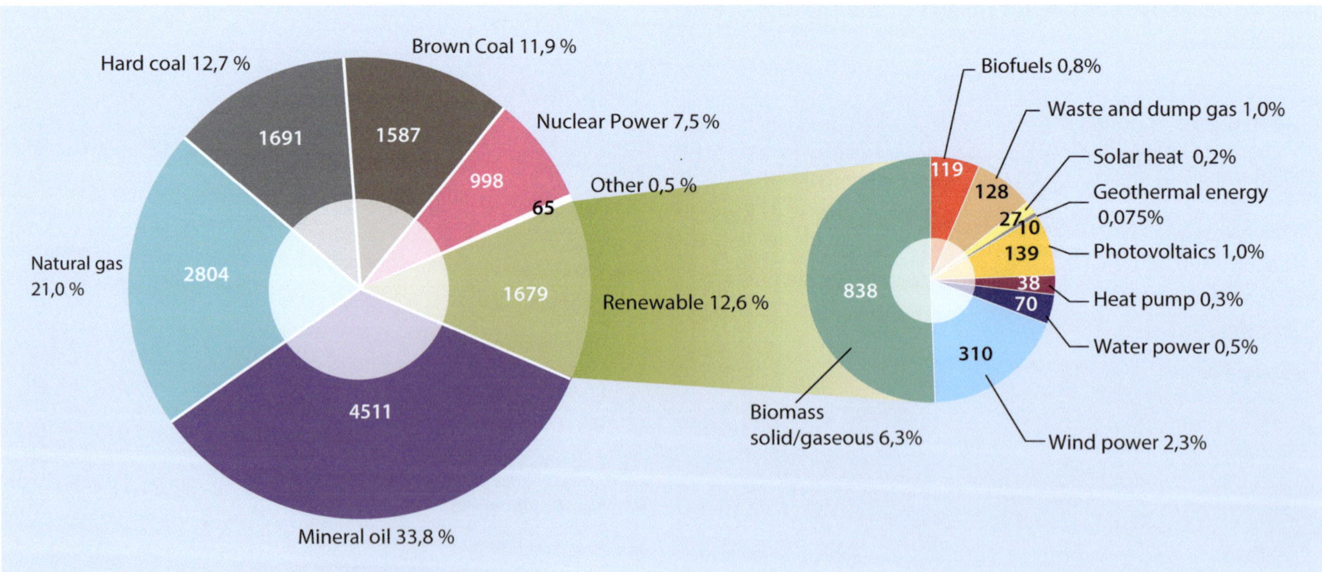

▫ **Fig. 8.3** Primary energy consumption in Germany by raw material source. (Source BMWi 2012)

8.2.3.6 Diversity and Efficiency of Raw Materials

If the current conventional fossil-based ethylene production in Europe of 24.5 million tonnes per year was to be substituted with bioethylene – given the relatively low yield described above – at least 79 million tonnes of sugar would be needed, a quantity that significantly exceeds the current annual sugar production of around 17 million tonnes in the European Union. Obviously, agriculture, as it is practiced today, can only cover part of the bioeconomic demand for raw materials. A greater diversity of raw materials and better raw material efficiency offer ways out of this dilemma. The needed biomass must be processed with a higher yield than before, and so must previously unused biomass fractions and waste streams also be made accessible for processing. Lignocellulose, the saccharification of which has been used in industrial practice for a few years, offers great potential. In Germany alone, the volume of wheat straw produced annually as a potential raw material is estimated at 8–13 million tonnes (DBFZ 2012). In Malaysia, leaves, trunks, empty fruit stalks and fruit peels with a mass of 70 million tonnes per year remain unused on palm oil farms and in oil mills. In Canada, the province of Alberta produces a third of Canada's wheat, rapeseed, flax, hemp and sugar beet. Every year, 64 million tonnes of previously unused agricultural waste and 2 million tonnes of forest residues accumulate there. In Russia, 35 million tonnes of residual materials from wheat processing and 150 million tonnes from wood processing could be used annually (Kircher 2012).

These few examples show that agriculture and forestry still have large reserves of biomass to be exploited. Marine resources, which cannot be discussed here in greater detail, also have potential.

Moreover, gas fermentation has the potential to produce further raw materials by using synthesis gas based on biomass or municipal waste and gaseous carbon sources such as CO emissions from steel mills and CO_2-emissions from power and cement plants. If it were possible to expand the application spectrum of gas fermentation accordingly (▶ Chaps. 5 and 7), then:

- the raw material efficiency of lignocellulose, in particular, would be improved by up to 30%,
- it would represent an entry into the carbon circular economy, in which industrially processed carbon is recycled directly as a part of the technical process, rather than via plant photosynthesis after its emission into the atmosphere,
- it would relieve agriculture of the burden of producing raw industrial materials,
- both fossil and bio-based carbon could be used industrially. This is an advantage that should not be underestimated in the long transition phase from the fossil-based economy to the bioeconomy, because it would also attract companies that still deal with raw fossil materials today as investors in processes that will be applied in the bioeconomy in the long term. In the transition phase that is beginning, processes that can work with fossil carbon sources today and with bio-based carbon sources tomorrow should be particularly promoted.

8.2.3.7 Land Use, Biodiversity and Environmental Protection

The increasing use of bio-based raw materials goes hand in hand with an intensification of agricultural and forestry activities that trigger land use, as well as ecological sustainability conflicts. The environment must not be further polluted by the various forms of agricultural production. Agriculture accounts for up to 30% of global greenhouse gas emissions (BMBF 2009) and biodiversity must be preserved (▶ Chap. 9). The following examples show how bioeconomic products and processes can contribute to relieving land use conflicts:

A total of 33% of the world's agricultural land is used for the production of animal feed (Steinfeld et al. 2006). Increasing the efficiency of its use by the animals thus has a direct influence on the land requirements. The essential amino acids play a key role here, because farm animals cannot synthesise these amino acids themselves, but have to ingest them with their food. The amino acid profile of the vegetable feed does not correspond to that of the animals. They have to absorb food until their need for the mostly limiting amino acid is covered. The resulting problem is illustrated in the picture of the Liebig barrel (◘ Fig. 8.5).

The animals inevitably absorb an abundance of other amino acids, but are unable to utilize them in their metabolism and excrete them. This leads to a high nitrogen load in the slurry. Above all, the feed is only incompletely converted into animal biomass. This reduced raw material efficiency can be improved by supplementing the feed with limiting amino acids, because, in this way, the feed is adapted to the amino acid requirements of the animal. The most limiting amino acid is L-methionine. It can be fed with fish meal, which has a high proportion of this amino acid. Alternatively, DL-methionine is used, which saves 54 kg of fish meal per kilogram. DL-methionine is chemically synthesized, and is therefore produced as a racemate (mixture of the D and L forms). Using the racemate as feed is possible because animals can convert the D-form into the required L-form using their own body's racemase. However, such a racemase is missing for other amino acids. Worldwide, 750,000 tonnes of DL-methionine are used as feed annually. The essential amino acids L-lysine, L-threonine, L-tryptophan and others, which are essential for the limitation to methionine, must be given in the L-form. Only 50% of the racemate is recycled. Since only the biotechnological production process provides the L-form, microbial fermentation remains the process of choice. Approximately 1.5 million tonnes of L-lysine are produced annually (IBVT/TU Braunschweig 2016), which, as an animal feed additive, saves 17 kg of soy meal per kilogram (Evonik 2014). If this quantity had to be fed exclusively in the form of soy protein, the theoretical area under cultivation

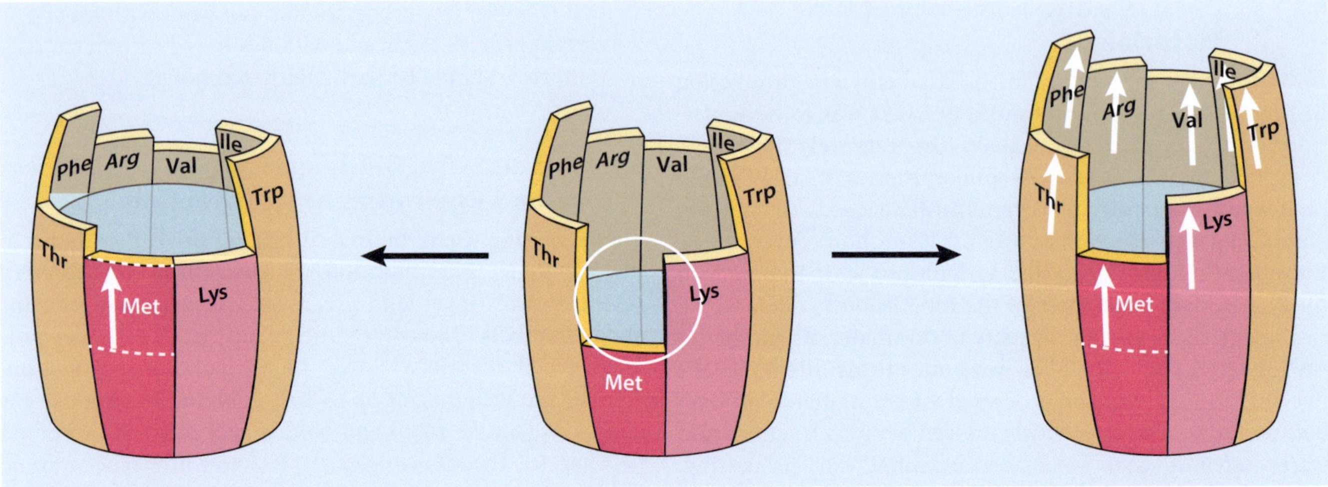

Fig. 8.5 Liebig's Barrel and the Law of Minimums. The length of the staves of the barrel in the middle corresponds to the average amino acid content of vegetable animal feed. For recovery, symbolized by the water level, L-methionine is limiting, followed by L-lysine. On the left, methionine is added, and the utilization of all amino acids increases. On the right, more feed protein is fed. The amino acid supply increases, but the utilization of the feed is not improved. (Source Evonik 2014)

8

would be around 70 million hectares. By way of comparison, the global soybean cultivation area is 110 million hectares (WWF Germany 2016). Another biotechnological product that increases the raw material efficiency of animal feed and reduces the environmental impact of animal husbandry is the enzyme phytase (▶ Excursus 8.1), which enables the utilisation of vegetable phosphate.

In summary, it can be stated that the competitiveness of the bioeconomy at different levels (technology, product, value chain) meets complex challenges and depends on many factors. A key factor for competitiveness at the product level – i.e., the success of bio-based products – is their acceptance at the level of customers, consumers and, ultimately, society as a whole.

8.3 Customer and Consumer Acceptance

The bioeconomy thrives on innovative activity and the capacity to develop innovations that can be linked together to foster more sustainable material and energy cycles. Nonetheless, for the full benefits of these innovations to be reaped, one crucial, though frequently overlooked, factor is the acceptance and adoption of innovations by manufacturers, firms, and consumers alike. "Acceptance" therefore signifies a paramount criterion for the success of the bioeconomy, and, moreover, one that must be fulfilled at all stages of a value chain. According to Rogers' oft-cited theory of diffusion, acceptance is also a crucial precursor for the adoption or implementation of innovative products and technologies – indeed, the extent to which acceptance takes place serves as a determinant not only of the degree of diffusion in a particular sector, but also the more general establishment (i.e., as a dominant design; see ▶ Sect. 8.2.1) of particular product applications or the technology itself on the market (Rogers 1983). Underlying technologies and their product applications are thus closely linked, as is especially evident from the questions and concerns surrounding "green" genetic engineering – that is, the application of biotechnologies for food and agricultural production and, as a result, potential restrictions relating to biomass production or even the bioeconomy as a whole. If consumers and the general public are broadly disposed against a technology (e.g., biotechnology) used for production, this can be expected to negatively impact the acceptance of the product itself, not to mention its potential sales and market share. If, however, consumers, as a result of their deliberations, both public and private, come to see the overall benefits of a technology as outweighing the risks, perhaps because the resulting products are easier to use or have superior environmental performance, then their evaluations of these products will likely be more positive and the likelihood of acceptance and adoption higher. Depending on the extent of adoption and acceptance at the micro-level, we can, moreover, see these distributed decisions represented with regard to the degree of diffusion at the broader level of economies and societies. In other words, as more individuals opt to purchase or use a product or technology, each on the basis of a process of reasoning that is likely to be idiosyncratic, we may see it become more diffuse and potentially dominant.

8.3.1 Foundations of the Acceptance and Adoption of Innovations

The adoption of innovative technologies can initially be easily clarified with the help of the notion of *degrees of innovativeness*. According to this scheme, *continuous innovations* refer to those that make incremental improvements to existing systems, for instance, by focusing on certain features or components of products. *Disruptive innovations*, on the other hand, are characterised by the utilization of novel insights, e.g., from research and development, in order to cultivate novel types of systems. However, owing to their more systems-level consequences and the greater overall potential for "disruption," the latter are likely to be attended by conflicts with existing value chains and technology platforms, namely, those that they seek to replace, not to mention questions of a more broadly political and societal nature. While the adoption of continuous innovations such as rapeseed press cake is therefore usually high, disruptive innovations are likely to be subject to the same scrutiny as genetic engineering. From the perspective of users and consumers, this is rather intuitive, given that products that are modified only slightly from those with which individuals are familiar, and therefore remain easy to understand, do not then require an extensive amount of evaluation, nor, indeed, do they have to be tried out and experimented with. In sum, the process and practice of acceptance and adoption is necessarily more complex for radical innovations than for incremental ones.

And yet, in spite of the substantial diversity and complexity of these processes, the majority of all studies on technology acceptance have tended to focus only on the last stage of the value chain, i.e., the end-user or consumer. As a result, there is a general understanding of the disparate factors that explain and influence the decision-making processes of consumers related to acceptance and adoption, e.g., benefits, risks, cost, trust, involvement, and subjective norms. By integrating these factors into a common framework or model, we are thereby able to better predict the conditions in which a particular behaviour (i.e., acceptance or adoption) is likely to occur, or, for that matter, the likelihood that consumers with certain attributes or characteristics will choose one option over another. For example, trust in political and social institutions is likely to be of particular importance for decisions regarding the use of biotechnology in food production, especially given the more controversial tenor of the surrounding discussions. It can be further supposed, on a more general level, that broad disparities in one's life circumstances, as reflected by such socio-demographic factors as gender and age, will also be impactful for consumer decision-making.

Many innovative products and production processes in the context of the bioeconomy have, however, not yet been introduced to the market. As a result, research into the acceptance of technology in such situations is necessarily hypothetical, which not only means that novel approaches are required, but also that the actual decisions and behavior of consumers may differ once these products and processes do come to market. Accordingly, in such hypothetical situations, the first thing to note is that observations of adoption and acceptance behaviour need not necessarily correspond to actions in the real world. Rather, these better reflect the promise or desire of consumers to engage in a given behavior under certain circumstances, i.e., a plan that may or may not be realized once other individual and contextual factors have had the chance to intervene. The concepts most often used to characterize and measure the likelihood that an individual behaves in a certain way consist of attitudes, preferences, and intentions (Kahneman et al. 1999):

- **Preferences** are evaluations about the relative desirability of two (or more) potential alternatives, and thus express the value an individual attaches to a specific object or event. In a real-world shopping situation, such alternatives correspond to actual products that the individual can decide to buy or not buy. It is also possible, however, for a preference to simply be 'stated' if a person is asked to identify a product that they would like or intend to buy. Stated preferences are increasingly utilized to explore 'hypothetical' choice situations, i.e., when products are not yet to market.
- **Attitudes** express the degree of favor or disfavor that is associated with particular objects of events (Eagly and Chaiken 1993). Unlike preferences, attitudes do not make a direct comparison among different alternatives, but rather reflect the degree of emotional valence (affective component) and the level of cognitive belief (cognitive component) that immediately emerge in regard to an object (Ajzen 2001).
- **Intentions** describe the existence of one's motivation in the sense of a conscious plan or the decision to make an effort to behave in a planned manner (Conner and Armitage 1998). Intentions are thus often used in consumer research as an intermediate link between attitudes and (stated) preferences on the one hand and actual behavior on the other (◘ Fig. 8.6).

One of the most frequently used models in the literature, as well as an example that features many of the variables outlined above, is the **Theory of Planned Behaviour**. This conceptual framework was first introduced by Icek Ajzen (1985, 1991) for the purpose of clarifying and predicting individual intentions, as well as the way in which these relate to specific purposeful behaviors. To do so, it utilizes three relevant factors: attitudes, subjective norms, and perceived behavioural control. The central idea here is that, by taking together a person's overall evaluation of a product (*attitude*), the felt perception of social standards (*subjective norms*), and their perceived capacity to behave in a desired way (*perceived behavioral control*), we can predict the likelihood that a certain behavior will occur, specifically, by being able to better

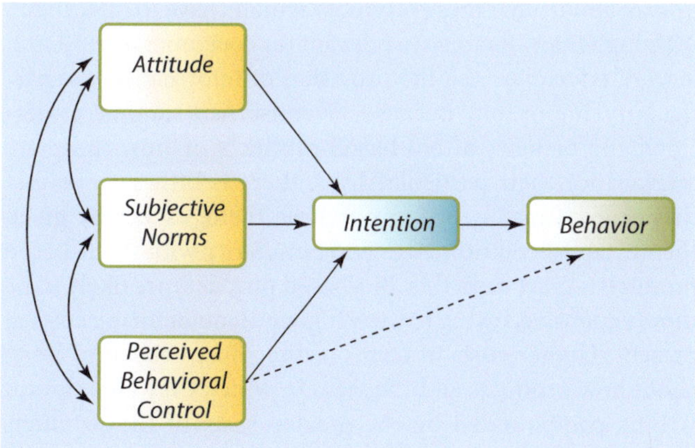

◘ **Fig. 8.6** Theory of planned behaviour. (Based on Ajzen 1991)

predict the strength of the conscious desire or aim (*intention*) to behave in a particular fashion.

Indeed, the Theory of Planned Behaviour has proven to be applicable across a variety of decision contexts, including health-related and self-improvement decisions (Ajzen and Madden 1986; Brewer et al. 1999) and those tied to sustainability and environmental protection (Bamberg and Schmidt 2003; Harland et al. 1999). In addition, this framework has been used to explain the adoption of decisions by agricultural producers (Beedell and Rehman 1999; Lynne et al. 1995). Neither attitudes nor intentions alone are sufficient, however, to predict the likelihood of behaviours or behavioural change. Notably, despite the increasingly positive attitude of the general public towards the protection of our environment, as well as the (stated) willingness of many individuals to undertake relevant action, the level of environmentally harmful emissions continues to increase (Bamberg and Möser 2007). This persistent absence of behavioural change, at least to the extent that might have been expected from the changes in individual attitudes and intentions, has been conceptualized as the *attitude-behavior gap* (Kollmuss and Agyeman 2002). Demonstrating the scope of this puzzle, as well as motivating further research to explain why such a disparity occurs, such gaps are evident both throughout the consumer acceptance literature and in relation to the adoption and implementation of new technologies by firms, organisations, and institutions.

8.3.2 Factors Influencing Consumer Acceptance

Consumer behaviour represents the ultimate outcome of a dynamic process that not only reflects the interplay of emotions, cognitive processes, and intentions but is also influenced by a range of individual and contextual factors. This section thus aims to clarify and provide insight into the factors that are important for consumer acceptance of novel products and technologies.

8.3.2.1 Product Attributes

Bio-based products are innovative in many ways. It is precisely for this reason that consumers are likely to be initially unsure about what these products actually have to offer them. Although many factors are relevant for consumer acceptance, level of education and the provision of information are particularly important. Because, if consumers cannot understand the benefits of bio-based products or how these are relevant for their particular lives, there is little chance that they will ultimately opt to purchase them, especially given their established knowledge and familiarity with fossil-based products. Apart from this, bio-based products are likely to be more expensive, owing to their higher manufacturing costs at present. Higher costs in terms of the effort needed to learn about new products and the ways in which they are relevant is thus compounded by the greater costs of the products themselves. In order to increase consumers' willingness to pay premium prices for bio-based products, one potential strategy might be to employ labels in order to better distinguish them. As demonstrated by their successful use in the growing market for environmentally-friendly products, such labels are able both to provide consumers with additional information about product quality and signal compliance with established quality standards (Carus et al. 2014).

However, information provision alone is not sufficient to convince consumers of the advantages of bio-based products. It may happen, for instance, that individuals actually find the promise of greater sustainability to have a burdensome impact on the fulfillment of other aspects of product quality. Taking the case of cleaning detergents, if the perceived "performance" or strength of a product is adversely affected by the fact of its being marketed as sustainable, then we might expect, somewhat paradoxically, that such products will be seen to be less valuable than more conventional detergents (Luchs et al. 2010). Generally speaking, this illustrates the consequences of those product benefits, e.g., environmental and health benefits, that cannot be directly verified for consumers. Whether a cleaning detergent that promotes itself as more sustainable or bio-based products that aim to make more efficient use of existing resources, the prominence of these credence attributes requires further consideration (Darby and Karni 1973; ◘ Table 8.5).

In contrast to the search and experience attributes (Nelson 1970, 1974) that consumers can directly evaluate, this is not possible for credence attributes, especially if they cannot directly access or evaluate the production process themselves. Rather, the capacity to determine whether or not such attributes are present would require a level of education and expertise that they usually would not have. However, as consumers come to place growing importance on health and sustainability, it is the existence and implications of exactly these kinds of credence attributes that must be taken into account (Cuthbertson and Marks 2007; Moser et al. 2011). Companies are thus making increasing use of quality labels, among other strategies, in order to communicate and persuade individuals that these attributes are present. These

◘ **Table 8.5** Consumer evaluation of product attributes

Search attributes	Experience attributes	Credence attributes
Evaluation *possible* before purchase or consumption (e.g., ripeness of fruit)	Evaluation *possible* on the basis of one's experience after product consumption (e.g., taste)	Direct evaluation by the consumer *not possible* (e.g., sustainability of overall production process, CO_2 emission during production)

Own creation based on Nelson (1970), Darby and Karni (1973)

include, for example, the EU Ecolabel, the EU organic logo and, in Germany, the Green Dot symbol ("Der Grüne Punkt") to designate products manufactured by companies participating in the recycling programme of the same name.

Insofar as consumers determine the information they receive through quality labels to be credible, it is then possible for these credence attributes to be translated into search attributes – in other words, one only need look for a specific label. In markets in which credence attributes proliferate, it is therefore possible for these types of labels, as well as the relative strengths of one labeling scheme over another, to serve as the basis for differentiating between products and firms in order to make purchasing decisions (Boehlje 2016; Cuthbertson and Marks 2007; Löbnitz and Bröring 2015). For this reason, the communication strategies of companies within the bioeconomy should seek clearly to communicate to potential customers the relative advantages, and thereby justify the premium prices of their products by explicitly exploring how to make the relevant credence attributes more visible so that they can be transformed into search attributes.

8.3.2.2 Perceived Risks and Perceived Benefits

Fundamentally, consumer acceptance has been found to depend on the ratio of expected benefits to perceived risks for a given product or technology (Cardello 2003; Lusk et al. 2004). As a result, one frequently asserted hurdle for promoting acceptance more generally is the degree of knowledge about the benefits of new technologies and, more specifically, how limited knowledge of this kind can create room for individuals to be more concerned about prospective risks and dangers. For this reason, one of the essential insights of the acceptance literature is that risk perceptions are not therefore independent from the other factors (Ueland et al. 2012).

Green genetic engineering, i.e., the application of these technologies to plant production, has often been referred to as the "black sheep" of biotechnology. In a representative survey by the European Commission, technology in general is viewed rather positively by the European population, while biotechnology and genetic engineering are also viewed in a positive fashion by 53% (Gaskell et al. 2010). In fact, only 20% of individuals expect this technology to have any negative effects on the standard of living in 20 years. However, when the topic turns to the explicit application of genetic engineering for purposes of food production, the picture changes quite a bit. At this point, a total of 61% of the population think that genetic engineering should not be used for such purposes, whereas only 23% would or could support such applications. As a further cause for skepticism, such attitudes have become slightly more reinforced since 2005. However, some perspective is required here, given that, if we set this attitude alongside other types of risk in the context of food production (such as pesticide contamination), genetic engineering is actually seen as less risky. What is more, if genetically modified foods are promoted and demonstrated to offer concrete benefits, the level of support has the potential to be even higher, perhaps even surpassing the degree of concerns in this context (Desaint and Varbanova 2013). So, it thus seems that decisions related to consumer acceptance ultimately depend on the interplay and exchange among the different constituent elements, not only in regard to perceptions of risks and benefits, but also the specific application area and prevailing societal and political context.

Further insight has therefore been gained into some of the other factors that matter for consumer acceptance. For instance, risk perceptions of some individuals are also influenced by their more general willingness to experiment with new types of food (Cox and Evans 2008; Pliner and Salvy 2006). Indeed, the broad relevance of this so-called neophobia has been demonstrated in regard to reservations about the use of nanotechnology (Matin et al. 2012; Schnettler et al. 2013) and genetically modified foods (Vidigal et al. 2015). In a similar vein, deeply rooted notions of naturalness are also shown to trigger the great suspicion that some people have of foods that have not been produced in a "traditional" way (Rozin 2005; Tenbült et al. 2005).

Taking all of these disparate factors into account, increasing attempts are being made to influence the risk perceptions of individuals in the general public. As one prominent example, Protection Motivation Theory offers a general description of how individuals behave in threat-related situations (▶ Excursus 8.2). Any hazards to which they are exposed without their consent (e.g., the production of nuclear energy) are, for instance, perceived by people to be riskier than those that they believe that they themselves can control (e.g., smoking). It is the sense of control and being in control that therefore matters, even if the objective risks of the latter are significantly higher (Slovic 1987; Leikas et al. 2009). In order to confront not only the perceived riskiness of a situation, but also the more general complexity of making decisions amidst uncertainty, people are shown to lean heavily on so-called heuristics (Kahneman and Tversky 1974, 1979). These types of "if-then" rules based on experience are useful for simplifying the decision-making process, notably, by focusing on specific types of characteristics or giving priority to certain criteria more than others. In the case of a new and unknown product, such a rule might, for example, be: "I never buy new products if I do not know anyone who has previously had a good experience with them." By means of such heuristics, individuals are able to screen out certain alternatives from the choice set or better focus on particular evidence that past experience has taught them to pay closer attention to.

Excursus 8.2 Protection Motivation Theory

Protection Motivation Theory offers a conceptual framework for dealing with one's feelings of fear (Rogers 1975; Maddux and Rogers 1983). Within this framework, the ability to cope with threats of a health-related nature is depicted at the level of individuals. Specifically, the test persons or patients are asked to assess, on the one hand, their health threat and, on the other, their ability to cope with this threat. The resulting *threat appraisal* then measures the joint importance of the *perceived severity* of the health threat and the respondents' perceptions of their particular *vulnerability*. With regard to *coping appraisal*, meanwhile, the effectiveness that the respondents attribute to a given action, i.e., *response efficacy*, is explored alongside the extent to which they feel competent to perform the actions required in order to promote greater health, i.e., *self-efficacy*. By taking the results of both assessments together, it is then possible to deduce the level of *protection motivation* that exists – or, in other words, how strong the intention is to protect one's health through undertaking appropriate or recommended behaviours. To this point, this framework has mainly been used in health-related contexts, e.g., to reduce alcohol consumption. However, it is also increasingly being applied to a range of other contexts in which the motivation to protect one's health is also important, e.g., in order to investigate consumer behaviour towards biofortified pulses that have been fortified with iodine (Mogendi et al. 2016). Because consumers tend to perceive new, and especially genetically modified, foods as a threat to their health, Protecton Motivation Theory (◘ Fig. 8.7) offers significant opportunities with regard to the further investigation and understanding of the potential acceptance of new food products and new food technologies.

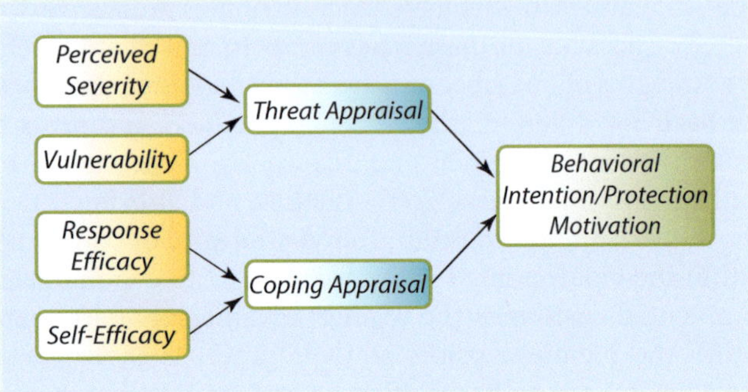

◘ **Fig. 8.7** Protection Motivation Theory. (Based on Maddux and Rogers 1983)

8

8.3.2.3 Trust

The more a person knows about novel complex products or technologies, the more likely it is that they will perceive them to be less risky – at least so long as the actual danger is not objectively high. However, the average consumer usually does not have detailed knowledge of the risks and benefits of novel technologies, making it necessary to rely on expert knowledge. Thus, there is research that illustrates, for instance, how higher levels of consumer trust in the companies and institutions working with genetic engineering can foster greater acceptance of the technology itself (Siegrist 2000, 2008). On a general level, it is therefore notable that the German public has high confidence in science and scientific actors. Nonetheless, there are particular research areas and technologies, such as synthetic biology, for which the general public holds largely negative views (Hacker and Köcher 2015). This therefore renders the introduction and implementation of these technologies, as well as their prospective applications, quite difficult – no matter the potential advantages that may exist.

However, taking synthetic biology as an example, this can have substantial consequences for society. Synthetic biology involves the increasing use of genetic and genomic knowledge for the purpose of making fundamental changes to biological systems and constructing synthetic organisms that can serve a variety of purposes, including as new pharmaceuticals (e.g., for cancer), plant-protection compounds free from fossil fuels, and more sustainable of flavorings for food production (► Sect. 5.2). What is more, the transformation of specialized molecules into microscopically small, self-contained factories is of particular importance for the bioeconomy, in view of the potential to thereby produce cleaner types of fuel or even filter carbon dioxide directly from the atmosphere (Specter 2009) . Nonetheless, the fact that the public interest in synthetic biology is so low echoes the low level of knowledge about this domain in general, not to mention the likelihood that consumers simply do not see any of the (prospective) applications as having any bearing on their everyday concerns. Indeed, simple reference to the potential advantages of particular applications of synthetic biology is enough to make people more enthusiastic and positive about the development of these technologies (Hacker and Köcher 2015; Pauwels 2013). In this way, we can observe the "double-edged" nature of information provision: depending on how and by whom the information is presented, it is possible to either build greater support for new technologies or, instead, to endow them with a richer sense of danger or riskiness. In such situations, trust-building initiatives, especially those engaging with individuals on an emotional level, are often more important for increasing acceptance than the dissemination of knowledge or information to the general public. Indeed, if the consumer considers those actors undertaking the development and introduction of a technology or novel product application to be trustworthy, responsible, and knowledgeable, then this trust can help to reduce the overall complexity of the topic (Lusk et al. 2014). Even if he/she does not explicitly understand how the technology works, this greater "accessibility" to the topic can then facilitate agreement or acceptance, or even simply make her/him more receptive to new information that is relevant. As such, trust and confidence in scientific and political actors, as well as the belief that these parties have the broad interests of society in mind, are a crucial consideration for overcoming difficulties and potential hurdles for the public acceptance of novel technologies.

8.3.3 Determinants of Technology Adoption

While acceptance is mostly measured in terms of behavior or purchasing, adoption instead assumes the form of the decision that is actually made. A distinction is thus required between products, processes, and technologies in order to make sense of technology adoption. Notably, consumer acceptance usually refers only to the end product, perhaps contingent on specific features or novel attributes. If this product is manufactured using a novel technology, consumer evaluations thus focus on the product itself, though potentially also taking into account all of the risks and benefits associated with the underlying technology. As a result, the nature of this evaluation is quite similar across a range of adoption situations, including both the adoption of novel technologies, i.e., at the level of companies, and the introduction of products on the market. Among the most notable frameworks, the Technology Acceptance Model (TAM) from Davis (1989) outlines the factors and processes specifically relevant to the adoption of information technology (◘ Fig. 8.8).

This model centers on two particular factors: the *perceived ease of use* (E) of a technology and its *perceived usefulness* (U). In turn, each of these factors is influenced by a range of external variables, such as age and gender, before the joint influence of the two then determines the *attitude toward using* (A) the new technology. Together with perceived usefulness, which again features as a direct predictor here, this attitude then specifies the strength of the *behavioral intention to use* (BI) the technology in question and, ultimately, its actual use or adoption.

If we apply the Technology Acceptance Model to the context of the bioeconomy, a further distinction between two already distinct types of relationship is useful: namely, that between producers and their business partners – i.e., business-to-business relationships (B2B) – and another between producers and the consumers who are the ultimate users of the manufactured goods available on the market – i.e., business-to-consumer relationships (B2C).

From the point of view of businesses, competitiveness serves as a key prerequisite for taking the initial step towards devoting more resources to the bioeconomy. On the one hand, this certainly presupposes that the related technologies offer a higher usefulness (U) compared to the previous approach. One example would be biotechnology, which could increase the U factor by, for example, enabling the cheaper, more resource-efficient manufacture of new pharmaceutical compounds. Furthermore, we again note that both the usefulness and the attitude towards using the technology are influenced by its perceived ease of use. As a result, for implementation to be successful, it is crucial to avoid creating insurmountable obstacles that make using the technology difficult (E). To remain on the example of biotechnology, perceived ease of use from the company's point of view could then be satisfied if, among other things, there were to exist a transparent mechanism to conduct safety assessments and clear rules to be followed for marketing. If such conditions were in place, the process of switching to the new technology would then be rendered more cost-effective and predictable. If the risks for firms could be placed in a more favourable relation to the expected benefits, this would encourage a more positive attitude towards the introduction of the new technology (A).

From the consumer's point of view, the story remains broadly similar, though the focus here is situated at the very end of the value chain, i.e., on the product available for purchase. Consequently, the crucial consideration is whether the product itself is easy to purchase, use, or consume. Biofuels can be cited here as an example. These goods are available at a range of petrol stations, just like the more conventional fuel offerings, and able to be used in a range of car models and makes (E). It therefore makes sense that the product was accepted and adopted relatively quickly on the market. Nonetheless, this leaves unanswered the question of the potential impact on the perceived benefits (U). This factor is integrated into consumer evaluations in two ways. On the one hand, it is crucial that the product offers some kind of advantage vis-à-vis its established counterparts, in this case, a lower price, and on the other, such advantages must not be attended by excessive disadvantages that might, for instance, even outweigh the advantages on offer, such as the unexpected consequences of fostering greater competition between feedstocks and foodstuffs. In sum, consumer evaluations with respect to the benefits, risks, and usability of the product ultimately come together to form the attitude (A) toward its use.

As a general rule, the main implications of the TAM framework are that (a) the benefits of novel products and

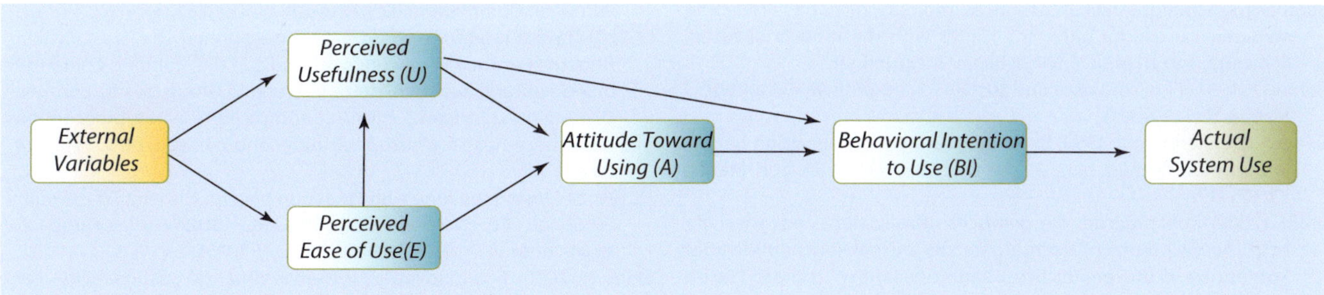

◘ **Fig. 8.8** Technology acceptance model. (Based on Davis 1989)

technologies must be quite high (in case, e.g., the risks presented by established products or technologies are also perceived to be low) and (b) consuming the innovation or integrating it into one's daily life must, as much as possible, avoid any difficulties or inconvenience that might risk that the prospective benefits do not materialize. In so doing, according to the model, the attitudes and intentions of both producers and consumers would be positively influenced, so that the acceptance and adoption potential of the product or technology will be dramatically increased. Such implications from the Technology Acceptance Model thereby enable us to better understand the conditions and prerequisites for the success of the bioeconomy, notably, by focusing more on the decisions and evaluations of consumers and producers. And yet, these assorted insights represent a proverbial drop in the bucket of the research that is needed, both because many models of technology acceptance research have only been sparsely applied to the context of the bioeconomy and, what is more, the growing appreciation of attitude-behavior gaps denotes that changing attitudes and intentions is far from sufficient on its own. As a result, the overall message is that research is required not only to better understand the acceptance of consumers, as individual actors and members of larger societies, but also to include upstream actors across the supply chain, especially given their important role as both technology adopters and sources of information for consumers. Only through greater insights and understanding in these manifold variables will it be possible to identify and then address the potential challenges and opportunities related to the emergence of new bio-based value chains and, indeed, the success of the bioeconomy as a whole.

References

Ajzen I (1985) From intentions to actions: a theory of planned behavior. In: Kuhl J, Beckmann J (eds) Action control: from cognition to behavior. Springer, Berlin, pp 11–39

Ajzen I (1991) Theory of planned behaviour. Organ Behav Hum Decis Process 50(2):179–211

Ajzen I (2001) Nature and operation of attitudes. Annu Rev Psychol 52(1):27–58

Ajzen I, Madden TJ (1986) Prediction of goal-directed behavior: attitudes, intentions, and perceived behavioral control. J Exp Soc Psychol 22:453–474

Bamberg S, Möser G (2007) Twenty years after Hines, Hungerford, and Tomera: a new meta-analysis of psycho-social determinants of pro-environmental behaviour. J Environ Psychol 27(1):14–25

Bamberg S, Schmidt P (2003) Incentives, morality, or habit? Predicting students' car use for university routes with the models of Ajzen, Schwartz, and Triandis. Environ Behav 35(2):264–285

Barney J (1991) Firm resources and sustained competitive advantage. J Manag 17(1):99–120

Beedell JDC, Rehman T (1999) Explaining farmers' conservation behavior: why do farmers behave the way they do? J Environ Manag 57:165–176

BMBF (2009) Kohlendioxid: die gemischte Bilanz der Landwirtschaft. http://www.pflanzenforschung.de/de/journal/journalbeitrage/kohlendioxid-die-gemischte-bilanz-der-landwirtschaft-10011. Accessed: 11.09.2016

BMU und BMELV (2010) Nationaler Biomasseaktionsplan für Deutschland: Beitrag der Biomasse für eine nachhaltige Energieversorgung. Bundesministerium für Umwelt, Naturschutz und Reaktorsicherheit (BMU) und Bundesministerium für Ernährung, Landwirtschaft und Verbraucherschutz (BMELV), Berlin

BMWi (2012) Energiestatistiken: Energiegewinnung und Energieverbrauch. Bundesministerium für Wirtschaft und Energie (BMWi), Berlin. https://www.bmwi.de/BMWi/Redaktion/PDF/E/energiestatistiken-energiegewinnung-energieverbrauch,property=pdf,bereich=bmwi2012,sprache=de,rwb=true.pdf

Boehlje M (2016) How might big data impact industry structure and enhance margins? Int Food Agribus Manage Rev 19(A):13–16

Brewer JL, Blake AJ, Rankin SA, Douglass LW (1999) Theory of reasoned action predicts milk consumption in women. J Am Diet Assoc 99(1):39–44

Cardello AV (2003) Consumer concerns and expectations about novel food processing technologies: effects on product liking. Appetite 40(3):217–233

Carus M, Eder A, Beckmann J (2014) GreenPremium prices along the value chain of bio-based products. Nova paper #3 on bio-based economy. Hürth 04.2014. Download at www.bio-based.eu/novapapers

Christensen CM (1997) The innovator's dilemma: when new technologies cause great firms to fail. Harvard Business School Press, Brighton

Conner M, Armitage CJ (1998) Theory of planned behavior: a review and avenues for further research. J Appl Soc Psychol 28:1429–1464

Corbion (2016) Carbon footprint: Emissions from production of common polymers (kg CO_2 eq per kg polymer – cradle to gate). http://www.corbion.com/bioplastics/about-bioplastics/sustainability

Cox DN, Evans G (2008) Construction and validation of a psychometric scale to measure consumers' fears of novel food technologies: the food technology neophobia scale. Food Qual Prefer 19(8):704–710

Cuthbertson B, Marks N (2007) Beyond credence: emerging consumer trends in international markets. Melbourne, Australia State of Victoria Department of Primary Industries. http://ageconsearch.umn.edu/bitstream/5980/2/cp08cu01.pdf

Darby MR, Karni E (1973) Free competition and the optimal amount of fraud. J Law Econ 16(1):67–88

Davis FD (1989) Perceived usefulness, perceived ease of use, and user acceptance of information technology. MIS Q 13:319–339

DBFZ (2012) Basisinformationen für eine nachhaltige Nutzung von landwirtschaftlichen Reststoffen zur Bioenergiebereitstellung. DBFZ Report Nr. 13, Deutsches Biomasseforschungszentrum (DBFZ), Leipzig. https://www.dbfz.de/fileadmin/user_upload/DBFZ_Reports/DBFZ_Report_13.pdf

Desaint N, Varbanova M (2013) The use and value of polling to determine public opinion on GMOs in Europe: limitations and ways forward. GM Crops Food 4(3):183–194

Eagly AH, Chaiken S (1993) The psychology of attitudes. Harcourt Brace Jovanovich, Fort Worth

EBP (2014) Case studies of market-making in the bioeconomy. European Bioeconomy Panel (EBP), Brussels. https://ec.europa.eu/research/bioeconomy/pdf/13-case-studies-0809102014_en.pdf

EC (2012) Innovating for sustainable growth: a bioeconomy for Europe. European Commission (EC), Brussels

EC (2014) What next for the European bioeconomy? The latest thinking from the European Bioeconomy Panel and the Standing Committee on Agricultural Research Strategic Working Group (SCAR). European Commission, Brussels. http://ec.europa.eu/research/bioeconomy/pdf/where-next-for-european-bioeconomy-report-0809102014_en.pdf

EC (2015) Closing the loop – an EU action plan for the circular economy. European Commission (EC), Brussels. http://eur-lex.europa.eu/legal-content/EN/TXT/?uri=CELEX%3A52015DC

Ekins P (2010) Eco-innovation for environmental sustainability: concepts, progress and policies. IEEP 7(2):267–290

Evonik Industries, Health & Nutrition (2014) What if...; How amino acids from Evonik contribute to sustainable food production
FNR (2015) Massebezogener Substrateinsatz in Biogasanlagen 2014. Fachagentur Nachwachsende Rohstoffe (FNR), Gülzow-Prüzen, Deutschland. https://mediathek.fnr.de/catalog/product/gallery/id/93/image/1424/
Gaskell G, Stares S, Allansdottir A, Allum N et al (2010) Europeans and biotechnology in 2010: winds of change? European Commission, Brussels. http://ec.europa.eu/public_opinion/archives/ebs/ebs_341_winds_en.pdf
Hacker J, Köcher R (2015) The public understanding of synthetic biology: considerations in the context of science-based policy advice to policy-makers and the public. 01.2015/Discussion No.3, IfD Allensbach und Leopoldina National Akademie der Wissenschaften, Halle (Saale), Deutschland
Harland P, Staats H, Wilke HAM (1999) Explaining pro environmental behavior by personal norms and the theory of planned behavior. J Appl Soc Psychol 29:2505–2528
Hasler K, Olfs H-W, Omta O, Bröring S (2016) Drivers for the adoption of eco-innovations in the German fertilizer supply chain. Sustainability 8:682–699. Journal of Sustainability (in press)
Henderson BD, Gälweiler A (1984) Die Erfahrungskurve in der Unternehmensstrategie. Campus Verlag, Frankfurt
IBVT/TU Braunschweig (2016) Biobasierte Produktion von L-Lysin mit Corynebacterium glutamicum – Maßgeschneiderte Zellfabriken und Bioprozesse. Institut für Bioverfahrenstechnik (IBVT), Technische Universität Braunschweig. http://www.ibvt.de/DE/Forschung/Abgeschlossene_Projekte/Lysin.php
Junqueira TL, Cavalett O, Bonomi A (2016) The virtual sugarcane biorefinery – a simulation tool to support public policies formulation in bioenergy. Ind Biotechnol 12(1):62–67
Kahneman D, Tversky A (1974) Judgment under uncertainty: heuristics and biases. Science 185(4157):1124–1131
Kahneman D, Tversky A (1979) Prospect theory: an analysis of decision under risk. Econometrica 47(2):263–291
Kahneman D, Ritov I, Schkade D (1999) Economic preferences or attitude expressions? An analysis of dollar responses to public issues. J Risk Uncertain 19:220–242
Kemp R, Pearson P (2007) Final report MEI project about measuring eco-innovation. Deliverable 15 of Measuring Eco-Innovation (MEI) project (D15), UM Merit, Maastricht
Kircher M (2012) The transition to a bio-economy: emerging from the oil age. Biofuels Bioprod Biorefin 5:369–375
Kircher M (2015) Sustainability of biofuels and renewable chemicals production from biomass. Curr Opin Chem Biol 29: 26–31
Kircher M (2016) KADIB Unternehmensdatenbank. www.kadib.de
Kollmuss A, Agyeman J (2002) Mind the gap: why do people act environmentally and what are the barriers to pro-environmental behavior? Environ Educ Res 8(3):239–260
Leikas S, Lindeman M, Roininen K, Lähteenmäki L (2009) Who is responsible for food risks? The influence of risk type and risk characteristics. Appetite 53(1):123–126
Löbnitz N, Bröring S (2015) Consumer acceptance of new food technologies for different product categories: the relative importance of experience versus credence attributes. J Int Consum Mark 27: 307–317
Luchs MG, Naylor RW, Irwin JR, Raghunathan R (2010) The sustainability liability: potential negative effects of ethicality on product preference. J Mark 74(5):18–31
Lusk JL, House LO, Valli C, Jaeger SR, Moore M, Morrow JL, Traill WB (2004) Effect of information about benefits of biotechnology on consumer acceptance of genetically modified food: evidence from experimental auctions in the United States, England, and France. Eur Rev Agric Econ 31(2):179–204
Lusk JL, Roosen J, Bieberstein A (2014) Consumer acceptance of new food technologies: causes and roots of controversies. Ann Rev Resour Econ 6(1):381–405
Lynne GD, Casey CF, Hodges A, Rahmani M (1995) Conservation technology adoption decisions and the theory of planned behavior. J Econ Psychol 16(4):581–598
Maddux JE, Rogers RW (1983) Protection motivation and self-efficacy: a revised theory of fear appeals and attitude change. J Exp Soc Psychol 19(5):469–479
Matin AH, Goddard E, Vandermoere F, Blanchemanche S, Bieberstein A, Marette S, Roosen J (2012) Do environmental attitudes and food technology neophobia affect perceptions of the benefits of nanotechnology? Int J Consum Stud 36(2):149–157
Mogendi JB, De Steur H, Gellynck X, Makokha A (2016) Consumer evaluation of food with nutritional benefits: a systematic review and narrative synthesis. Int J Food Sci Nutr 67(4):355–371
Moser R, Raffaelli R, Thilmany-McFadden D (2011) Consumer preferences for fruit and vegetables with credence-based attributes: a review. Int Food Agribus Manage Rev 14(2):121–142
Nameroff TJ, Garant RJ, Albert MB (2004) Adoption of green chemistry: an analysisbased on US patents. Res Policy 33:959–974
Nelson P (1970) Information and consumer behavior. J Polit Econ 78(2):311–329
Nelson P (1974) Advertising as information. J Polit Econ 82(4):729–754
OECD (2009) The bioeconomy to 2030: designing a policy agenda, main findings. Organisation for Economic Cooperation and Development (OECD), Paris. http://www.oecd.org/futures/bioeconomy/2030
Pauwels E (2013) Public understanding of synthetic biology. BioScience 63(2):79–89
Penrose E (1959) The theory of the firm. Wiley, New York
Pliner P, Salvy S (2006) Food neophobia in humans. In: Shepherd R, Raats M (eds) The psychology of food choice. Frontiers in nutritional science 3. CAB International, Wallingford, pp 75–92
Porter ME (1980) Competitive strategy: techniques for analyzing industries and competitors. The Free Press, New York
Prahalad CK, Hamel G (1990) The core competence of the corporation. Harv Bus Rev 68(3):79–91
Rogers RW (1975) A protection motivation theory of fear appeals and attitude change. J Psychol Interdiscip Appl 91(1):93–114
Rogers EM (1983) Diffusion of innovation: a cross-cultural approach. The Free Press, New York
Rozin P (2005) The meaning of "natural" process more important than content. Psychol Sci 16(8):652–658
SCAR (2015) Sustainable agriculture, forestry and fisheries in the bioeconomy: a challenge for Europe. 4th SCAR Foresight Exercise, Standing Committee on Agricultural Research (SCAR), Brussels
Schnettler B, Crisóstomo G, Sepúlveda J, Mora M, Lobos G, Miranda H, Grunert KG (2013) Food neophobia, nanotechnology and satisfaction with life. Appetite 69:71–79
Siegrist M (2000) The influence of trust and perceptions of risks and benefits on the acceptance of gene technology. Risk Anal 20(2): 195–204
Siegrist M (2008) Factors influencing public acceptance of innovative food technologies and products. Trends Food Sci Technol 19(11):603–608
Slovic P (1987) Perception of risk. Science 236:280–285
Specter M (2009) A life of its own: where will synthetic biology lead us? New Yorker (28 September 2009). http://www.newyorker.com/magazine/2009/09/28/
Steinfeld H, Gerber P, Wassenaar T, Castel V, Rosales M, Haan CD (2006) Livestock's long shadow: environmental issues and options. Food and Agriculture Organization of the United Nations(FAO), Rome
Teece DJ, Pisano G, Shuen A (1997) Dynamic capabilities and strategic management. Strateg Manag J 18(7):509–533
Teece DJ (2016) Dynamic capabilities and entrepreneurial management in large organizations: Toward a theory of the (entrepreneurial) firm. European Economic Review 86:202–216
Tenbült P, De Vries NK, Dreezens E, Martijn C (2005) Perceived naturalness and acceptance of genetically modified food. Appetite 45(1): 47–50

Ueland Ø, Gunnlaugsdottir H, Holm F, Kalogeras N, Leino O, Luteijn JM, Verhagen H (2012) State of the art in benefit–risk analysis: consumer perception. Food Chem Toxicol 50(1):67–76

Utterback JM, Abernathy WJ (1975) A dynamic model of process and product innovation. Omega 3(6):639–656

Vidigal MC, Minim VP, Simiqueli AA, Souza PH, Balbino DF, Minim LA (2015) Food technology neophobia and consumer attitudes toward foods produced by new and conventional technologies: a case study in Brazil. LWT-Food Sci Technol 60(2):832–840

Wernerfelt B (1984) A resource-based view of the firm. Strateg Manag J 5(2):171–180

World Energy Council; London, UK (2011) Oil recoverable reserves by region. https://www.worldenergy.org/data/resources/resource/oil/. Accessed July 2016

WWF Deutschland (2016) Soja: Wunderbohne mit riskanten Nebenwirkungen. WWF Deutschland, Berlin. http://www.wwf.de/themen-projekte/landwirtschaft/produkte-aus-der-landwirtschaft/soja/

8

The Conditions of a Sustainable Bioeconomy

Uwe Fritsche and Christine Rösch

Rice terraces in the Philippines, where the European Legato project is investigating the sustainable development of ecosystems in rice cultivation. (© André Künzelmann, Helmholtz Centre for Environmental Research – UFZ)

J. Pietzsch (ed.), *Bioeconomy for Beginners*, https://doi.org/10.1007/978-3-662-60390-1_9

9.1 Goals of Sustainable Development

The conditions of a sustainable bioeconomy refer to the concept of sustainability, which originates from forestry. In the 18th century, this was already used to describe the fact that the amount of wood that may be cut is only as much as can grow again. The basis for the much more comprehensive concept of sustainable development used today as a guiding political principle was created by the report of the *World Commission on Environment and Development* (WCED 1987), chaired by the Norwegian Prime Minister Gro Harlem Brundtland, that defined sustainable development as "development that meets the needs of the present generation without compromising the ability of future generations to meet their own needs." In 1992, at the UN Conference on Environment and Development in Rio de Janeiro, the international community of states agreed on the mission statement of sustainable development. The Rio Declaration expresses the joint responsibility to use the earth's resources in a way that all countries of the world have fair development opportunities without compromising the development opportunities of future generations (UNCED 1992). The 2030 Agenda for Sustainable Development, adopted by all United Nations Member States in 2015, provides a shared blueprint for peace and prosperity for people and the planet, now and into the future. At its heart are the 17 Sustainable Development Goals (SDGs), which are an urgent call for action by all countries – developed and developing – in a global partnership (UN 2015a). They recognize that ending poverty and other deprivations must go hand-in-hand with strategies that improve health and education, reduce inequality, and spur economic growth – all while tackling climate change and working to preserve our oceans and forests (▣ Fig. 9.1).

Several of the 17 SDGs have direct relevance for the bioeconomy, in particular:

- SDG 2: End hunger, achieve food security and improved nutrition, and promote sustainable agriculture
- SDG 6: Ensure availability and sustainable management of water and sanitation for all
- SDG 7: Ensure access to affordable, reliable, sustainable and modern energy for all
- SDG 12: Ensure sustainable consumption and production patterns
- SDG 13: Take urgent action to combat climate change and its impacts by regulating emissions and promoting developments in renewable energy
- SDG 15: Protect, restore and promote sustainable use of terrestrial ecosystems, sustainably manage forests, combat desertification, and halt and reverse land degradation and halt biodiversity loss

These goals also have interconnections with each other and with the other 17 SDGs (Fritsche and Iriarte 2016).

The guiding principle of sustainable development has established itself both in global politics and, more specifically, in European and German politics. In 2002, Germany enacted its national sustainability strategy (BuReg 2002), which includes 21 indicators for the goals of intergenerational justice, quality of life, social cohesion and international responsibility. This set of indicators does not explicitly include the objectives of bioeconomy, but it does comprise indicators on land management. Agriculture is of central importance for the bioeconomy, since it is – besides forests – the most important producer of raw materials for food, as well as for the supply of biological material, i.e., for construction, and renewable energy (▶ Chap. 2). Indicators relevant to agriculture include the nitrogen surplus and the area under organic farming. Two further sustainability goals of the German government that interact with the bioeconomy are land consumption for settlement and transport and the development of biodiversity and landscape quality.

The bioeconomy is regarded as an important component of the German sustainability strategy (BuReg 2016, 2018). It should contribute to climate protection (▶ Sect. 9.7), diversification of the source of raw material and the development of a sustainable and resource efficient economic system (▶ Sect. 9.6). Besides, the bioeconomy is expected to improve competitiveness, add value and create employment opportunities in rural areas (▶ Sect. 9.8). At the same time, the aim is to prevent a growing bioeconomy from competing with food production (▶ Sect. 9.1), endangering soil fertility and biodiversity (▶ Sect. 9.5) or worsening air and water quality (▶ Sect. 9.4).

These goals and the sustainability concepts upon which they are based are still insufficiently differentiated (Pfau et al. 2014), but there are already proposals for relevant criteria and indicators (Fritsche and Iriarte 2016) and for corresponding monitoring activities (DBFZ 2015a; O'Brien et al. 2015). Implicit conflicts of objectives on the path to sustainability are, however, obvious: "Bioeconomy is not sustainable per se" (BÖR 2014a). Irrespective of this challenge, the German government is supporting the development of the bioeconomy through various strategies and research programmes, including the "National Research Strategy Bioeconomy 2030" (NFSB 2030) and the "National Policy Strategy Bioeconomy" (BMBF 2010; BMELV 2013; BMEL 2014), and is currently working on merging the research and policy sides into one cohesive strategy. The bioeconomy is also seen as a key strategy for sustainable development at the European level. It is intended to reduce dependence on fossil fuels, promote economic growth, create jobs in rural areas and improve the sustainability of primary production and the manufacturing industry (EC 2012, 2016, 2018). The bioeconomy is being integrated into a complex field of challenges in many policy fields, which requires an explicit handling of the corresponding conflicts of objectives (▶ Sect. 9.9). The successful handling of these conflicting goals on the way to a sustainable bioeconomy will also depend on the establishment of global bodies of governance (▶ Sect. 9.10).

Fig. 9.1 UN goals for sustainable development *(Sustainable Development Goals)*. (UN 2015b)

9.2 Food Security

To the extent that it uses raw materials from agriculture and, in the future, also from fisheries and aquaculture, the bioeconomy must take into account the effects of this use on traditional agricultural products and their markets. This demands not only the SDG 2 ("Zero Hunger"), but also the right to nutrition enshrined in the UN Charter of Human Rights. Food security (▶ Chap. 3) can, however, be impaired if agricultural products are increasingly used as energy and raw materials or if agricultural production areas are used to a greater extent for the cultivation of crops for energy (▶ Sect. 9.3). This can reduce the availability of food and increase its prices. Although the latter can have positive effects on farmers' incomes, even in poorer countries, and thus on their food security (Mirzabaev et al. 2014), these gains are offset by the negative consequences for many non-producers and landless people (FAO 2008; Kalkuhl 2014, 2015). The bioeconomy can also intensify competition for scarce land resources (▶ Sect. 9.3), which leads to higher rents for arable land and, in turn, to higher food prices, and may have crowding-out effects for smaller, lower-yielding farms (FAO 2010). These effects have so far been investigated primarily for biofuels. However, they apply equally to uses of biomass for other energy carriers and for biomaterials (Bardhan et al. 2015).

Yet, these effects do not follow a simple cause-and-effect chain: Higher agricultural product and land prices can lead to more investments in agriculture, and thus increase its efficiency (Zeddies et al. 2014), as well as induce changes on the demand side, e.g., reduce the comparatively expensive consumption of animal products (Searchinger et al. 2015).

Food security is also measured against the criterion of the stability of food availability and prices (FAO et al. 2014), so that increased material and energy use in the context of the bioeconomy can trigger further positive effects in the medium and longer terms if no "shocks" are induced by short-term demand policies. This role as a "stabiliser" requires a flexibility in political support instruments such as biofuel quotas, which would have to be reduced in times of high food prices and increased again in times of low prices (HLPE 2011; FAO and OECD 2011; Kline et al. 2016).

In order to operationalize the primacy of the "right to food" in the bioeconomy, various approaches have been developed (Mohr et al. 2015; Schneider 2014, among others), but these still need to be implemented in practical policy. For bioenergy, the *Global Bioenergy Partnership* agreed on indicators for the national level (GBEP 2011), which include food security and are now applicable at the project level (FAO 2012; Maltsoglou et al. 2015). In addition, the *Voluntary Guidelines on the Responsible Governance of Tenure of Land, Fisheries and Forests in the Context of National Food Security* (VGGT) of the *Committee on World Food Security* (CFS 2012) is another important way to integrate this issue into national policies (▶ Sect. 9.10).

To the extent that the bioeconomy also relies on an increased role for forests in the provision of raw lignocellulosic materials, many do not see any direct influence on food security (BÖR 2016). For North European and North American forests, this is certainly true to a large extent. However, it does not apply to forests in Latin America, Oceania, Southeast Asia and sub-Saharan Africa that are used productively by the indigenous population and make an important contribution to local nutrition (CIFOR 2013; FAO

2013a, b; IUFRO 2015). In these regions, also because of a greater use of wood for the bioeconomy may therefore only take place with appropriate "safeguards," of the often-weak state institutions (▶ Sect. 9.10). Lignocellulose from perennial cultures (short rotation with poplars and willows, energy grasses) is seen as another raw material basis of the bioeconomy (▶ Sect. 2.2) (BÖR 2016). These crops can have positive effects on biodiversity and climate balance, but both the nature and the previous use of the land on which such crops are planted are crucial (▶ Sect. 9.3).

Food security is influenced not least by dietary patterns: the consumption of animal products implies feed requirements whose cultivation occupies corresponding areas, and thus tends to raise prices for agricultural products. However, it should be borne in mind that the sustainable use of grassland by pasture farmers does not compete with arable crops.

In this context, it is important that global food security, which also creates scope for biogenic raw materials, indeed is possible without further deforestation: the combination of moderate increases in yields with sustainable intensification of pasture management and sustainable diets offers sufficient space for a bioeconomy that integrates food and feed production with biogenic raw material provision, even as the population continues to grow (Erb et al. 2016). Examples show that these are not simply model calculations: A holistic view of food security and energy supply problems offers solutions that focus on greater integration of different land uses with a positive balance for food (de Laurentiis et al. 2016; Mirzabaev et al. 2014). This integrating perspective is important for the bioeconomy in order to relativize the polarizing food-versus-fuel discussion about the so-called first-generation biofuels (von Braun 2014) and to recognize the important opportunities of bio-based products for improved food security (Kline et al. 2016). First generation biofuels are produced from food crops such as rapeseed, corn and soybeans, which led to food shortages on world markets from 2007 to 2008, with corresponding price increases.

The use of biofuels is controlled by political measures. According to the phased plan of the EU Renewable Energies Directive (EU 2009), all member states would need to cover 5.75% of their fuel consumption in the transport sector with renewable energies (including e-mobility) by 2010, with an increase to 10% by 2020. Since the introduction of super petrol with 10% bioethanol (E10) and diesel fuel with up to 7% biodiesel (B7), these requirements have been the subject of controversy due to the food-fuel dilemma. To mitigate the conflict between food and energy crop cultivation, the EU made a paradigm shift in 2015 and is now increasingly relying on so-called second-generation biofuels produced from waste or non-food biomass such as straw, forest residues and third-generation biofuels from algae. In April 2015, the European Parliament decided to amend the Renewable Energy Directive (RED) and the Fuel Quality Directive (FQD) to include various fuels in the greenhouse gas reduction quota. While the share of first generation biofuels (biodiesel and bioethanol) will be limited to 7%, second and third generation biofuels will be supported by multiple crediting. EU Member States were allowed to introduce a sub-target of 0.5% for second generation biofuels. More recently, the recast of the RED (RED-II) calls for a gradual phase-out of first-generation biofuels by 2030 (EU 2018).

Despite all of the controversy, the debates on biofuels show that the bioeconomy can provide two important impulses for food security (Osseweijer et al. 2015):

- On the one hand, investments in agriculture increase its productivity, and thus the amount of available food, while maintaining or even reducing land use (Zeddies et al. 2014; Woods et al. 2015).
- On the other hand, the use of coupled products and by-products, e.g., from biorefineries (EC 2016), offers the opportunity to provide feed without additional land use, and thus have a dampening effect on future food prices.

The bioeconomy can therefore promote necessary innovations in agriculture – including in the EU (SCAR 2015, 2016; EC 2018) (▶ Chap. 8). From the point of view of sustainability, it is important that this is done with consideration for biodiversity (▶ Sect. 9.5), and that climate protection (▶ Sect. 9.7) and the social aspects of land use are taken into account.

9.3 Land Resources

People have always influenced land use, mainly through deforestation for the purpose of creating arable land, animal husbandry, resource extraction and settlement. However, the scale and speed of land use change has increased sharply since the 18th century, due to high population growth and changes in agricultural practices. Since the 1950s, further grassland and forests have been converted into arable land in order to meet the needs of a growing population and changes in dietary habits and to enable the use of biomass for energy and material purposes (Fritsche et al. 2015).

Today, agriculture uses 12% of the global land area for arable land and 26% for pastures (grassland). 31% of the global land area is covered by forest (◘ Fig. 9.2).

Only about 35% of global crop production is used directly as food, while 62% is used for animal feed, which indirectly contributes to food production. Bioenergy, biomaterials and seed production account for only 3%.

Population growth and urbanization will remain fundamental drivers of future land use, especially in Asia and sub-Saharan Africa. By 2050, the population of cities will increase to more than two thirds of the world's population (UNDESA 2014). Urban and infrastructure areas will grow, at the very least, in proportion to the population, especially in developing countries (WBGU 2016). This contrasts with a significant decline in the rural population, with corresponding consequences for the work and life opportunities of the population remaining there – including in Germany. Here, the bioeconomy offers important opportunities for rural employment

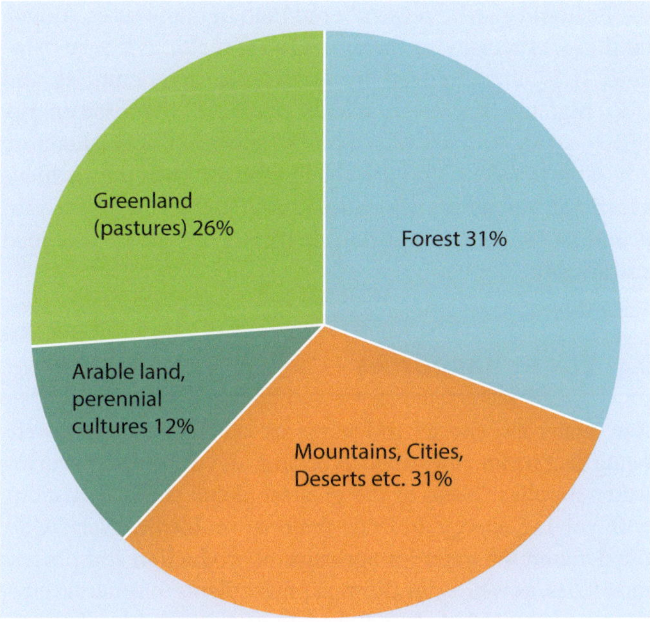

Fig. 9.2 Global land use and biomass. (Own presentation according to Fritsche et al. 2015)

and regional added value – but mainly when decentralised concepts are pursued (▶ Sect. 9.8).

As far as the question of sustainable land use is concerned, it should generally be noted that, compared to other energy and raw material sources, biomass requires significantly larger areas of production, due to lower energy and raw material densities. For example, for the provision of comparable quantities of electricity, the land requirement of bio-based technology paths is five to ten times larger than those of wind farms or photovoltaic systems, and more than 100 times larger than for electricity from natural gas, oil, coal or nuclear power (EEA 2013), but it can also be rather low when the biomass comes from organic residues, or in cases of intercropping and degraded land (Fritsche et al. 2017).

Many studies have investigated whether there is any long-term potential for the cultivation of biomass for energy and material use against the background of changing land use and the high specific land requirements, both at the global level (e.g., IEA 2017; WBA 2016; Woods et al. 2015), in the EU (EEA 2013, 2016a; Panoutsou et al. 2016) and for Germany (Thrän et al. 2015). If sustainability restrictions such as biodiversity conservation and climate protection are taken into account, there is considerable potential for sustainability in the medium to long term (2030–2050) (Piotrowski et al. 2016; IEA 2017). Here, however, land-use competition (Fritsche et al. 2015; Thrän et al. 2011) must be considered, which, in addition to biodiversity, primarily concerns cultivation for food and feed: If biomass is cultivated as a raw material for bioenergy or biomaterials, it may displace agricultural land products, which then are no longer available for food or other purposes. These indirect effects are particularly noticeable at the global level and have a strong impact on greenhouse gas balances if the "displaced" food is cultivated on other land, leading, for example, to deforestation there. This effect is described as indirect land use change (ILUC) (▶ Sect. 9.7).

In addition to the environmental aspects of sustainability, land use is strongly influenced by its social dimension: Food security is closely linked to access to land as the basis for securing the livelihood of many small farmers. However, there is a fundamental conflict potential here if financially strong companies or investors acquire land through purchase or displacement in the course of biomass production (BMZ 2012). This so-called *land grabbing* effect (Cotula et al. 2009) is not exclusively linked to the production of biogenic raw materials, but is rather a general problem of agriculture and forestry in countries where no or few formal land titles exist and governments are often too weak to protect the rights of the local population (PANGEA 2011).

Special protection against *land grabbing* is necessary for indigenous peoples and traditional municipal land use without a title of ownership (ILC 2015) – initial proposals have been made for a corresponding global inventory that identifies areas in need of protection (RRI 2015).

Secure land rights are a key factor in implementing the human right to food. The increase in investments in agriculture in emerging and developing countries since 2007 has made the recognition and protection of existing land rights even more explosive and topical. In response, more than 100 member states of the UN Committee on World Food Security (CFS) unanimously adopted the VGGT (▶ Sect. 9.2) (CFS 2012). The VGGT includes:

- Minimum standards for the recognition, transfer and management of rights of ownership, possession and use of land, fishing grounds and forests
- Regulations for expropriations, compensation processes and agricultural reform measures up to redistributive land reforms
- Procedural norms and standards of good governance for land management, from pricing and evaluation to land administration

The VGGT also describe how the participation of those affected should be ensured, discrimination and corruption in land access and land management should be avoided, traditional and informal rights of use should be respected, and the rights of indigenous peoples should be adequately taken into account.

In addition, the VGGT formulate minimum standards for investments in land, forest and fishery resources: For example, assessments are required of the consequences of investment projects on property rights and rights of use, as well as on the right of the local population to food.

The VGGT thus provide an important basis for national policies to prevent or at least minimise negative ecological and social impacts of land use.

In the context of bioeconomics, four basic strategies are suitable for reducing or avoiding such negative impacts:

- Focusing the cultivation of biogenic raw materials on "surplus" areas that are not used for food or feed production due to low soil quality and profitability, as well as low precipitation (Zeddies et al. 2014). This strategy also offers potential for developing countries (Ostwald et al. 2015; Rahman et al. 2014; Wicke 2011). It should be noted, however, that many landless farmers – e.g., through pasture farming – are engaged in subsistence agriculture on such marginal areas and that commercial biomass cultivation can endanger their livelihoods (BMZ 2012; Baka 2014) (▶ Sect. 2.1). Marginal areas are also often characterized by a high level of biodiversity (▶ Sect. 9.5).

- Use of land that has been degraded, for example, by overgrazing, but on which biogenic raw materials can certainly grow using perennial crops (Gelfand et al. 2013; Wicke 2011), as well as contaminated, flooded or saline areas (IEA and GBEP 2016) that can be used with special perennial crops. This allows for the cultivation of raw materials for the bioeconomy that will contribute to achieving SDG 15 (▶ Sect. 9.1).
- Integration of biomass cultivation into "underused" conventional crop rotations through intercrops, double-cropping systems and mixed arable/tree systems *(agroforestry)* that provide additional yields, as biogenic raw materials are grown on the same land (▶ Sect. 2.2) (Rahman et al. 2014).
- Use of waste and residual materials that still have considerable unused potential globally (Woods et al. 2015), in the EU (Panoutsou et al. 2016) and in Germany (DBFZ 2015b).

Sustainability restrictions must be observed in all forms of biogenic raw material supply – from biodiversity and climate protection to the availability of water and access rights to land, e.g., for herding nomads. If these restrictions are observed, the additional use of biomass can lead to overall positive effects (Gerssen-Gondelach 2016), including in countries of sub-Saharan Africa (Karlberg et al. 2015). It is essential to pursue a holistic approach to sustainable land use that transcends sectoral boundaries and actively involves stakeholders (Kline et al. 2016; Fritsche et al. 2015).

It should also be remembered that, in many cases, today's land use, especially in the agricultural sector, is not sustainable. It is characterized by large-scale monocultures and (too) high use of water, nutrients, pesticides and fossil energy (SRU 2016). As part of a holistic approach, it is therefore advisable to assess land use changes positively if they achieve greater sustainability (Berndes and Fritsche 2016). The bioeconomy can be an important driver for this if it is designed sustainably.

9

9.4 Water Resources

Raw materials, energy and water are crucial for human well-being and sustainable socio-economic development. Securing water supplies is one of the greatest global challenges, along with solving energy and raw material problems. Worldwide, the demand for water for agricultural production and process industries, as well as for thermal power plants to generate electricity, is increasing. It will lead to ever greater local supply problems in the future, especially in developing countries. Already today, around 2 billion people have no access to clean and permanently available drinking water (UN 2013 in WWAP 2014). More than 70% of the anthropogenic use of freshwater resources takes place in agriculture. Due to changes in the climate and the increasing global demand for plant products and animal foodstuffs, irrigated agriculture will become even more important in the future and will require significantly more fresh water than it does today (Gerbens-Leenes and Nonhebel 2002, 2004; Rockström et al. 2007). With continued population and economic growth and changing consumption patterns, including an increase in the consumption of animal food, global water abstraction is expected to increase by 55% by 2050. By then, more than 40% of the world's population will live in regions with severe water shortages (Halstead et al. 2014; WWAP 2014). The bioeconomy has to tackle this challenge in order to gain further importance. This is possible, e.g., through the cultivation of plants with rather low water requirements, such as perennial crops.

Excursus 9.1 The high water footprint of biomass

The further implementation of the bioeconomy is closely linked to water use, as the provision of biomass is associated with a significantly higher water demand than that of raw fossil materials. This is what calculations of the water footprint show. It indicates the quantity of water required for the production of goods and services. A distinction is made between groundwater and surface water, evaporation by vegetation and the amount of water polluted by production processes (Hoekstra 2008). Approximately 92% of the average water footprint is associated with the production of agricultural goods, mainly cereals, meat and milk (Mekonnen and Hoekstra 2011). Energy production from biomass has a water footprint (WF) that is 70–400 times greater than energy production from a mix of non-renewable energy sources (Gerbens-Leenes et al. 2008). The wide range is based on plant physiology, climatic differences and the availability of groundwater. An example: The WF of maize in the Netherlands is 9, while that of winter rape is 67. In Brazil, the WF of maize is 39, and in Zimbabwe, 200. The substitution of fossil fuels with biogenic energy sources in industrialized countries would mean that the WF per capita for food and energy supply would be the same (Hoekstra and Chapagain 2007, 2008). According to O'Brien et al. (2015), the global water footprint of the bioeconomy is one of the key questions for assessing its sustainability and resource base. To date, only a few research projects have investigated the water needs that bioeconomy will entail both in the regional context and globally, and the changes in the water availability that this will trigger. It is also unclear as to how the water demand in the countries in which biomass originates is to be met if undernourishment and malnutrition are to be overcome at the same time and biomass is to be exported to industrialised countries on a large scale.

The increasing consumption of fresh water can lead to a decrease in groundwater resources and, ultimately, to their overuse. Already today, 20% of aquifers are overused (WWAP 2014). The implementation of the bioeconomy concept may, due to its water demand, lead to further intensification of water stress, both in the dry regions of Germany and in biomass source countries, which are already suffering from water shortages (▶ Excursus 9.1). In Germany, however, irrigation with groundwater has so far played only a minor role in agricultural production. Only 6% of farms use this option. But, in certain regions, such as eastern Lower Saxony, Hesse, Brandenburg and Bavaria, it may be of greater importance (LAWA 2014).

Through biomass cultivation, the bioeconomy not only has an influence on the water supply, but also on the quality of ground and surface waters. Nutrient and pesticide inputs from agricultural land use represent a major problem on the way to a sustainable water supply. Nitrogen, phosphorus and eroded soil particles, as well as pesticides and their metabolites, enter groundwater and surface waters from agricultural land via different input paths. The European Nitrates Directive (91/676/EEC) and the Fertiliser Ordinance form the legal framework for water-conserving fertilisation, in particular, with organic nutrient carriers, as well as adapted land management. Despite the existing regulations, 27% of all groundwater bodies in Germany have a poor status due to excessive nitrate levels alone (LAWA 2014). Of the 72 bodies of coastal water, 71 are in moderate to poor condition due to excessive nutrient concentrations (nitrogen and phosphorus). The sustainability indicator "nitrogen surplus" selected by the German government (as of 2012: 101 kg N/ha) is still far from its target value (80 kg N/ha). The achievement of this goal is hampered by a noticeable change in agricultural structure and production, as well as by the increasing demand for biomass for food production and energy use. The resulting production concentration at selected locations and the intensification of agriculture vary from region to region. For example, maize cultivation for biogas production has increased strongly in some federal states and, due to current cultivation practices, leads to an increase rather than a decrease in nitrogen emissions into the soil, groundwater and, from there, into surface waters. At the same time, there is an increase in the input of soil particles, phosphorus and possibly also pesticides into surface waters. These changing and evolving framework conditions hamper the sustainable development of water resources and the achievement of the objectives of the Water Framework Directive (WFD).

Sustainability certifications contribute to assessing the water management relevance of bioeconomic pathways and *water grabbing*, water scarcity and water pollution from biomass cultivation. The international draft standard DIN EN ISO 14046:2015-11 defines principles, requirements and guidelines for the determination of the water footprint of products and processes on the basis of a life cycle assessment. However, these auditing instruments will not work if so-called energy crop landscapes are established to implement the bioeconomy and the water balance in river basins, for example, is changed as a result.

On the other hand, the supply of raw materials for the bioeconomy can lead to reduced water use and substance inputs if perennial crops (short rotation with poplars and willows, energy grasses) are integrated into the landscape, as examples from all over the world show (IEA and GBEP 2016; Neary 2015). Other measures for ensuring the sustainable use of water resources include

- Efficiency increases through modern irrigation technologies,
- the use of treated wastewater for irrigation,
- the sustainable consumption of food.

The last point, in particular, contains significant potential for conserving water resources. It includes the reduction of the consumption of meat and dairy products, as well as the amount of food waste produced (▶ Sect. 2.4; ▶ Chap. 3). It is estimated that one third of the world's food is lost. This corresponds to 1.3 billion tonnes per year (Gustavsson et al. 2011). Food waste in the EU-27 amounts to 89 million tonnes per year or 179 kg per capita (Monier et al. 2010).

9.5 Biodiversity

In addition to climate change, the loss of biodiversity is one of the major global environmental threats and one of the greatest challenges facing humankind. Species loss affects the environment, the economy and society as a whole, as genetic resources represent an important natural asset. Preserving biodiversity is therefore a central goal of sustainable development. The protection of biological diversity is intensively discussed and promoted at the global, European and national levels in scientific, social and political terms. The number of political agreements on biodiversity is correspondingly diverse. The relevant international agreement is the *UN Convention on Biological Diversity* (CBD). This goes beyond pure nature conservation, also including the economic potential of natural resources, and social concerns. The three main objectives of the CBD are to

- protect biological diversity,
- sustainably use the components of biodiversity, and
- equitably share the benefits arising from the use of genetic resources.

The CBD is a framework agreement under international law that lays down guidelines and principles that are implemented by the signatory states in national strategies. Every two years, a Conference of the Parties to the Convention (COP) takes place. The main results of these conferences are protocols that serve to regulate specific topics. The Nagoya Protocol of the COP 2010 and the "Strategic Plan 2011–2020" adopted there regulate access to genetic resources and the equitable sharing of benefits. The so-called Aichi targets contained in the plan specify strategic benchmarks and 20 targets to be achieved by 2020. This includes, in particular,

- greater integration of the biodiversity problem into governments and society,
- the sustainable use of natural resources,
- the protection of ecosystems, different species and genetic diversity,
- a fair and equitable distribution of the benefits arising from the use of biodiversity,
- better accessibility and use of existing knowledge.

In addition, the *Access and Benefit Sharing* ensures that genetic resources and their diversity are used and developed into patents only with the consent of the countries of origin and with profit-sharing included. The economic incentives are intended to promote the conservation of biodiversity in developing countries and (◘ Fig. 9.3) will be given a permanent higher status.

The protection of biodiversity is also included in the *Sustainable Development Goals:* SDG 15 calls for halting the loss of biodiversity and protecting, restoring and promoting the sustainable use of ecosystems. By 2020, the aim is to stop deforestation, reforest forests and prevent the extinction of species.

Both the EU and Germany have formulated their own strategies and goals for the protection of biodiversity in accordance with the CBD. In 2007, the National Strategy on Biological Diversity (NBS) was developed and embedded in the EU's National Sustainability Strategy and Biodiversity Strategy. The strategy extends to 2020 and comprises around 330 targets and 430 measures and indicators for monitoring and improving them.

Despite isolated advances, there is no doubt that the success record of biodiversity policy has so far been rather low – including as measured against its own objectives. Obviously, classical concepts of nature conservation, which mainly rely on regulatory requirements and prohibitions, are reaching their limits. Although a greater number of areas are protected worldwide, it has not yet been possible to halt the loss of biological diversity. The majority of habitats and species in Germany and Europe are also in an unfavourable-to-alarming conservation status (EEA 2015). The UN Strategic Plan 2011–2020 and the EU Biodiversity Strategy 2020 therefore no longer pursue the classical protected area approach, as is pursued with Natura 2000 sites, but instead focus on economically-oriented strategies. One example of this is the financial support for organic farming as an agri-environmental measure to preserve and promote biodiversity within the framework of European agricultural policy. The promotion of the entry into or changeover to an organic farming method takes place in the first 2–3 years, during which the products may not yet be sold as organic goods with correspondingly higher prices. A second example is the support and revitalisation of beekeeping as an economic activity, e.g., by promoting the marketing of bee products

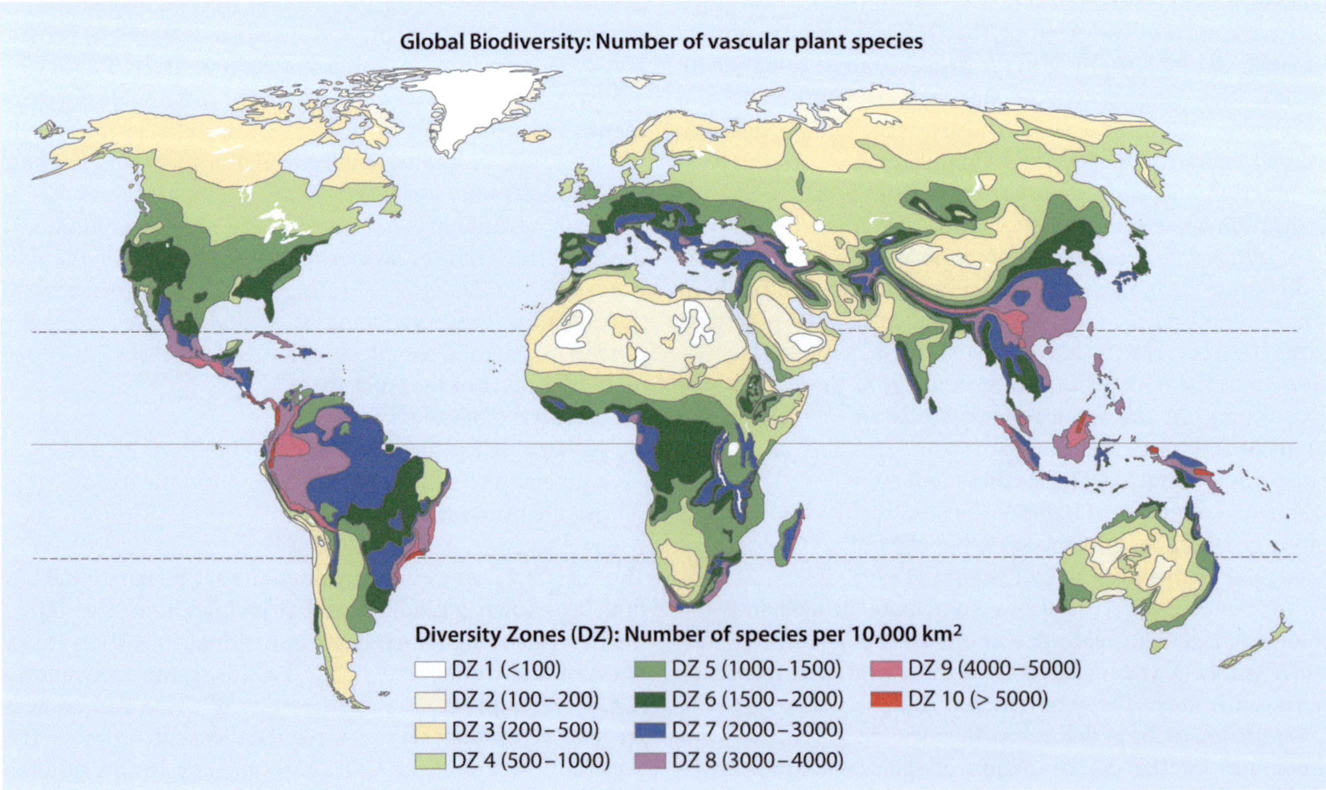

◘ **Fig. 9.3** Global distribution of biodiversity. (Barthlott et al. 2016)

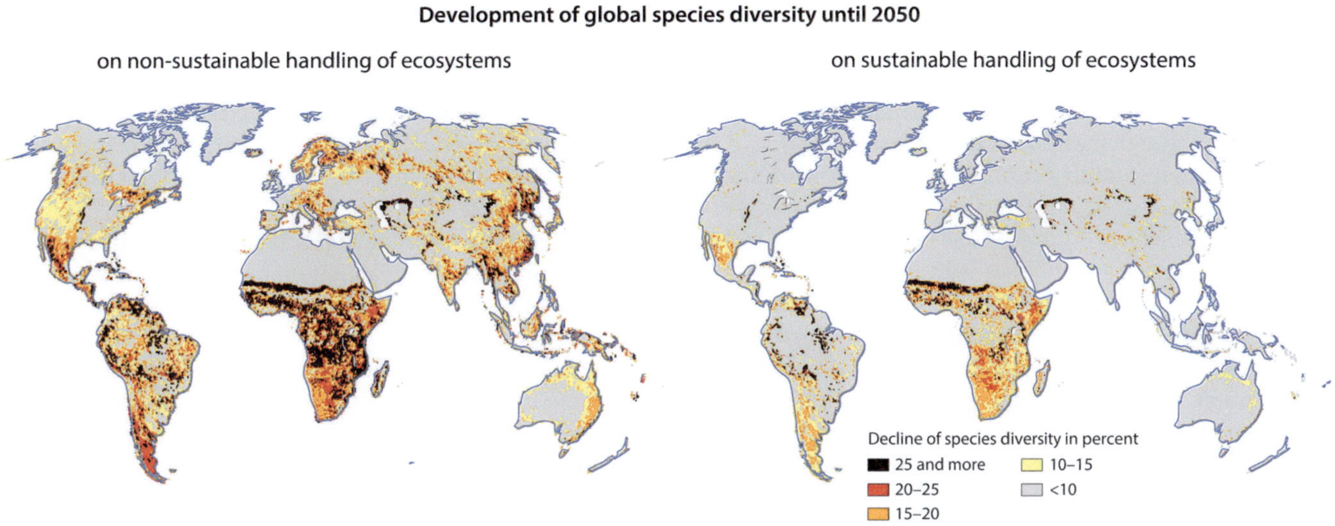

Fig. 9.4 Development of global biodiversity by the year 2050. (© Hugo Ahlenius, UNEP/GRID-Arendal, ▶ http://www.grida.no/graphics-lib/detail/biodiversity-loss-state-and-scenarios-2006-and-2050_d944)

and improving the framework conditions for training and information. This is because bees ensure the survival of many plant species, and thus biodiversity, through their pollination function.

The main causes of species decline (Fig. 9.4) include intensive forms of land cultivation. Further hazards lie in forestry, hydraulic engineering and water maintenance, construction measures and sports and leisure activities (EEA 2015). Obviously, such hazardous causes are dominant, and are associated with an intensification of the use of nature and landscape and the associated changes in habitats.

For many wild animal and plant species, agricultural landscapes are habitats, food sources and breeding and retreat areas. At the same time, they are production areas for biomass, the raw material of the bioeconomy. Intensive plant cultivation aimed at maximum yields considerably limits species diversity and forces the loss of biological diversity. Tight crop rotation reduces habitat diversity, requires significant use of fertilisers and pesticides and displaces the natural vegetation adapted to the location. The use of efficient technology leads to an increase in field felling and to the disappearance of field shrubs, natural landscape elements, such as hedges, or flowering strips.

A further increase in the demand for production areas and biomass by the bioeconomy can lead to a further intensification or change in land use, or even to an expansion of agricultural production areas, and to habitat loss, thus reducing the ecological compensatory function of the landscape. This is illustrated by the example of energy crop cultivation, which has led to the cultivation of set-aside land and the conversion of grassland into arable land. These areas are among the most species-rich habitats. More than half of the species population in Germany are located on grassland sites, which are among the most species-rich biotopes in Central Europe. The ploughing up of grassland for the cultivation of energy and raw material crops for the bioeconomy is therefore usually accompanied by a loss of species (Rösch et al. 2009). Other critical developments in agriculture, such as short breaks in cultivation, tight crop rotation, concentration on a few crop species and the increasing intensity of cultivation, are also seen as being the result, above all, of quantity-oriented energy crop cultivation.

The primacy of food security and the resulting demand to grow raw materials for the bioeconomy primarily on land that is either not suitable or only of limited suitability for food and feed production (▶ Sects. 9.2 and 9.3) leads to conflicting objectives with biodiversity, as these “inferior” areas are often particularly species-rich. In order to solve this dilemma, manufacturers of biofuels and electricity from liquid bioenergy sources have been obliged, since 2011, to prove the sustainable production of the bioenergy source. The sustainability ordinances for biofuels and biomass electricity apply to both biomass from Germany and that from other countries if it is to be credited against the biofuel quota, and thus granted a tax reduction or remuneration under the Renewable Energy Sources Act (EEG) in Germany. According to these regulations, no biomass may be used that comes from areas with a high nature conservation value or from carbon stocks.

When assessing the impacts of the bioeconomy on biodiversity, a distinction must be made between direct local effects that can be clearly assigned geographically and causally and indirect effects that result from an overall increase in demand for biomass (Delzeit et al. 2014). The biodiversity effects of indirect land use changes (ILUC) and the resulting climate effects are difficult to measure (Edwards et al. 2010; Delzeit et al. 2014). It is controversial as to whether the agricultural area of around 1.5 billion ha worldwide can be expanded at all without endangering biodiversity conservation. Due to its land use abroad, Germany also has an influence on the global reduction of biodiversity. Converted into agricultural area, Germany has a net import of about 4 million ha (FNR 2014a), which corresponds to about one third of the German cultivated area (► Sect. 9.2). The demand for raw materials to achieve the goals of the bioeconomy cannot be met solely by domestic production, but must be provided by biomass imports. Thus, the problem of indirect land use effects due to bioeconomy (► Sect. 9.3) is of high relevance.

A way out of this dilemma could be the sustainable intensification of agricultural production. With this form of agricultural production, yield increases can be achieved, for example, through more efficient use of nutrients, improved nutrient dynamics or modified soil properties, without causing negative effects on the environment or taking up additional land for cultivation (The Royal Society London 2009). What 'sustainable intensification' means in concrete terms, and to what extent it differs from current agricultural practices, remains largely unclear in both science and practice (Petersen and Snapp 2015). The concept continues to be the subject of controversy. For some, it is a key element of a more sustainable agriculture. Others recognise the need for more resource-efficient agricultural production, but criticise the technocratic orientation and insufficient attention to biological processes and doubt the potential of this concept to achieve sustainable biomass production (Garnett et al. 2013). In Germany and Europe, according to scientific surveys, 45% and 43%, respectively, of the areas are not regarded as suitable for sustainable intensification (◘ Fig. 9.5)

An alternative to sustainable intensification is the concept of ecologising biomass production. Its aim is to increase biodiversity through the cultivation of energy crops. In addition to the creation of habitats for numerous species, the targeted promotion of ecosystem services in the open agricultural landscape is also important. Diversification of energy crop cultivation can, for example, promote soil regulation and the attractiveness of the landscape. The production alternatives for energy and raw material production are not limited to traditional food crops such as corn and rapeseed. Rather, there are various annual and perennial plants and a wide range of cultivation systems to choose from. For example, targeted mixtures of wild plants have been developed that provide biomass for energy production in biogas plants, as well as habitats and food resources for birds and insects (◘ Fig. 9.6).

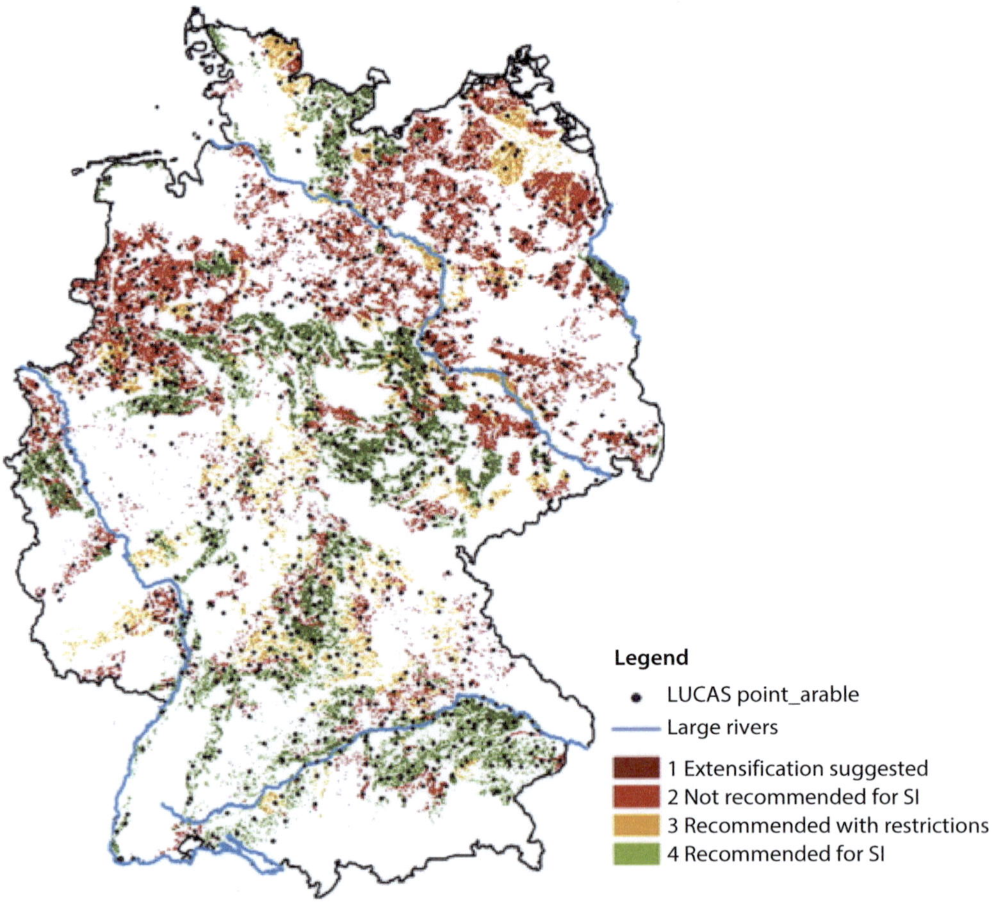

◘ **Fig. 9.5** Suitable land for sustainable intensification in Germany. (Source Schiefer et al. 2015)

Fig. 9.6 Species-rich grassland as a raw material supplier for the bioeconomy. (© Christine Rösch)

The bioeconomy can thus contribute to the conservation and promotion of agrobiodiversity and stop the decline of crop varieties. In addition to species diversity, the protection of variety diversity is of particular importance for necessary breeding progress in view of the future challenges in the context of the bioeconomy and climate change. Biological diversity and genetic diversity are the raw materials for breeding more efficient and adaptable plant varieties, and thus for a sustainable increase in the production of biogenic raw materials (BÖR 2014). New concepts for the design and extension of agrobiodiversity can facilitate the sustainable development of agricultural production. This includes, in particular, concepts for the optimal design of crop rotations with a significant increase in the number of crop species and biological production systems (▶ Excursus 9.2). Concepts of optimal crop rotation and fertilisation management adapted to the location are also advantageous, and include both the soil-internal nutrient supply potential provided by biodiversity and the nutrient appropriation potential of the crop rotation members that can close nutrient cycles as far as possible. So far, however, the inclusion of such questions in the sustainability criteria, especially for solid biomass, has only been realised to a limited extent (Fritsche and Seyfert 2013).

Excursus 9.2 Crop rotations

Multiple crop rotations increase the stability of agricultural ecosystems, have economic advantages in regard to labour and also reduce the business risk. Although these advantages are widely known, the design and implementation of multiple crop rotations is often very difficult, due to economic constraints. To a certain extent, disadvantages of monotonous crop rotations can be corrected through the use of chemical pesticides. Increasing problems, e.g., due to the spread of resistant populations of arable foxtail and common wind stalk, have, in recent times, clearly shown the limits of this repair-oriented production method. The repeated cultivation of winter wheat in particular has a negative effect on disease pressure, soil structure and nitrogen dynamics. Even if soil characteristics and climatic conditions limit crop rotation, technical and breeding advances have improved the suitability of certain crops for cultivation, and thus the possibilities for crop rotation (example: late sowing of winter wheat). The increasing importance of energy crops can also have a positive impact on the versatility of crop rotations. The aim should be to avoid cereal-based crop rotations (cereal content above 65%), breaks in cultivation or long fallow periods and high work peaks through the cultivation of similar crops, in order to reduce the risk of soil-damaging and non-situational measures. The change from summering to wintering should be encouraged, possibly in combination with intercropping and legume cultivation.

The priority use of organic wastes and residues is often regarded as a solution to the conflict between biomass production for the purpose of covering the raw material requirements of the bioeconomy and the preservation of biodiversity. The fact is often overlooked that residual materials can also be important for maintaining soil fertility and biodiversity. This is demonstrated by the examples of deadwood and its functions for the ecosystem services of forests, as well as straw and its significance for the fertility of arable land. If these restrictions are respected, Germany, for example, will still show relevant biomass potential for the bioeconomy (DBFZ 2015b).

Overall, biodiversity as a very important aspect of sustainability has so far been given too little consideration in the discussion about the bioeconomy (Larsen 2012). That must change in the future. The current indicator report on the National Strategy on Biological Diversity (BMUB 2015b) should be the reason for this. It shows that species diversity decreased significantly between 2001 and 2011, and is now only 63% of the target value (DESTATIS 2016).

9.6 Resource Efficiency

Natural resources, especially in the form of raw materials, are essential production factors. They represent the foundations of our social productive potential and prosperity. However, the increasingly intensive use of natural resources by humans can exceed the load limits of ecosystems and exacerbate global environmental problems. In the past 30 years, global raw material extraction has doubled to around 70 billion tonnes per year. Already today, this is clearly exceeding the earth's ability to regenerate and endangers the development opportunities for future generations.

Natural resources should be used more intelligently and more efficiently in order to achieve the same production result or the same service with a lower consumption of resources, and thus accrue more benefit and more prosperity (EEA 2016b). The drastic increases in resource efficiency (synonym: resource productivity) indicate that there is enormous potential in many areas of production and consumption that has so far gone largely unused and is, in some cases, not yet sufficiently understood. In order to tap this potential, far greater efforts are needed in politics, science and society than has previously been the case (EEA 2016b). The strategy of increased resource productivity is an important component in the concept of the *green economy* of the United Nations Environment Programme (UNEP 2011). At the national and European levels, strategic concepts for increasing resource productivity were adopted in 2012 with the German resource efficiency programme ProgRess (BMUB 2015a, c, d) and the flagship initiative Resource Conserving Europe (EC 2011). ProgRess is part of the National Sustainability Strategy of 2012 and its update (BuReg 2016, 2018). Despite its limited availability and resource-intensive production, biomass as a resource is excluded from these strategies (BMUB 2015a).

In its policy recommendations on resource efficiency, the OECD advises a mix of different strategies consisting of the elements of reduction, reuse and recycling (OECD 2016), which could be applied to the bioeconomy, too. This is also true for the concept of the circular economy, which is regarded as central to increasing resource efficiency (EEA 2016b). Its implementation is a complex process that requires fundamental changes at the different levels of the production-consumption system and technological, economic and social innovation (EEA 2016b) (◘ Fig. 9.7).

According to UNEP estimates, the possible contributions of the resource efficiency strategy to climate protection are enormous. In view of the existing trends of a growing world population, an expanding global middle class and further urbanisation, all of which could be accompanied by an increase in raw material extraction from 85 to 186 billion tonnes by 2050, such a strategy is not only important, but imperative (UNEP-IRP 2016). Demand for biotic and abiotic raw materials is growing, especially as emerging countries catch up in unsustainable development. This has economic, ecological and social consequences. The scarcity of important raw materials, rising raw material prices and supply risks are weighing on the economy and leading to environmental damage (cf., e.g., Fritsche 2013). These can range from the release of greenhouse gases and pollutants into the air, water and soil to the degradation of ecosystems and a threat to biodiversity.

In view of the long-term necessity of transforming fossil-based economic systems into more sustainable systems based on renewable resources, the bioeconomy also faces the challenge of taking into account the limited and increasingly scarce raw material base. It must develop strategies to reduce the specific demand for biomass and the resources required to produce it, respectively, land, water and nutrients, including reuse and recycling concepts. An essential element for increasing the efficiency of the biomass resource is the concept of **cascading use** (◘ Fig. 9.8) (► Chap. 7). The biomass is used as long, as frequently and as comprehensively as possible, only being used for energy at the end of the product life cycle (UBA 2013). Ideally, the cascade use should follow successive stages, from the highest possible level of value creation to the lower levels. A prominent example of this is the use of wood: It is first used materially (e.g., for furniture), then recycled (e.g., chipboard production from waste wood), and then burned to generate electricity and heat. In order to implement the concept of cascading use, it is necessary to create not only structures and processes, but also legal regulations to promote cascading. In Germany, for example, the legal basis for cascading use of waste is already laid down in the so-called Wasted Management Act (► Sect. 2.4).

The concept of the **circular economy** is another essential strategy for increasing resource efficiency in addition to cascading use (► Chap. 7). The use of production residues and biowastes reduces the extraction of primary raw materials and the need for mineral fertilizers. At the same time, there is less waste to be disposed of, which can lead to cost savings and resource conservation (◘ Fig. 9.9) On the other hand, the use of resources and energy required for the processing of residual

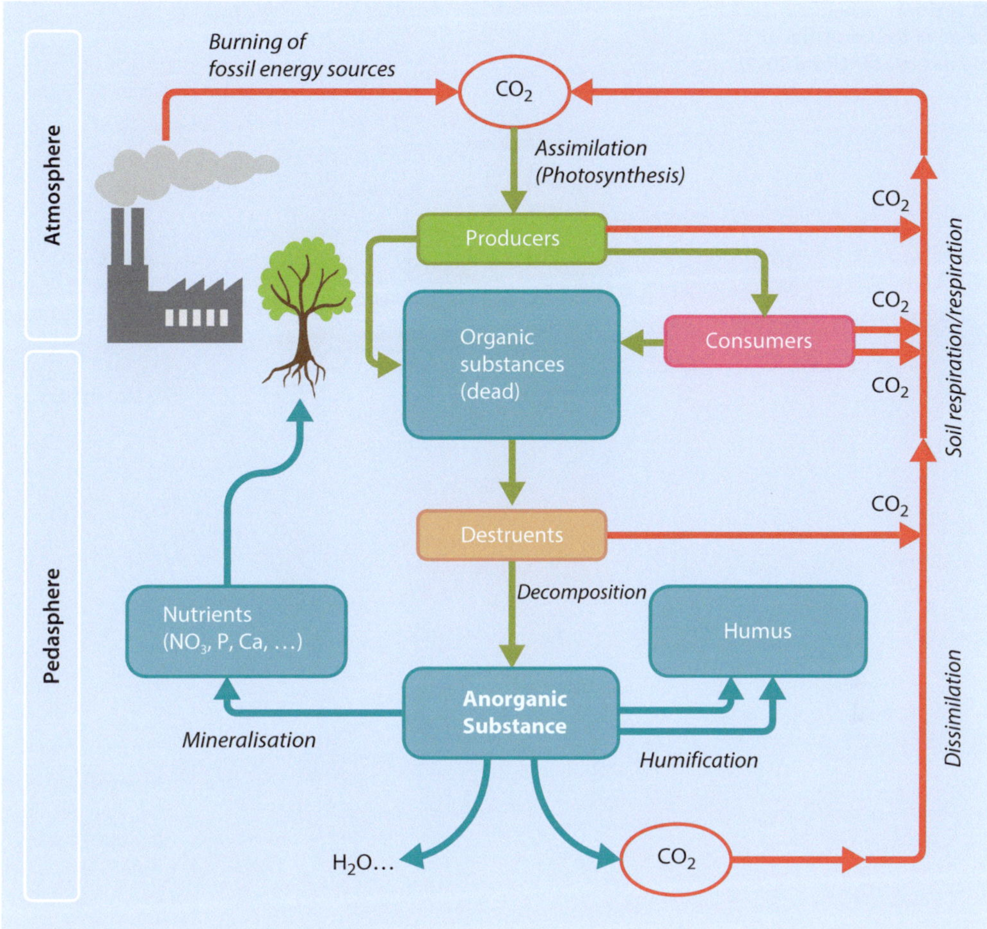

Fig. 9.7 Carbon cycle. (Hypersoil Uni Münster 2016)

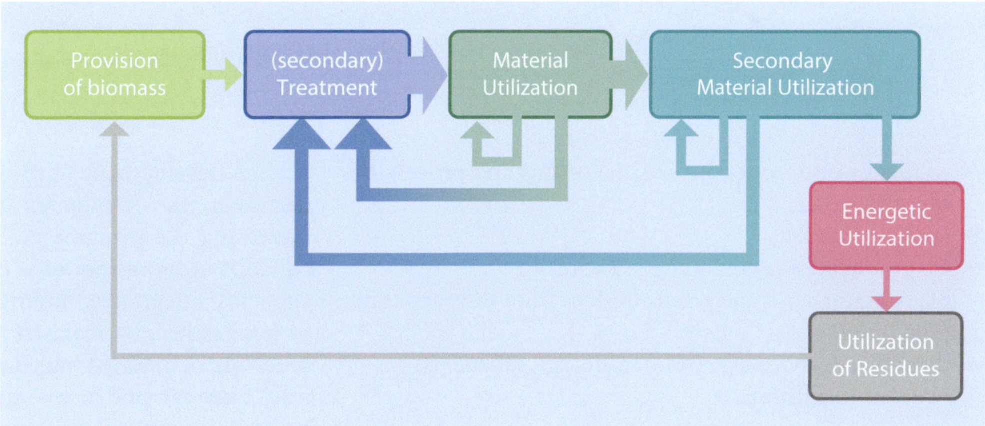

Fig. 9.8 Concept of cascade use of biomass

materials can cause environmental pollution, for example, through the release of greenhouse gas emissions from composting biological waste. A further restriction results from the legal regulations of the circular economy, which, in the interest of an orderly recovery and disposal of wastes, defines requirements for recycling that may hinder the use of biogenic secondary raw materials. Nevertheless, residual materials and wastes from agriculture, forestry and the food sector have a significant potential for a circular economy and, by this, can increase resource efficiency in the bioeconomy, a fact that has not yet been sufficiently exploited. Residues and wastes as feedstock for the bioeconomy include the use of

- animal excrements (e.g., slurry, manure, faeces),
- organic and green wastes (e.g., food waste, herb-like wastes),
- straw and crop residues (e.g., beet leaves, straw),
- sewage sludge from sewage treatment plants,
- organic household waste, municipal waste and by-products of food production (e.g., used plant oils, animal fats, potato peels),

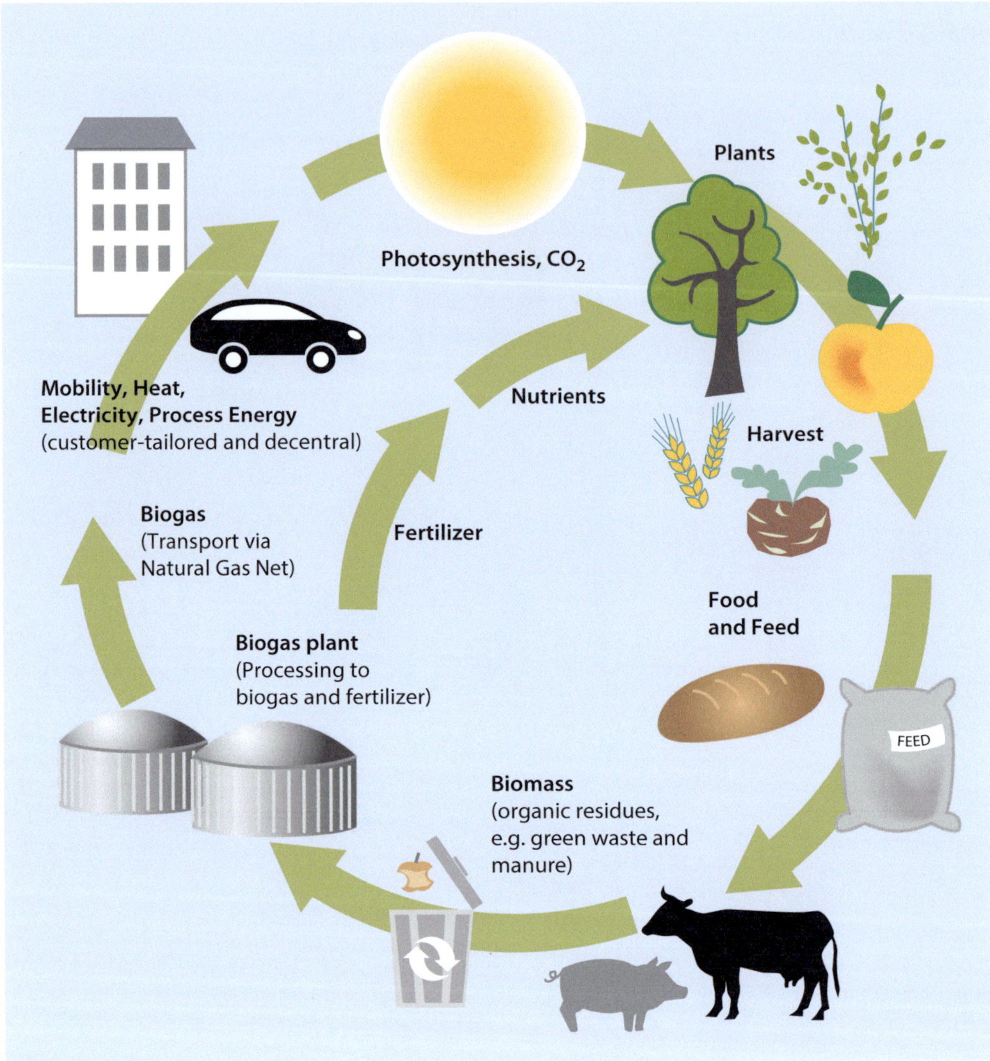

Fig. 9.9 Schematic representation of the circulation of biomass. (© SFPI Inwil 2012)

9

- residual wood from thinning, harvesting and processing of forest wood,
- landscape wood,
- by-products of sawmills (e.g., sawdust),
- black liquor, bark and other residues from the paper and pulp industry,
- demolition wood (e.g., wooden storage pallets, old wooden furniture).

It is estimated that between 8% and 13% of future energy requirements could be covered by these residual and waste materials (BMU 2012).

However, the use of biogenic residues and waste materials has its limits, because biomass also has a variety of ecological functions, for example, as a habitat and as a carbon store. More efficient use of residual materials can therefore further aggravate the critical state of global ecosystems and their productive and regulatory functions. For example, the removal of residual wood from the forest and straw from the field can lead to a situation in which an insufficient amount of organic substances and minerals remain.

The development of processes and technologies for the thorough use of biomass is another central element for increasing the resource efficiency of the bioeconomy. The **concept of biorefineries** is of outstanding importance. This explicitly integrative, multifunctional concept uses biomass as a versatile raw material source for the production of a spectrum of different intermediate products, as well as chemicals, materials and bioenergy, using all raw material components as completely as possible. Food and feed may be produced as by-products of biorefineries. The biorefineries operate on various raw material bases. They use cereals and sugar beet as suppliers of sugar and starch, grasses, wood and straw as suppliers of lignocellulose, or algae. Most of them are still in the development phase and require further research before they can be commercialised in practice (Fig. 9.10) (► Chap. 4).

The biorefinery concept focuses on the integration of different processes into coherent technical processes and the upscaling of the concept from the experimental and pilot to the industrial scale. In addition, there is the need for optimisation of the raw material supply and primary and secondary

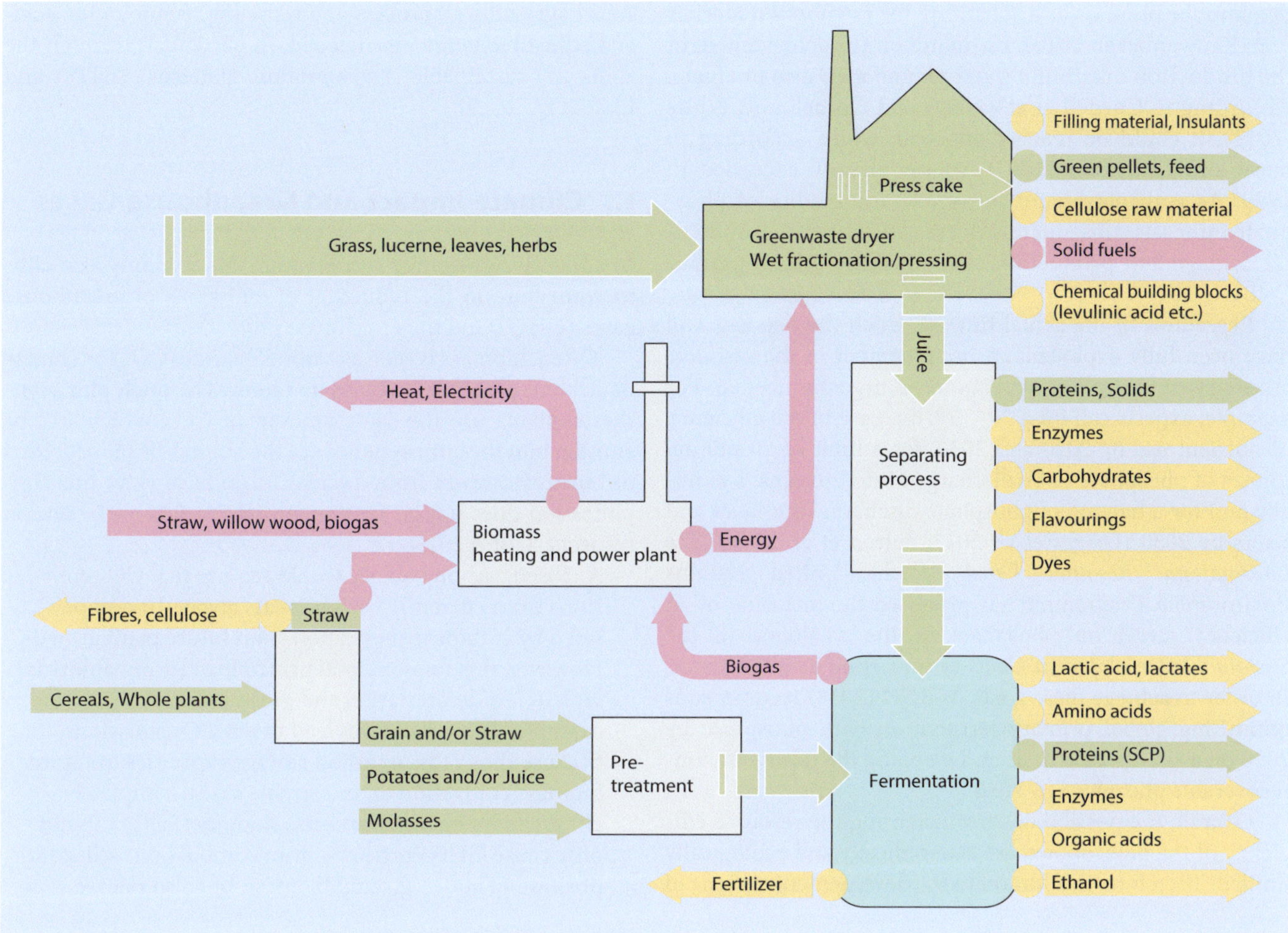

Fig. 9.10 Design of a biorefinery. (Biorefinery concept 2016)

refining processes (Arnold et al. 2011; Fritsche et al. 2012; Maga 2015; Wolf et al. 2016). The implementation of the integrative and multifunctional approach of the biorefinery must be supported by political framework conditions. Otherwise, there is a risk that a biorefinery will, in practice, be more materially oriented or energy-related for economic reasons, thereby significantly reducing product diversification and resource efficiency.

The concept of the efficient use of resources does not only include biomass, but also the resources required for biomass production, such as soils, water, and nutrients. Besides nitrogen and potassium, phosphorus, in the form of phosphate, is one of the three main plant nutrients and components of mineral fertilizers. The availability of phosphate as a plant nutrient is indispensable for increasing land productivity and resource efficiency (Bouwman et al. 2009; WBGU 2011). An increase in the efficiency of the use of phosphate is needed in agricultural production, as this substance cannot be replaced by other substances or artificially produced, and its occurrence in highly concentrated, degradable phosphate rock is limited (WBGU 2011) (Fig. 9.11).

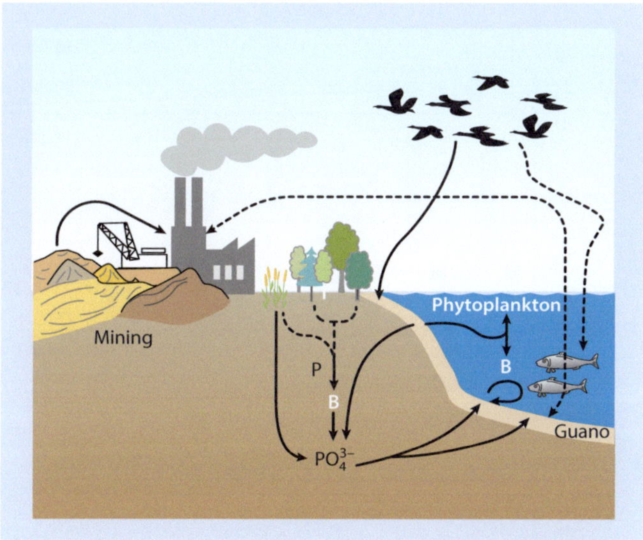

Fig. 9.11 The path of phosphorus. (Source: dtv-Atlas Ecology 1998)

However, the data on phosphate rock reserves, as well as on the duration of their complete extraction, are subject to considerable uncertainties. They fluctuate between 61 and 400 years, with the lowest estimates assuming the highest increases in demand due to a growing world population, changed consumption patterns and an increasing demand for biomass as a substitute for raw fossil materials (van Kauwenbergh 2010). Estimates on the achievement of the production maximum *(peak phosphorus)* also fluctuate. According to Cordell et al. (2009) and Cordell and White (2011), it could be reached in 2030, while, according to Déry and Anderson (2007), it was already exceeded in 1989. As with petroleum *(peak oil)*, the quality of phosphate minerals subsequently decreases, while environmental damage and production costs increase ceteris paribus (WBGU 2011).

Regardless of the actual time at which the deposits will have been fully exploited, an improvement in the resource efficiency of phosphate appears to be urgently needed. For example, experts call for a 20–30% increase in the efficiency of nutrient use by 2020 and 2030, for a limit of 10 million tonnes of phosphate to be discharged into the oceans annually and for a halving of phosphate discharge into lakes and rivers by 2030 (Griggs et al. 2013; Sutton et al. 2013). The *International Resource Panel* of the United Nations Environment Programme has proposed the inclusion of an efficiency target on phosphate in the catalogue of the *Sustainable Development Goals* to underline its importance. In order to achieve these goals, WBGU (2014) recommends optimising global primary fertilisation with phosphate by 2030 on a site-specific basis and stopping the release of non-recoverable phosphate by 2050.

Overall, the possibilities for increasing the resource efficiency of the bioeconomy are economically and ecologically limited. Therefore, in addition to the development and use of smart agricultural, process and recycling technologies, societal adjustments are also needed, in particular, through the shift to sustainable consumption patterns (ETPs and EUFETEC 2015).

9.7 Climate Impact and Greenhouse Gases

In addition to resource conservation, bioeconomy also aims to contribute to the reduction of emissions of greenhouse gases (GHG), and thus to climate protection.

Often, biomass is seen as carbon dioxide (CO_2) or climate neutral. In principle, to build up biomass through photosynthesis, plants use the same amount of CO_2, which will be emitted into the atmosphere once the biomass is biodegraded or burnt. However, from a scientific point of view, this simplification does not go far enough (Fig. 9.12) – in several respects (DBFZ 2015c):

- A closed carbon balance would require that CO_2 emissions from currently used biomass always be compensated by carbon sequestration from future plant growth. However, this fundamental principle of sustainability is violated by deforestation and grassland conversion, for example. These measures lead to net CO_2 emissions, because they reduce carbon storage capacities in future biomass cultivation or, in extreme cases, completely eliminate them. Such land-use changes (LUC) can not only cause GHG emissions from combustion or decomposition of above-ground biomass, but also release

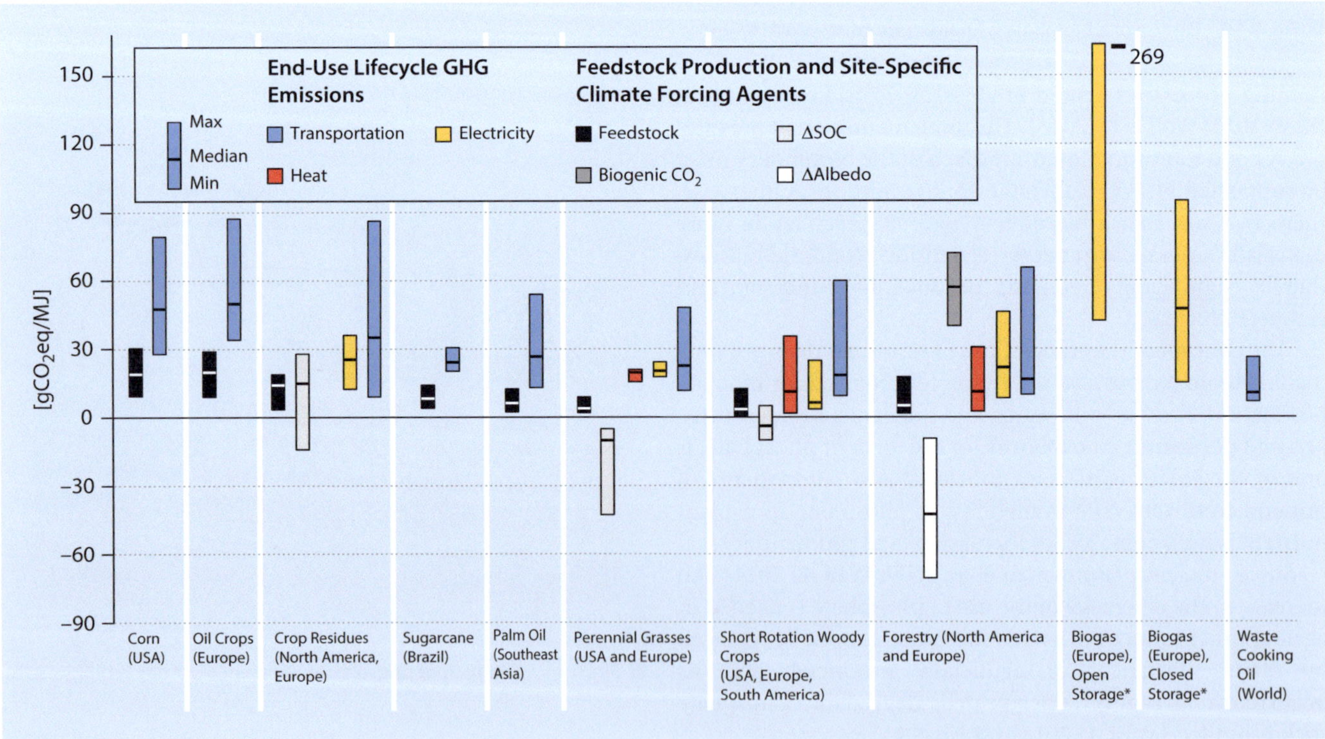

Fig. 9.12 Overview of the greenhouse gas emissions in the entire value chain of the most important bioenergy production systems, based on a meta-analysis of the technical literature. The median is marked with a dash. The lower and upper ends of the bars represent the minimum and maximum, respectively, of the values reported in the literature. (Creutzig et al. 2015)

organic carbon from soils. LUC-related GHG emissions can be very high (IPCC 2011).
- In addition to carbon dioxide, there are other greenhouse gases that are particularly relevant for agricultural biomass – methane (CH_4) from anaerobic decomposition of plant material and nitrous oxide (N_2O) from fertilization. The various greenhouse gases can be measured by their respective global warming potential (GWP) in CO_2 equivalents and, through that, taken into account in GHG balances in a weighted manner (IPCC 2013).
- In addition to direct GHG emissions from cultivation and combustion, there are usually further GHG releases from the life cycles of products made from biomass, e.g., through auxiliary energies and materials, transport and processing. These emissions are "upstream" of the use of biogenic products and must be taken into account in life cycle analysis, and assessments (DBFZ 2015c).
- Further GHG emissions may be caused by indirect land-use changes (ILUC), which can occur when the cultivation of biomass uses land that was previously used for food or feed production (Fritsche et al. 2010a, b; Edwards et al. 2010) – but corresponding model calculations show high bandwidths and uncertainties (Delzeit et al. 2016; De Rosa et al. 2016; Plevin et al. 2015; Valin et al. 2015). ILUC impacts can also be reduced through appropriate measures, such as double- and intercropping, the use of degraded land or cultivation on "surplus land" (RSB 2015; Wicke et al. 2015).
- In addition to the purely numerical balancing of GHG emissions, their temporal dynamics are also important for the net climate impact, i.e., changes in global temperature. This dynamic is referred to as the *carbon debt*, especially for biomass from forests: If, for example, thinnings and logs are burned to provide bioenergy, the CO_2 released is re-absorbed in the long term through the growth of replanted trees, but, depending on the type of forest, this can take up to several hundred years. Bioenergy "owes" its climate neutrality during this time, because, although it may have replaced coal, for example, there is temporarily more CO_2 in the atmosphere, leading to more global warming compared to a scenario without bioenergy (JRC et al. 2014; Matthews et al. 2015). It should be noted, however, that the use of residual forest wood typically only takes a few years to offset this "debt," because, if it were not used, these "harvest wastes" would remain in the forest and quickly decompose, releasing CO_2, and CH_4. Also, energy use of wood from short rotation forestry (SRF) and of lignocellulose from energy grasses has almost no carbon debt, because biomass will grow again in a few years, for example.
- Whether the bioeconomy is climate-friendly is ultimately determined through a comparison with a case in which no biomass is used for energy and/or materials. The choice of this reference scenario representing the non-bio ("counterfactual") case strongly influences the results, especially the possible carbon debt of bioenergy from forests (Matthews et al. 2015).

This list shows that the question of the climate impact of the bioeconomy is not easy to answer. The complexity of the question is further increased by the variability of biomass uses. Depending on where, when, how and which biomass is used, there are different GHG emissions and net effects in relation to also different reference systems. This must be taken into account in the following examples – they are not suitable for making general statements about the GHG balance of the bioeconomy, but show the range of such balances.

- In the case of biofuels, GHG emissions vary greatly, depending on the raw material, its origin and processing paths compared with fossil fuels (Edwards et al. 2014; Elshout et al. 2015), which is valid also if ILUC effects are included (Valin et al. 2015).
- In the case of bioenergy, a distinction must be made between electricity and heat and their combined provision (cogeneration) (IFEU 2016; van Hilst et al. 2015; JRC 2015; Strengers et al. 2015). Respective balances for bioenergy systems are shown in ◘ Fig. 9.12.
- In the vast majority of cases, GHG emissions from biomass use for the production of biomaterials are significantly lower than those from fossil reference products (Barth and Carus 2015; Carus et al. 2014; McKechnie et al. 2015; Tsiropoulos et al. 2015). However, ILUC effects must also be taken into account here. If the supply of raw materials leads to a displacement of food production (Bardhan et al. 2015), the GHG reduction may be lower or even negative.
- Depending on their design, biorefineries provide a combination of energy and material products (e.g., chemicals, fertilizers), as well as animal feed (cf. ▸ Sect. 4.2). Their GHG balance, therefore, depends on the respective type, but generally shows lower emissions than conventional comparison systems (Arnold et al. 2009; Haro et al. 2015; Maes et al. 2015; Maga 2015; Parajuli et al. 2015).

9.8 Rural Development

The bioeconomy is seen as an opportunity for the development of rural areas, as it opens up new ways to use renewable resources in industrial processes. It thus can contribute to sustainable economic growth, employment opportunities and income in rural areas, because biomass is produced, stored, processed and – at least partially – sold locally. However, the use of renewable raw materials per se neither guarantees sustainable economic development nor does it necessarily generate added economic value compared with a conventional fossil economy. Yet, transforming a fossil once-through ("linear") economy into closed material cycles with "circular" value chains offers potential for sustainable rural development.

Supported by environmental and climate policy targets and fiscal measures, the bioeconomy has experienced a kind of boom in recent decades and has developed into an important economic sector (BÖR 2010; GBS 2015; Scarlat et al. 2015). In the EU, between 1,600 and 2,200 million tonnes of biomass are produced annually. At 70%, carbohydrates such as cellulose, sugar and starch account for the largest share of the biomass used. The European bioeconomy generates an annual turnover of around €2 trillion and employs around 18 million people. This corresponds to about 9% of all jobs in the EU (EC 2012; JRC 2015; Carus 2016). In Germany, too, the bioeconomy is a prospering economic field, employing almost 5 million people in 2007. This corresponds to 12.5% of total employment. 125,000 of these jobs are related to the production of bioenergy, the current core of the bioeconomy, with 25,000 employees for the provision of biofuels such as bioethanol and biodiesel (O'Sullivan et al. 2014). In 2012, the gross value added as a result of the bioeconomy in Germany amounted to € 160 billion or 7.6% of the total gross value added. Of this, 62% was accounted for by agriculture, food and horticulture, 33.5% by forestry and timber, 2.3% by bioenergy and 1.3% by the material use of raw agricultural materials (Efken et al. 2012). Technological progress and productivity gains are leading to ongoing structural changes in agriculture, which are reducing the number of farms and agricultural jobs. The bioeconomy can mitigate, albeit to a modest extent, the steady decline in agricultural employment by creating new employment opportunities. Although the bioethanol industry in the EU has created around 50,000 direct and indirect jobs, 4.8 million full-time agricultural jobs disappeared in the EU between 2002 and 2012. In Germany, there are still about 300,000 agricultural enterprises, with about 1.1 million employees and a turnover of about € 50 billion. However, only about 18% of these employees are permanently employed. More than half of them are family workers and around 30% are seasonal workers. Especially in small companies, the sideline plays an important role.

As in the agricultural sector as a whole, the employment effects of the bioeconomy are decisively based on subsidies, promotional measures and economic and regulatory frameworks set by policymakers. Through state compensatory measures and support for agriculture, biomass production has a special role compared to other sectors of the market economy. German farmers receive about € 5 billion in direct payments through the EU agricultural policy, allocated according to the area cultivated. In addition, there are extensive financial transfers from other policy areas, such as agricultural social policy, rural development and the promotion of bioenergy. The direct payments and subsidies to agricultural holdings in the main occupation amounted to more than € 30,000 in 2014/15 (Statista 2016).

In Germany, around 2.4 million hectares of renewable raw materials are cultivated, corresponding to 14.3% of Germany's total agricultural area (FNR 2014a; DESTATIS 2015). 88% of this area is devoted to the cultivation of energy crops. In the use of wood, bioenergy use now exceeds material use, too (◘ Fig. 9.13).

The figures show that bioenergy has the largest share of the bioeconomy in Germany and its material use has been comparatively low to date. This is due to the promotion of bioenergy value chains (◘ Fig. 9.14) by the Renewable Energy Sources Act (EEG), the Biofuel Quota Act and the technically well-managed use of bioenergy in various processes. However, due to the economic impact chains and fiscal circumstances, it is assumed that the benefits to public budgets from bioethanol production exceed the original production value for bioethanol (Schöpe and Britschkat 2006).

Due to their often small-scale structures, the cultivation of energy crops and the use of bioenergy create decentralised employment opportunities and can help secure jobs in rural areas and promote investment in structurally weak regions. Imports can also have an impact on employment in emerging and developing countries, provided the economic prospects and the right framework conditions are in place. According to FAO (2008), the use of unused agricultural potential can attract urgently needed investments in rural development. The resulting increase in the productivity of the agricultural sector can contribute to food security and poverty reduction.

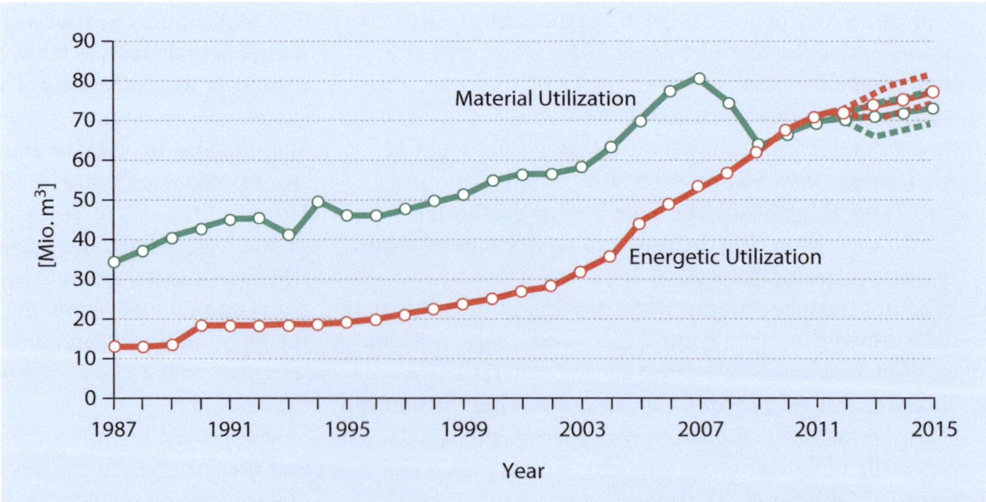

◘ **Fig. 9.13** Development of the material and energy-related use of wood in Germany. (Source Mantau 2012)

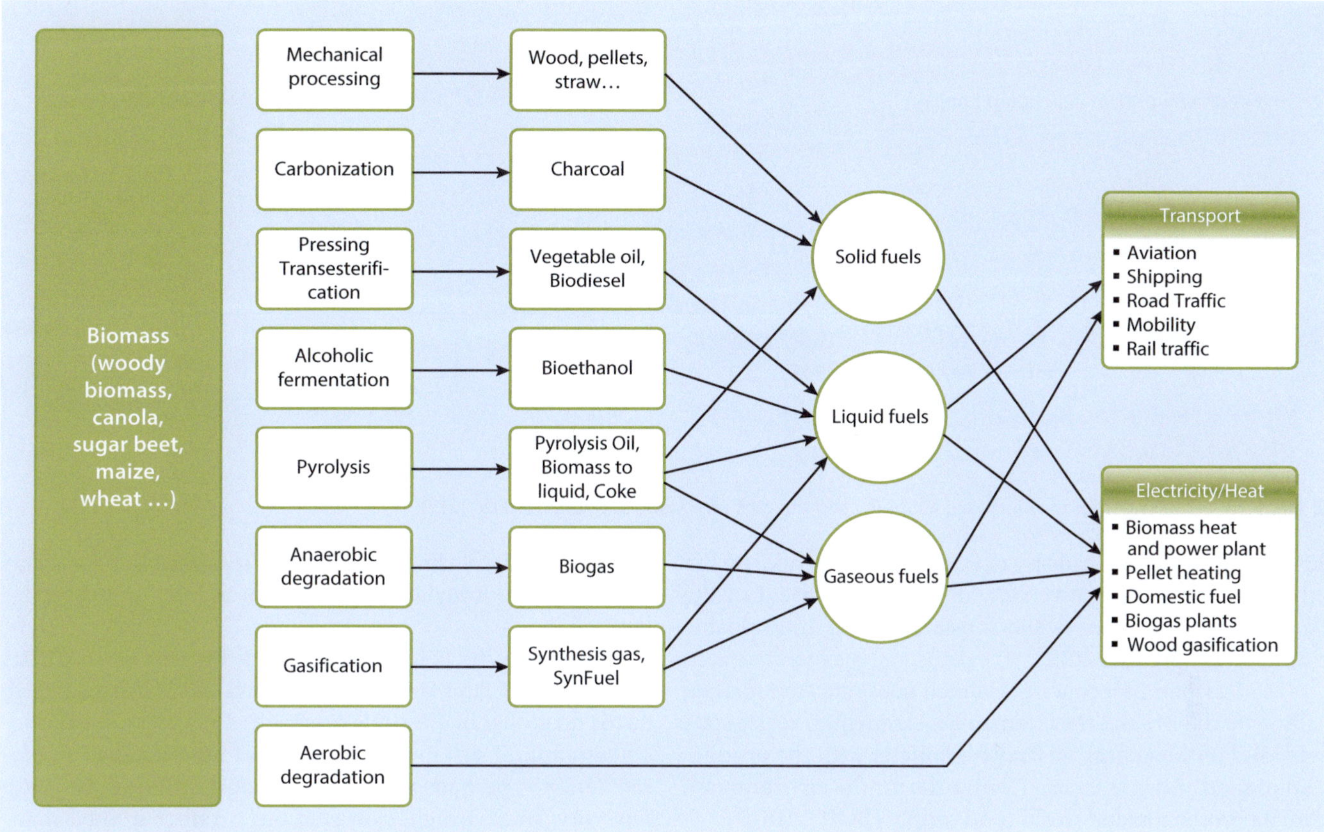

Fig. 9.14 Value chains of bioenergy. (FNR 2014b)

9.9 Conflicting Goals

The transformation of the economy from fossil- to bio-based has many impacts on humans and the environment. Climate protection, biodiversity conservation, the sustainable use of natural resources and the satisfaction of human needs are closely interlinked through a variety of interactions.

The transformation of the economic system into an age without fossil fuels is therefore more than simply a change in industry's carbon supply. It requires a variety of changes that are inevitably associated with conflicting objectives ("trade-offs"). The implementation of the bioeconomy can lead to interactions between economic, ecological and social aspects and to unintended conflicts of objectives and interests at different spatial levels (local, national and global) (Fig. 9.15). A systematic identification of these interdependencies and conflict potentials and a cross-scale, cross-sectoral and integrated consideration of the individual areas of the bioeconomy are, therefore, indispensable for the political steering of the process.

At the same time, the question arises as to an equitable distribution of resources and access to them, both within the present generations (intragenerational justice) and between the present and future generations (intergenerational justice). In the bioeconomy debate so far, in regard to its ecological and socio-economic impacts and possible conflicts, the primacy of food security *(Food first)* is conjured (▶ Chap. 3). However, it is largely unclear as to what social and economic effects the transformation will have on different regions and household types. It is also unclear how these can be controlled and who will be involved in the negotiation processes. To date, the participation of social groups in shaping the transformation process is largely missing (e.g., NABU 2014; Lahl 2014; IÖW 2011). In most decision-making bodies, nature conservation and environmental protection are represented neither by civil society nor by science. The bioeconomic strategy is primarily developed by industry and technologically-oriented science representatives in the interest of industry.

Most analyses agree on the conflicting objectives that arise between food security, biodiversity conservation and the transition of the economy to renewable raw materials. However, the options for action and recommendations for implementation derived from this perception drift far apart. The European Commission and the German government regard investments in research and innovation, the development of modern and sustainable agriculture and the use of residual and waste materials as the preferred means of resolving these conflicting land use objectives. However, concrete regulatory proposals for implementation are lacking in the research strategy.

The reports of the German Advisory Council on the Environment (SRU) make clearer statements, proposing to limit biomass use to sectors in which biomass use is indispensable (SRU 2007, 2011). The German Advisory Council on Global Change (WBGU) highlights the aspects of responsible consumption and increased efficiency (WBGU 2008,

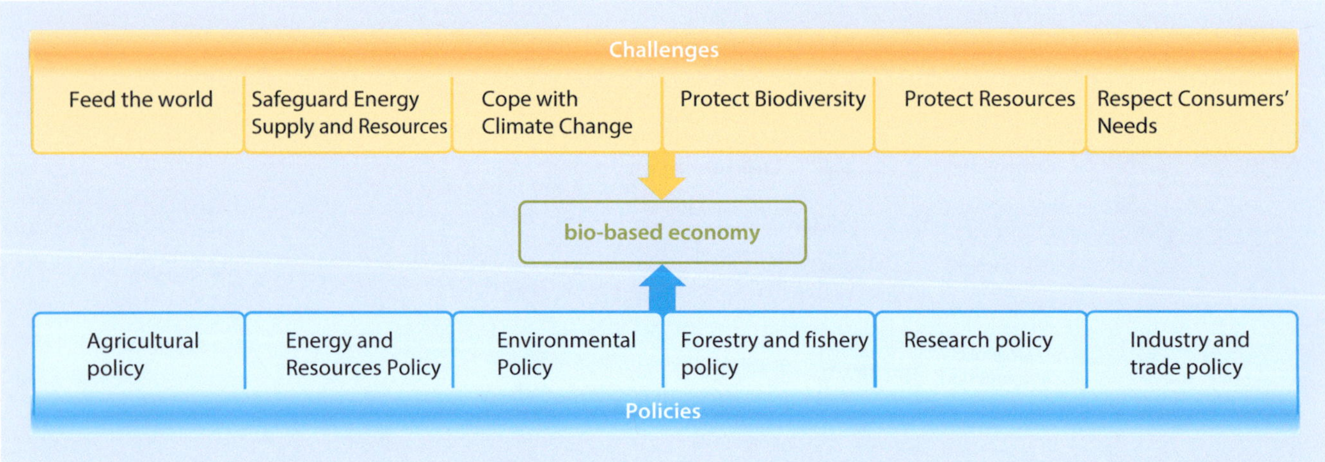

Fig. 9.15 Challenges for the bio-based economy and relevant policy areas. (Source BMELV 2012)

2011). The National Academy of Sciences Leopoldina, on the other hand, has a negative view of the development of the bioeconomy and stresses the foreseeable and unavoidable conflicts (Leopoldina 2012).

The German Bioeconomy Council goes one step further, calling on the EU and the German government to revise state subsidies for bioenergy, as its use competes with the production of food and is associated with risks for the environment, climate protection and world food supply (BÖR 2015).

9.10 Good Global Governance

Good Governance is a term from the international discussion on sustainable development – public discourse, participation and transparency are important aspects. *Good global governance* concerns the regulation of cross-border problems.

The bioeconomy is not just a German or European concept, but is pursued by many countries (► Chap. 1). It connects, via international value chains (► Chap. 7) and worldwide trade, raw material-producing countries with those in which intermediate products are manufactured or end products are consumed, i.e., it has a global reach. Bearing this in mind, and in view of its global potential (Piotrowski et al. 2016), the bioeconomy is a challenge for global sustainability.

Good global governance is necessary for the bioeconomy, because its opportunities and risks must be understood across borders, both in the geographical sense and in relation to "planetary borders" (Rockström et al. 2009a, b). This is one of the reasons why the WBGU, in its reports on global land use (WBGU 2008) and on the Great Transition (WBGU 2011), identified the sustainable use of biomass as one of the central policy tasks of this century. With the *Sustainable Development Goals* (SDGs), the countries represented in the United Nations General Assembly set themselves ambitious goals for globally sustainable development. Biomass as the basis of the bioeconomy is linked to many SDGs – both positively and negatively (► Sect. 9.1). So far, however, there is a lack of suitable bodies to assume responsibility for shaping the bioeconomy at the global – or at least multilateral – level.

Neither in the field of biomass (cf. Fritsche et al. 2010a) nor in that of land use (cf. Boudreaux 2015; Fritsche et al. 2015) do global institutions exist with a mandate to draw up corresponding sustainability rules and monitor their implementation – perhaps with the exception of the *Global Crop Diversity Trust*, which, however, has a rather limited mandate. The global governance of the bioeconomy has so far been a blank page.

There are a number of UN organisations, such as CFS, FAO and UNEP, that deal with aspects of a sustainable bioeconomy with regard to voluntary commitments and standards (such as the VGGT, cf. CFS 2012). They are, however, literally powerless against the interests of states or economic actors that do not accede to such agreements or that effectively ignore their implementation. In addition, the International Free Trade Agreement (GATT) and the World Trade Organization (WTO) largely exclude social and ecological aspects. Only the UN environmental conventions (biological diversity, climate, desertification) have established certain protective rights. Social regulations, such as the ILO Convention on Occupational Safety and Health, on the other hand, are seen as barriers to trade. The extent to which the implementation of the SDGs will improve this situation remains to be seen. In any case, they are an important milestone on the way to a cooperative, multilateral shaping of globalization and, therefore, also point the way forward for the bioeconomy.

After all, the lack of global governance has led to the establishment of intergovernmental partnerships such as the *Global Bioenergy Partnership* (GBEP) and the *Global Soil Partnership* (GSP). They include a large number of countries and international organisations and are actively working on – again, voluntary – regulations for the sustainable management of biomass (GBEP 2011; GSP 2016). These initiatives at least partially provide the basis for global governance.

The attempt to create a comprehensive international platform for the bioeconomy is still in its infancy. In this respect, the German Government has initiated the *Global Bioeconomy Summits*, which took place in 2015 and 2018 (GBS 2015, 2018) (▶ Chap. 1). The EU "Bioeconomy Manifesto" called for the broad participation of stakeholders in the discussion of the bioeconomy (EU 2016), and the recent update of the EU Bioeconomy Strategy argues in the same direction (EU 2018).

A further form of governance is sustainability-related standards for biomass, which do address producers, processors and consumers and make use of voluntary certification and control procedures. Such biomass-related sustainability standards have been developed in many different ways in the interim (Stupak et al. 2016; Thrän and Fritsche 2016). However, sufficient regulations are still lacking, especially for biodiversity (Fritsche and Seyfert 2013).

Also, most standards are only related to certain segments of the bioeconomy (e.g., energy: GBEP 2011; biomaterials: INRO 2013), and thus are fragmented (van Dam 2015). However, initial proposals have been made for overarching sustainability criteria and corresponding indicators (Iriarte et al. 2015).

Whether the transition to a sustainable bioeconomy will succeed will depend on the extent to which the currently very weak global governance is strengthened and resilient sustainability criteria can be implemented. It remains to be seen whether this will be achieved through multilateral agreements, through cooperation between national governments and transnational corporations or in hybrid forms – but this strengthening of global forces of law and order must take place if the bioeconomy is to live up to its claim of being a sustainable, post-fossil alternative.

We must not wait for a global solution. As other transformation processes, such as the German energy transition ("Energiewende"), show, central discussions on sustainability are not conducted merely at the global level, but also at the local, regional and national levels. There, concrete examples can be demonstrated, approaches to action tested and alliances forged between different groups of actors. This decentralised, bottom-up dynamic is no substitute for internationally accepted and enforceable rules for the bioeconomy – but it paves the way for them.

References

Arnold K et al (2009) Klimaschutz und optimierter Ausbau erneuerbarer Energien durch Kaskadennutzung von Biomasse – Potenziale, Entwicklungen und Chancen einer integrierten Strategie zur stofflichen und energetischen Nutzung von Biomasse. Wuppertal Institut. Wuppertal. https://epub.wupperinst.org/files/4448/WR5.pdf. Accessed 7 June 2016

Arnold K et al (2011) BioCouple – Kopplung der stofflich/energetischen Nutzung von Biomasse – Analyse und Bewertung der Konzepte und der Einbindung in bestehende Bereitstellungs- und Nutzungsszenarien. Endbericht zum Verbundprojekt gefördert vom BMU (FKZ-Nr. 03 KB 006 A-C). Wuppertal, Oberhausen, Darmstadt. http://www.energetische-biomassenutzung.de/fileadmin/user_upload/Downloads/Endberichte/Endbericht_BioCouple.pdf. Accessed 7 June 2016

Baka J (2014) What wastelands? A critique of biofuel policy discourse in South India. Geoforum 54:315–323

Bardhan S et al (2015) Biorenewable chemicals: feedstocks, technologies and the conflict with food production. Renew Sustain Energy Rev 51:506–520. Accessed 6 June 2016

Barth M, Carus M (2015) Carbon footprint and sustainability of different natural fibres for biocomposites and insulation material. Nova-Institute, Huerth. http://bio-based.eu/?did=14089&vp_edd_act=show_download. Accessed 7 June 2016

Barthlott W, Rafiqpoor MD, Erdelen WR (2016) Chapter 3: Bionics and biodiversity – bio-inspired technical innovation for a sustainable future. In: Knippers J, Nickel K, Speck T (eds) Biomimetic research for architecture and building construction: biological design and integrative structures. Springer, Berlin Heidelberg

Berndes G, Fritsche U (2016) May we have some more land use change, please? Biofuel Bioprod Bior 10:195–197

BMBF (2010) Nationale Forschungsstrategie BioÖkonomie 2030 – Unser Weg zu einer biobasierten Wirtschaft. Bundesministerium für Bildung und Forschung. Bonn, Berlin. http://www.bmbf.de/pub/biooekonimie.pdf. Accessed 25 May 2016

BMEL (2014) Nationale Politikstrategie Bioökonomie –Nachwachsende Ressourcen und biotechnologische Verfahren als Basis für Ernährung, Industrie und Energie. Berlin. http://www.bmbf.de/pubRD/Politikstrategie_Biooekonomie_barrierefrei.pdf. Accessed 25 May 2016

BMELV (2012) Strategie für eine biobasierte Wirtschaft. Präsentation beim Workshop im Bundesministerium für Ernährung, Landwirtschaft und Verbraucherschutz Berlin, 25. Oktober 2012

BMELV (2013) Politikstrategie Bioökonomie. Bundesministerium für Ernährung, Landwirtschaft und Verbraucherschutz. Berlin. http://www.bmelv.de/SharedDocs/Downloads/Broschueren/BioOekonomiestrategie.pdf?__blob=publicationFile. Accessed 5 Oct 2016

BMU (2012) Langfristszenarien und Strategien für den Ausbau der erneuerbaren Energien in Deutschland bei Berücksichtigung der Entwicklung in Europa und global. Stuttgart/ Kassel/Teltow, März 2012 (Leitszenario 2008–2012)

BMUB (2015a) Deutsches Ressourceneffizienzprogramm (ProgRess). Programm zur nachhaltigen Nutzung und zum Schutz der natürlichen Ressourcen. Berlin

BMUB (2015b) Indikatorenbericht 2014 zur Nationalen Strategie zur biologischen Vielfalt. http://www.bmub.bund.de/fileadmin/Daten_BMU/Pools/Broschueren/indikatorenbericht_biologische_vielfalt_2014_bf.pdf. Accessed 25 May 2016

BMUB (2015c) Deutsches Ressourceneffizienzprogramm (ProgRess): II: Fortschrittsbericht 2012–2015 und Fortschreibung 2016–2019. Berlin. http://www.neress.de/fileadmin/media/files/pdf/2015/ProgRessII_RA_pdf.pdf. Accessed 25 May 2016

BMUB (2015d) Deutsches Ressourceneffizienzprogramm (ProgRess). Programm zur nachhaltigen Nutzung und zum Schutz der natürlichen Ressourcen (2. Aufl.). Berlin. http://www.bmub.bund.de/fileadmin/Daten_BMU/Pools/Broschueren/progress_broschuere_de_bf.pdf

BMZ (2012) Investitionen in Land und das Phänomen des "Land Grabbing". BMZ-Strategiepapier 2/2012. Bonn. http://www.bmz.de/de/mediathek/publikationen/reihen/strategiepapiere/Strategiepapier316_2_2012.pdf. Accessed 27 May 2016

BÖR (2010) Gutachten des Bioökonomierats 2010. Innovation Bioökonomie. Forschung und Technologieentwicklung für Ernährungssicherung, nachhaltige Ressourcennutzung und Wettbewerbsfähigkeit. Forschungs- und Technologierat Bioökonomie. BioÖkonomierat. Berlin

BÖR (2014) Nachhaltige Bereitstellung von biobasierten agrarischen Rohstoffen. BioÖkonomierat. Berlin. http://www.biooekonomierat.de/fileadmin/Publikationen/berichte/030714_RUN_Nachhaltige_Bereitstellung.pdf. Accessed 1 June 2016

BÖR (2014a) Positionen und Strategien des Bioökonomierates. Berlin. http://www.biooekonomierat.de/fileadmin/Publikationen/empfehlungen/Strategiepapier.pdf. Accessed 1 June 2016

BÖR (2015) Landwirtschaft in Deutschland ihre Rolle für die Wettbewerbsfähigkeit der Bioökonomie. BÖRMEMO 01. BioÖkonomierat. Berlin. http://www.biooekonomierat.de/fileadmin/Publikationen/berichte/BOERMEMO_Landwirtschaft_final.pdf. Accessed 1 June 2016

BÖR (2016) Holz in der Bioökonomie – Chancen und Grenzen. Bioökonomierat BÖRMEMO 05. Berlin. http://www.biooekonomierat.de/fileadmin/Publikationen/berichte/BOERMEMO_Holz_vorla__ufig_final.pdf. Accessed 2 June 2016

Boudreaux K (2015) Land governance as a corporate performance standard: opportunities, challenges and recommended next steps. Prepared for the Global Donor Platform for Rural Development. https://www.scribd.com/document_downloads/299546579?extension=pdf&from=embed&source=embed. Accessed 27 May 2016

Bouwman A et al (2009) Human alteration of the global nitrogen and phosphorus soil balances for the period 1970–2050. Glob Biogeochem Cycles 23. https://doi.org/10.1029/2009GB003576

BuReg (2002) Perspektiven für Deutschland – Unsere Strategie für eine nachhaltige Entwicklung. Berlin. http://www.bundesregierung.de/Content/DE/_Anlagen/Nachhaltigkeit-wiederhergestellt/perspektiven-fuer-deutschland-langfassung.pdf?__blob=publicationFile. Accessed 24 May 2016

BuReg (2016) Deutsche Nachhaltigkeitsstrategie Neuauflage 2016. Entwurf Stand: 30. Mai 2016. Berlin

BuReg (2018) Deutsche Nachhaltigkeitsstrategie – Aktualisierung 2018. Presse- und Informationsamt der Bundesregierung (Hrsg.). Berlin. https://www.bundesregierung.de/resource/blob/975274/1546450/65089964ed4a2ab07ca8a4919e09e0af/2018-11-07-aktualisierung-dns-2018-data.pdf. Accessed 8 Mar 2019

Carus M (2016) 2,1 Billionen Umsatz und 18,3 Millionen Beschäftigte in der europäischen Bioökonomie: Pressemitteilung nova-Institut GmbH. Hürth

Carus M et al (2014) Ökologische Innovationspolitik – Mehr Ressourceneffizienz und Klimaschutz durch nachhaltige stoffliche Nutzungen von Biomasse. UBA-Texte 01/2014 (FZK 3710 93 109), Dessau. http://bio-based.eu/news/media/2014/02/14-02-%C3%96kologische-Innovationspolitik-%E2%80%93-Mehr-Ressourceneffizienz-und-Klimaschutz-durch-nachhaltige-stoffliche-Nutzungen-von-Biomasse-Langfassung-nova.pdf. Accessed 9 June 2016

CFS (2012) Voluntary Guidelines on the Responsible Governance of Tenure of Land, Fisheries and Forests in the Context of National Food Security. Committee on World Food Security. Rome. http://www.fao.org/fileadmin/user_upload/nr/land_tenure/pdf/VG_Final_May_2012.pdf. Accessed 10 June 2016

CIFOR (2013) Food security and nutrition: the role of forests. Jakarta. http://www.cifor.org/publications/pdf_files/WPapers/DPSunderland1301.pdf. Accessed 30 May 2016

Cordell D, White S (2011) Peak phosphorus: clarifying the key issue of a vigorous debate about long-term phosphorus security. Sustainability 3:2027–2049

Cordell D et al (2009) The story of phosphorus: global food security and food for thought. Glob Environ Chang 19:292–305

Cotula L et al (2009) Land grab or development opportunity? Agricultural investment and international land deals in Africa. FAO, IIED & IFAD. London, Rome. http://www.ifad.org/pub/land/land_grab.pdf. Accessed 27 May 2016

Creutzig F et al (2015) Bioenergy and climate change mitigation: an assessment. GCB Bioenergy 7(5):916–944

DBFZ (2015a) Sachstandsbericht über vorhandene Grundlagen für ein Monitoring der Bioökonomie: Nachhaltigkeit und Ressourcenbasis der Bioökonomie. Leipzig. https://www.dbfz.de/fileadmin/user_upload/Referenzen/studien/P3310038_Sachstandsbericht_Nachhaltigkeit_und_Ressourcenbasis.pdf. Accessed 8 June 2016

DBFZ (2015b) Biomassepotenziale von Rest- und Abfallstoffen – Status Quo in Deutschland. Schriftenreihe Nachwachsende Rohstoffe Band 36 (Hrsg. FNR). Gülzow. http://mediathek.fnr.de/media/downloadable/files/samples/s/c/schriftenreihe_band_36_web_01_09_15.pdf. Accessed 8 June 2016

DBFZ (2015c) Method handbook material flow-oriented assessment of greenhouse gas effects. In: Thrän D, Pfeiffer D (eds) Series of the funding programme "Biomass energy use", vol 4, Leipzig. https://www.energetische-biomassenutzung.de/fileadmin/user_upload/Downloads/Ver%C3%B6ffentlichungen/04_MHB_en_web.pd

de Laurentiis V et al (2016) Overcoming food security challenges within an Energy/Water/Food Nexus (EWFN) approach. Sustainability 8:95

De Rosa M et al (2016) A comparison of land use change models: challenges and future developments. J Clean Prod 113:183–193

Delzeit R et al (2014) Indirect land use change (iLUC) revisited: An evaluation of approaches for quantifying iLUC and related policy proposals. Kiel Working Paper 1768, Institute for the World Economy. Kiel

Delzeit R et al (2016) An evaluation of approaches for quantifying emissions from indirect land use change. Kiel Working Paper No. 2035. Kiel. http://hdl.handle.net/10419/130757. Accessed 27 May 2016

Déry P, Anderson B (2007) Peak phosphorus. Energy Bulletin. energybulletin.net/node/33164. Accessed 30 Sept 2016

DESTATIS (2015) Landwirtschaftlich genutzte Fläche 2013: 71% sind Ackerland. Statistisches Bundesamt. Wiesbaden. https://www.destatis.de/DE/ZahlenFakten/Wirtschaftsbereiche/LandForstwirtschaftFischerei/FeldfruechteGruenland/AktuellFeldfruechte1.html. Accessed 2 June 2016

DESTATIS (2016) Umweltökonomische Gesamtrechnungen. Nachhaltige Entwicklung in Deutschland Indikatoren zu Umwelt und Ökonomie. Statistisches Bundesamt. Wiesbaden

DIN EN ISO 14046:2015–11: Umweltmanagement – Wasser-Fußabdruck – Grundsätze, Anforderungen und Leitlinien (ISO 14046:2014); Norm-Entwurf. FprEN ISO 14046:2015

Dtv (1998) dtv-Atlas Ökologie, Deutscher Taschenbuch Verlag, München, 4. Auflage

EC (2016) Report Workshop "Integrated biorefineries and innovations in the optimal use of biomass", Brussels, Dec 10, 2015. http://ec.europa.eu/research/bioeconomy/pdf/workshop_on_optimal_use_of_biomass-integrated_biorefineries_10Dec2015.pdf. Accessed 14 June 2016

EC (2018) A sustainable bioeconomy for Europe: strengthening the connection between economy, society and the environment – Updated Bioeconomy Strategy. European Commission DG Research and Innovation, Unit F – Bioeconomy. Brussels. https://ec.europa.eu/research/bioeconomy/pdf/ec_bioeconomy_strategy_2018.pdf. Accessed 8 Mar 2019

Edwards R et al (2010) Indirect land use change from increased biofuels demand: comparison of models and results for marginal biofuels production from different Feedstocks. JRC, Ispra

Edwards R et al (2014) Well-to-wheels analysis of future automotive fuels and powertrains in the European context. Ispra. http://iet.jrc.ec.europa.eu/about-jec/sites/iet.jrc.ec.europa.eu.about-jec/files/documents/wtw_report_v4a_march_2014_final.pdf. Accessed 10 June 2016

EEA (2013) EU bioenergy potential from a resource efficiency perspective. EEA Report No 6/2013. Copenhagen. http://www.eea.europa.eu/publications/eu-bioenergy-potential/at_download/file. Accessed 3 June 2016

EEA (2015) European ecosystem assessment – concept, data, and implementation. Contribution to Target 2 Action 5 Mapping and Assessment of Ecosystems and their Services (MAES) of the EU Biodiversity Strategy to 2020. EEA Technical report No 6/2015. European Environment Agency, Copenhagen

EEA (2016a) The direct and indirect impacts of EU policies on land. EEA Report No 8/2016. Copenhagen. http://www.eea.europa.eu/publications/impacts-of-eu-policies-on-land/at_download/file. Accessed 3 June 2016

EEA (2016b) Resource efficiency in Europe: benefits of doing more with less. EEA Report No 10/2016. Luxembourg. Publications Office of the European Union. http://www.eea.europa.eu/publications/more-from-less/at_download/file

Efken J et al (2012) Volkswirtschaftliche Bedeutung der biobasierten Wirtschaft in Deutschland. Arbeitsberichte aus der vTI-Agrarökonomie 07/2012. http://literatur.thuenen.de/digbib_extern/dn051397.pdf. Arbeitsberichte aus der vTI-Agrarökonomie 07/2012,54 Seiten

EC (2011) Ressourcenschonendes Europa – eine Leitinitiative innerhalb der Strategie Europa 2020. Mitteilung der Kommission an das Europäische Parlament, den Europäischen Wirtschafts- und Sozialausschuss und den Ausschuss der Regione, KOM(2011) 21

EC (2012) Innovation für nachhaltiges Wachstum: eine Bioökonomie für Europa. Mitteilung der Kommission an das europäische Parlament, den Rat, den europäischen Wirtschafts- und Sozialausschuss und den Ausschuss der Regionen. Kommission der Europäischen Union, Brüssel

Elshout P et al (2015) Greenhouse-gas payback times for crop-based biofuels. Nat Clim Chang 5:604–610

Erb K-H et al (2016) Exploring the biophysical option space for feeding the world without deforestation. Nat Commun 7:11382

ETPs & EUFETEC (2015) THE EUROPEAN BIOECONOMY IN 2030 – Delivering Sustainable Growth by addressing the Grand Societal Challenges. Brüssel. http://www.epsoweb.org/file/560. Accessed 24 May 2016

EU (2009) Directive 2009/28/EC of the European Parliament and of the Council of 23 April 2009 on the promotion of the use of energy from renewable sources amending and subsequently repealing Directives 2001/77/EC and 2003/30/EC. Official Journal of the European Union, L140/16 of 5.6.2009

EU (2016) European Bioeconomy Stakeholders Manifesto. http://www.bioeconomyutrecht2016.eu/Static/bioeconomyutrecht2016.eu/Site/Manifest-revisie-13-april-website-version.pdf. Accessed 25 May 2016

EU (2018) Directive (EU) 2018/2001 of the European Parliament and of the Council of 11 December 2018 on the promotion of the use of energy from renewable sources (recast). Official Journal of the European Union L 328/82-209. Accessed 8 Mar 2019

FAO (2008) The State of Agriculture 2008 – Biofuels: prospects, risks and opportunities. Rome. http://www.fao.org/3/a-i0100e.pdf. Accessed 27 May 2016

FAO (2010) Bioenergy and food security – The BEFS Analytical Framework. Environment and natural resources management working paper 16. Rome. http://www.fao.org/docrep/013/i1968e/i1968e.pdf. Accessed 27 May 2016

FAO (2012) Impacts of Bioenergy on Food Security – Guidance for Assessment and Response at National and Project Levels. Rome. http://www.fao.org/docrep/015/i2599e/i2599e00.pdf. Accessed 27 May 2016

FAO (2013a) Towards Food Security and Improved Nutrition: Increasing the Contribution of Forests and Trees. Rome. http://www.fao.org/docrep/018/i2969e/i2969e.pdf. Accessed 27 May 2016

FAO (2013b) Forests for Improved Nutrition and Food Security. Rome. http://www.fao.org/docrep/014/i2011e/i2011e00.pdf. Accessed 27 May 2016

FAO et al (2014) The State of Food Insecurity in the World 2014 – Strengthening the enabling environment for food security and nutrition. FAO, IFAD & WFP. Rome. http://www.fao.org/3/a-i4030e.pdf. Accessed 5 Oct 2016

FAO & OECD (2011) Price Volatility in Food and Agricultural Markets: Policy Responses. Policy Report including contributions by FAO, IFAD, IMF, OECD, UNCTAD, WFP, the World Bank, the WTO, IFPRI and the UN HLTF. Rome, Paris. http://ictsd.org/downloads/2011/05/finalg20report.pdf. Accessed 27 May 2016

FNR (2014a) Basisdaten biobasierte Produkte –Oktober 2014. Anbau, Rohstoffe, Produkte. Gülzow. http://mediathek.fnr.de/media/downloadable/files/samples/b/a/basisdaten-biooekonomie_web-v02.pdf. Accessed 26 May 2016

FNR (2014b) Marktanalyse Nachwachsende Rohstoffe. Schriftenreihe Nachwachsende Rohstoffe 34. Fachagentur Nachwachsende Rohstoffe, Gülzow

Fritsche U (2013) Global material flows and their environmental impacts: an overview. In: Angrick M, Burger A, Lehmann H (eds) Factor X resource – designing the recycling society, Eco-efficiency in industry and science, vol 30, pp 3–17

Fritsche U, Iriarte L (2016) The sustainable development goals (SDGs) and biomass in the EU: an issues brief. Report for IASS. International Institute for Sustainability Analysis and Strategy, Darmstadt, Madrid

Fritsche U, Seyfert U (2013) Aktueller Stand von nationalen und internationalen Zertifizierungsansätzen sowie Weiterentwicklung von Nachhaltigkeitskriterien für feste Biomassen. In: Biodiversitätsziele bei der energetischen Waldholznutzung als Beitrag der Nachhaltigkeit. BfN-Skripten 330. Bundesamt für Naturschutz, Bonn: 40–47 http://www.bfn.de/fileadmin/MDB/documents/service/skript_330.pdf. Accessed 14 June 2016

Fritsche U et al (2010a) Entwicklung von Strategien und Nachhaltigkeitsstandards zur Zertifizierung von Biomasse für den internationalen Handel. UBA-Texte 48/2010. Dessau. http://www.umweltbundesamt.de/sites/default/files/medien/461/publikationen/3960.pdf. Accessed 15 June 2016

Fritsche U et al (2010b) Direct and indirect land-use competition issues for energy crops and their sustainable production – an overview. Biofuel Bioprod Bior 4(6):692–704. https://doi.org/10.1002/bbb.258

Fritsche U et al (2012) Economic and ecological assessment of biorefineries – findings of the German biorefinery roadmap process. In: NWBC 2012 – 4th Nordic Wood Biorefinery Conference: 104–108 http://www.vtt.fi/inf/pdf/technology/2012/T53.pdf. Accessed 15 June 2016

Fritsche U et al (2015) Ressourceneffiziente Landnutzung – Wege zu einem Global Sustainable Land Use Standard (GLOBALANDS). IINAS in Kooperation mit Ecologic Institut, Öko-Institut und Leuphana Universität Lüneburg. Dessau. https://www.umweltbundesamt.de/sites/default/files/medien/378/publikationen/texte_82_2015_kurz_ressourceneffiziente_landnutzung.pdf. Accessed 16 June 2016

Fritsche U et al (2017) Energy and land use. Working Paper for the UNCCD Global Land Outlook. Darmstadt etc. https://global-land-outlook.squarespace.com/s/Energy-and-Land-Use__U_Fritsche-t9tw.pdf. Accessed 8 Mar 2019

Garnett T et al (2013) Sustainable intensification in agriculture: premises and policies. Science 341:33–34

GBEP (2011) The GBEP sustainability indicators for bioenergy. Rome. http://www.globalbioenergy.org/fileadmin/user_upload/gbep/docs/Indicators/The_GBEP_Sustainability_Indicators_for_Bioenergy_FINAL.pdf. Accessed 2 June 2016

GBS (2015) Global Bioeconomy Summit – Conference Report. Berlin. http://gbs2015.com/fileadmin/gbs2015/Downloads/GBS-2015_Report_final_neu_2.pdf. Accessed 14 June 2016

GBS (2018) Communiqué Global Bioeconomy Summit 2018. 19–20 April 2018, Berlin. http://gbs2018.com/fileadmin/gbs2018/Downloads/GBS_2018_Communique.pdf. Accessed 9 Mar 2019

Gelfand I et al (2013) Sustainable bioenergy production from marginal lands in the US Midwest. Nature 493:514–517

Gerbens-Leenes P, Nonhebel S (2002) Consumption patterns and their effects on land required for food. Ecol Econ 42:185–199

Gerbens-Leenes P, Nonhebel S (2004) Critical water requirements for food: methodology and policy consequences for food security. Food Policy 29:547–564

Gerbens-Leenes P et al (2008) Water Footprint of bioenergy and other primary energy carriers. Value of Water Research Reports Series No. 29. UNESCO-IHE Institute for Water Education. Delft

Gerssen-Gondelach S (2016) Yielding a fruitful harvest. Advanced methods and analysis of regional potentials for sustainable biomass value chains interlinked with environmental and land use impacts of agricultural intensification. PhD dissertation. Utrecht University. http://dspace.library.uu.nl/bitstream/handle/1874/325582/Gondelach.pdf?sequence=1. Accessed 27 May 2016

Griggs D et al (2013) Sustainable development goals for people and planet. Nature 495:305–307

GSP (2016) Voluntary Guidelines for Sustainable Soil Management (VGSSM). Global Soil Partnership. Rome. http://www.fao.org/3/a-bl813e.pdf. Accessed 26 May 2016
Gustavsson J et al (2011) Global food losses and food waste: extent, causes and prevention. FAO, Rome
Halstead M et al (2014) Understanding the Energy-Water Nexus. Policy Stud 2013 (2014), p. 2012 ff
Haro P et al (2015) Balance and saving of GHG emissions in thermochemical biorefineries. Appl Energy 147:444–455
HLPE (2011) Price volatility and food security. The High Level Panel of Experts on Food Security and Nutrition of the Committee on World Food Security. Rome. http://www.fao.org/3/a-mb737e.pdf. Accessed 6 June 2016
Hoekstra A (2008) Human appropriation of natural capital: a comparison of ecological footprint and water footprint analysis. Ecol Econ 68(7):1963–1974
Hoekstra A, Chapagain A (2007) Water footprints of nations: water use by people as a function of their consumption pattern. Water Resour Manag 21(1):35–48
Hoekstra A, Chapagain A (2008) Globalization of water: sharing the planet's freshwater resources. Wiley-Blackwell, Oxford. https://doi.org/10.1002/9780470696224
Hypersoil Uni Münster (2016) Kohlenstoffkreislauf. http://hypersoil.uni-muenster.de/0/05/14.htm. Accessed 30 Sept 2016
IEA (2017) Technology roadmap: delivering sustainable bioenergy. International Energy Agency and IEA Bioenergy TCP. Paris. http://www.iea.org/publications/freepublications/publication/Technology_Roadmap_Delivering_Sustainable_Bioenergy.pdf. Accessed 9 Mar 2019
IEA Bioenergy & GBEP (2016) Examples of positive bioenergy and water relationships. Rome. http://www.globalbioenergy.org/fileadmin/user_upload/gbep/docs/2015_events/AG6_workshop_25-26_August_2015/AG6_Examples_of_Positive_Bioenergy_and_Water_Relationships_Final.pdf. Accessed 28 June 2016
IFEU (2016) Aktualisierung der Eingangsdaten und Emissionsbilanzen wesentlicher biogener Energienutzungspfade (BioEm). Fehrenbach, Horst et al. UBA-Texte 09/2016. Dessau. https://www.umweltbundesamt.de/sites/default/files/medien/378/publikationen/texte_09_2016_aktualisierung_der_eingangsdaten_und_emissionsbilanzen_wesentlicher_biogener_energienutzungspfade_1.pdf. Accessed 5 Oct 2016
ILC (2015) Secure and equitable land rights in the Post-2015 Agenda. Rome. http://www.landcoalition.org/sites/default/files/documents/resources/LandRightsPost-2015Agenda.pdf Accessed 27 May 2016
INRO (2013) Nachhaltigkeitskriterien für die stoffliche Biomassenutzung. Initiative Nachhaltige Rohstoffbereitstellung für die stoffliche Biomassenutzung. Berlin. http://www.inro-biomasse.de/documents/Nachhaltigkeitskriterien.pdf. Accessed 3 June 2016
IÖW (2011) Bioökonomie – Können neue Technologien die Energieversorgung und die Welternährung retten? Stellungnahme im Auftrag des NABU. Institut für ökologische Wirtschaftsforschung, Berlin
IPCC (2011) IPCC special report on renewable energy sources and climate change mitigation. In: Edenhofer O et al (eds) IPCC working group III. https://www.ipcc.ch/pdf/special-reports/srren/SRREN_FD_SPM_final.pdf. Accessed 4 June 2016
IPCC (2013) Climate change 2013: the physical science basis. Contribution of WG I to the 5th Assessment Report of the IPCC. Cambridge, New York. http://www.climatechange2013.org/images/report/WG1AR5_ALL_FINAL.pdf. Accessed 5 Oct 2016
Iriarte L et al (2015) Consistent cross-sectoral sustainability criteria & indicators. Deliverable D5.4 of the S2Biom project. Darmstadt, Madrid. http://www.s2biom.eu/images/Publications/IINAS_2015_S2Biom_D5_4_Sustainability_C_I_proposal_Main_report_30_Mar.pdf. Accessed 2 June 2016
IUFRO (2015) Forests, trees and landscapes for food security and nutrition – a global assessment report. In: Vira B, Wildburger C, Mansourian (eds) IUFRO world series 33, Vienna. http://www.iufro.org/download/file/18901/5690/ws33_pdf/. Accessed 2 June 2016
JRC et al (2014) Forests, bioenergy and climate change mitigation Workshop May 19–20, 2014. Outcome statement. JRC in collaboration with EEA, IEA Bioenergy Tasks 38, 40, 43 & IINAS. Copenhagen. http://www.ieabioenergy-task38.org/workshops/copenhagen2014/CPH_Bioenergy_Workshop_Statement_2014.pdf. Accessed 5 Oct 2016
JRC (2015) The bioeconomy in the European Union in numbers – Facts and figures on biomass, turnover and employment. JRC 97789 Factsheet. Seville. https://ec.europa.eu/jrc/sites/default/files/JRC97789%20Factsheet_Bioeconomy_final.pdf. Accessed 27 May 2016
Kalkuhl M (2014) How strong do global commodity prices influence domestic food prices in developing countries? A global Price transmission and vulnerability mapping analysis. ZEF-Discussion Papers on Development Policy No. 191. Bonn. http://www.zef.de/uploads/tx_zefnews/zef_dp_191.pdf. Accessed 1 June 2016
Kalkuhl M (2015) Spekulation mit Nahrungsmitteln, Regulierung und Selbstregulierung. In: Aufderheide, D. & Dabrowski, M. (Hrsg.): Markt und Verantwortung – Wirtschaftsethische und moralökonomische Perspektiven. Volkswirtschaftliche Schriften Bd. 567 http://www.zef.de/fileadmin/webfiles/downloads/projects/volatility/downloads/Manuskript_Kalkuhl_final.pdf. Accessed 1 June 2016
Karlberg L et al (2015) Tackling biomass scarcity — from vicious to virtuous cycles in sub-Saharan Africa. Curr Opin Environ Sustain 15:1–8
Kline K et al (2016) Reconciling food security and bioenergy: priorities for action. GCB Bioenergy 9:557–576. https://doi.org/10.1111/gcbb.12366
Lahl U (2014) Bioökonomie für den Klima- und Ressourcenschutz – regulative Handlungskorridore. Studie im Auftrag des NABU. Berlin. https://www.nabu.de/imperia/md/content/nabude/gentechnik/studien/140821-nabu-biooekonomie-studie_2014.pdf. Accessed 7 June 2016
Larsen Y (2012) Bioökonomie – Gefahr oder Chance? Eine kritische Anmerkung zu den Prioritäten der Bioökonomieforschung in Bezug auf den Erhalt der biologischen Vielfalt. In: Feit, U. & Korn, H. (Hg.): Treffpunkt Biologische Vielfalt XI. BfN Bonn, pp. 145–148
LAWA (2014) Gewässerschutz und Landwirtschaft – Anforderungen an eine gewässerschonende Landbewirtschaftung aus der Sicht der Wasserwirtschaft. Bund/Länder-Arbeitsgemeinschaft Wasser. http://www.lawa.de/documents/Gewaesserschutz_und_Landwirtschaft_799.pdf
Leopoldina (2012) Bioenergie: Möglichkeiten und Grenzen – Empfehlungen. Halle. http://www.leopoldina.org/uploads/tx_leopublication/201207_Empfehlungen_Bioenergie_02.pdf. Accessed 2 June 2016
Maes D et al (2015) Assessment of the sustainability guidelines of EU Renewable Energy Directive: the case of biorefineries. J Clean Prod 88:61–70
Maga D (2015) A methodology to assess the contribution of biorefineries to a sustainable bio-based economy. PhD dissertation Ruhr-University Bochum. UMSICHT-Schriftenreihe 73. Fraunhofer-Institut UMSICHT, Oberhausen. http://publica.fraunhofer.de/eprints/urn_nbn_de_0011-n-3644316.pdf. Accessed 1 June 2016
Maltsoglou I et al (2015) Combining bioenergy and food security: an approach and rapid appraisal to guide bioenergy policy formulation. Biomass Bioenergy 79:80–95
Mantau U (2012) Holzrohstoffbilanz Deutschland – Entwicklungen und Szenarien des Holzaufkommens und der Holzverwendung von 1987 bis 2015. Universität Hamburg, Hamburg. http://literatur.ti.bund.de/digbib_extern/dn051281.pdf
Matthews R et al (2015) Carbon impacts of biomass consumed in the EU: quantitative assessment. Final report project: DG ENER/C1/427. Part A: Main Report. Forest Research. Farnham. https://ec.europa.eu/energy/sites/ener/files/documents/EU%20Carbon%20Impacts%20of%20Biomass%20Consumed%20in%20the%20EU%20final.pdf. Accessed 9 June 2016
McKechnie J et al (2015) Environmental and financial implications of ethanol as a bioethylene feedstock versus as a transportation fuel. Environ Res Lett 10:124018

9

Mekonnen M, Hoekstra A (2011) National water footprint accounts: the green, blue and grey water footprint of production and consumption, Value of Water Research Report Series No. 50. UNESCO--IHE. Delft

Mirzabaev A et al (2014) Bioenergy, Food Security and Poverty Reduction: Mitigating tradeoffs and promoting synergies along the Water-Energy-Food Security Nexus. ZEF Working Paper 135. Universität Bonn, Zentrum für Entwicklungsforschung, Bonn. http://www.zef.de/uploads/tx_zefportal/Publications/ZEF_WP_135_complete.pdf. Accessed 2 June 2016

Mohr A et al (2015) A rights-based food security principle for biomass sustainability standards and certification systems. ZEF Working Paper Series. Universität Bonn, Zentrum für Entwicklungsforschung, Bonn. http://www.zef.de/index.php?id=2213. Accessed 3 June 2016

Monier V et al (2010) Final report – preparatory study on food waste across EU 27. BIO Intelligence Service prepared for EC DG ENV. BIO Intelligent Service, Paris

NABU (2014) Nachhaltigkeit in der Bioökonomie. Zusammenfassung und Thesen als Ergebnis eines Workshops auf Vilm Dezember 2013. Naurschutzbund Deutschland e.V. Berlin

Neary G (2015) Best practices guidelines for managing water in bioenergy feedstock production. IEA Bioenergy Task 43 Report 2015:TR02 http://www.fs.fed.us/rm/pubs_journals/2015/rmrs_2015_neary_d001.pdf. Accessed 2 June 2016

O'Brien M et al (2015) Sachstandsbericht über vorhandene Grundlagen und Beiträge für ein Monitoring der Bioökonomie: Systemische Betrachtung und Modellierung der Bioökonomie. Wuppertal-Institut. Wuppertal. http://epub.wupperinst.org/files/5851/5851_Biooekonomie.pdf. Accessed 3 June 2016

O'Sullivan M et al (2014) Bruttobeschäftigung durch erneuerbare Energien in Deutschland im Jahr 2013. https://www.bmwi.de/BMWi/Redaktion/PDF/B/bericht-zur-bruttobeschaeftigung-durch-erneuerbare-energien-jahr-2013,property=pdf,bereich=bmwi2012,sprache=de,rwb=true.pdf. Accessed 30.9.2016

OECD (2016) Policy guidance on resource efficiency. Paris. http://www.oecd.org/environment/waste/Resource-Efficiency-G7-2016-Policy-Highlights-web.pdf. Accessed 30 Sept 2016

Osseweijer P et al (2015) Bioenergy and food security. In: Souza G et al (eds) Bioenergy & sustainability: bridging the gaps. Universidade de São Paulo, Sao Paulo, pp 90–136

Ostwald M et al (2015) Can India's wasteland be used for biomass plantation? Focali Brief 2015:02. Chalmers University, Gothenburg

PANGEA (2011) Land grab refocus – roots and possible demise of land grabbing. Brussels. http://pangealink.org/wp-content/uploads/2011/06/LandgrabStudy-Final.pdf. Accessed 27 May 2016

Panoutsou C et al (2016) Delivery of sustainable supply of non-food biomass to support a "resource-efficient" Bioeconomy in Europe. Final Report of the EU S2Biom project. London etc. www.s2biom.eu. Accessed 2 June 2016

Parajuli R et al (2015) Biorefining in the prevailing energy and materials crisis: a review of sustainable pathways for biorefinery value chains and sustainability assessment methodologies. Renew Sust Energ Rev 43:244–263

Petersen B, Snapp S (2015) What is sustainable intensification? Views from experts. Land Use Policy 46:1–10

Pfau S et al (2014) Visions of sustainability in bioeconomy research. Sustainability 6:1222–1249

Piotrowski S et al (2016) Sustainable biomass supply and demand: a scenario analysis. Open Agric 1:18–28

Plevin R et al (2015) Carbon accounting and economic model uncertainty of emissions from biofuels-induced land use change. Environ Sci Technol 49(5):2656–2664

Rahman M et al (2014) Extension of energy crops on surplus agricultural lands: a potentially viable option in developing countries while fossil fuel reserves are diminishing. Renew Sust Energ Rev 29:108–119

Rockström J, Lannerstad M, Falkenmark M (2007) Assessing the water challenge of a new green revolution in developing countries. PNAS 104(15):6253–6260

Rockström J et al (2009a) A safe operating space for humanity. Nature 461:472–475

Rockström J et al (2009b) Planetary boundaries: exploring the safe operating space for humanity. Ecol Soc 14(2):32. http://www.ecologyandsociety.org/vol14/iss2/art32/

Rösch C, Skarka J, Raab K, Stelzer V (2009) Energy production from grassland – assessing the sustainability of different process chains under German conditions. Biomass Bioenergy 33(4):689–700

Royal Society of London (2009) Reaping the benefits: science and the sustainable intensification of global agriculture. The Royal Society of London, London, UK, p. 72

RRI (2015) Who owns the world's land? A global baseline of formally recognized indigenous and community land rights. Rights & Resources Initiative. Washington, DC. http://www.rightsandresources.org/wp-content/themes/pushwp/download-pdf.php?link=%2Fwp-content%2Fuploads%2FGlobalBaseline1.pdf. Accessed 27 May 2016

RSB (2015) RSB low iLUC risk biomass criteria and compliance indicators. RSB, Chatelaine (Geneva)

SCAR (2015) Sustainable agriculture, forestry and fisheries in the bioeconomy: a challenge for Europe. 4th SCAR Foresight Exercise. Standing Committee on Agricultural Research, EC DG RTD. Luxembourg. https://ec.europa.eu/research/scar/pdf/ki-01-15-295-enn.pdf#view=fit&pagemode=none. Accessed 10 June 2016

SCAR (2016) Agricultural knowledge and innovation systems towards the future. A Foresight Paper. Standing Committee on Agricultural Research, EC DG RTD. Luxembourg. https://ec.europa.eu/research/scar/pdf/akis-3_end_report.pdf#view=fit&pagemode=none. Accessed 10 June 2016

Scarlat N et al (2015) The role of biomass and bioenergy in a future bioeconomy: policies and facts. Environ Dev 15:3–34

Schiefer J, Laira G, Blum W (2015) Indicators for the definition of land quality as a basis for the sustainable intensification of agricultural production. Int Soil Water Cons Res 3(1):42–49

Schneider R (2014) Tailoring the bioeconomy to food security. Rural 21–03/2014:19–21

Schöpe M, Britschkat G (2006) Volkswirtschaftliche Effekte der Erzeugung von Bioethanol zum Einsatz im Kraftstoffbereich. http://www.bdbe.de/application/files/7914/3566/9616/ifo-studie.pdf. Accessed 28 June 2016

Searchinger T et al (2015) Do biofuel policies seek to cut emissions by cutting food? Science 347:1420–1422

SFPI Inwil (2012) Biogas – ein geschlossener ökologischer Kreislauf. SwissFarmerPower Inwil AG. https://www.sfpinwil.ch/prozess/geschlossener-oekologischer-kreislauf/. Accessed 24 Feb 2017

SRU (2007) Klimaschutz durch Biomasse. Sondergutachten des Sachverständigenrats für Umweltfragen. Sachverständigenrat für Umweltfragen, Berlin. http://www.umweltrat.de/SharedDocs/Downloads/DE/02_Sondergutachten/2007_SG_Biomasse_Buc

SRU (2011) Wege zur 100% erneuerbaren Stromversorgung. Sondergutachten des Sachverständigenrats für Umweltfragen. Berlin. http://www.umweltrat.de/SharedDocs/Downloads/DE/02_Sondergutachten/2011_07_SG_Wege_zur_100_Prozent_erneuerbaren_Stromversorgung.pdf. Accessed 14 June 2016

SRU (2016) Umweltgutachten 2016 – Impulse für eine integrative Umweltpolitik. Sachverständigenrat für Umweltfragen, Berlin

Statista (2016) Direktzahlungen und Zuschüsse an landwirtschaftliche Betriebe in Deutschland in den Jahren 2005/06 bis 2014/15 (in Euro je Unternehmen). https://de.statista.com/statistik/daten/studie/207170/umfrage/unternehmensbezogene-direktzahlungen-in-den-landwirtschaftlichen-betrieben-in-deutschland/. Accessed 9 June 2016

Strengers B et al (2015) Greenhouse gas impact of bioenergy pathways. PBL study for IRENA. PBL, The Hague

Stupak I et al (2016) A global survey of stakeholder views and experiences for systems needed to effectively and efficiently govern sustainability of bioenergy. WIREs Energy Environ 5:89–118. https://doi.org/10.1002/wene.166

Sutton MA et al (2013) Our nutrient world: the challenge to produce more food and energy with less pollution. Global Overview of Nutrient Management. Centre for Ecology and Hydrology. Edinburgh

Thrän D, Fritsche U (2016) Standards for biobased fuels and resources – status and need. In: IEA Bioenergy Conference 2015 Proceedings: 148–158

Thrän D et al (2011) Identifizierung strategischer Hemmnisse und Entwicklung von Lösungsansätzen zur Reduzierung der Nutzungskonkurrenzen beim weiteren Ausbau der Biomassenutzung. DBFZ Report Nr. 4. DBFZ, Leipzig. https://www.erneuerbare-energien.de/EE/Redaktion/DE/Downloads/Berichte/identifizierung-hemmnisse-loesungsansaetze-reduzierung-nutzungskonkurenzen.pdf?__blob=publicationFile&v=2

Thrän D et al (2015) Endbericht Meilensteine 2030 – Elemente und Meilensteine für die Entwicklung einer tragfähigen und nachhaltigen Bioenergiestrategie. Schriftenreihe Förderprogramm "Energetische Biomassenutzung" Band 18. DBFZ, Leipzig. https://www.energetische-biomassenutzung.de/fileadmin/user_upload/Meilensteine/18_MS2030_final_max.pdf. Accessed 28 June 2016

Tsiropoulos I et al (2015) Life cycle impact assessment of bio-based plastics from sugarcane ethanol. J Clean Prod 90:114–127

UBA (2013) Globale Landflächen und Biomasse nachhaltig nutzen. Dessau. http://www.umweltbundesamt.de/sites/default/files/medien/479/publikationen/globalelandflaechenbiomassebfklein.pdf. Accessed 25 May 2016

UN (2015a) Transformation unserer Welt: die Agenda 2030 für nachhaltige Entwicklung. Resolution der Generalversammlung A/RES/70/1 verabschiedet am 25. September 2015. New York. http://www.un.org/depts/german/gv-70/a70-l1.pdf. Accessed 28 June 2016

UN (2015b) The global goals for sustainable development. http://www.globalgoals.org/de/. Accessed 5 Oct 2016

UNCED (1992) The earth summit. United Nations Conference on Environment and Development. Rio de Janeiro. http://www.un.org/geninfo/bp/enviro.html. Accessed 28 June 2016

UNDESA (2014) World urbanization prospects: the 2014 revision. Highlights (ST/ESA/SER.A/352). New York. http://esa.un.org/unpd/wup/Highlights/WUP2014-Highlights.pdf. Accessed 1 June 2016

UNEP (2011) Towards a green economy: pathways to sustainable development and poverty eradication. www.unep.org/greeneconomy. Accessed 30 Sept 2016

UNEP-IRP (2016) Global material flows and resource productivity. An Assessment Study of the UNEP International Resource Panel. Schandl, H. et al. Paris. http://unep.org/documents/irp/16-00169_LW_GlobalMaterialFlowsUNEReport_FINAL_160701.pdf. Accessed 5 Oct 2016

Valin H et al. (2015) The land use change impact of biofuels consumed in the EU – Quantification of area and greenhouse gas impacts. Study of Ecofys, IIASA and E4tech for EC DG ENER. Utrecht. https://ec.europa.eu/energy/sites/ener/files/documents/Final%20Report_GLOBIOM_publication.pdf. Accessed 26 May 2016

van Dam J (2015) Inventory trends sustainability of biomass for various end-uses. Study commissioned by the Dutch Commission Corbey. http://www.corbey.nl/includes/download.asp?mediald=338&md5=1F6AB8AABBC73AEA6149A0648AE59E34. Accessed 8 June 2016

van Hilst F et al (2015) Sustainable biomass for energy and materials: a greenhouse gas emission perspective. Working Paper. Utrecht University. http://www.uu.nl/sites/default/files/sustainable_biomass_for_energy_and_materials.pdf. Accessed 27 May 2016

van Kauwenbergh S (2010) World phosphate rock reserves and resources. International Fertilizer Development Center (IFDC), Muscle Shoals. http://firt.org/sites/default/files/SteveVanKauwenbergh_World_Phosphate_Rock_Reserve.pdf

von Braun J (2014) Bioeconomy and sustainable development – dimensions. Rural 21–03/2014:6–9. http://www.rural21.com/uploads/media/rural2014_03-S06-09.pdf. Accessed 5 Oct 2016

WBA (2016) Global biomass potential towards 2035. World Biomass Association Fact Sheet. WBA, Stockholm

WBGU (2008) Welt im Wandel: Zukunftsfähige Bioenergie und nachhaltige Landnutzung. Wissenschaftlicher Beirat der Bundesregierung Globale Umweltveränderungen. Berlin. http://www.wbgu.de/fileadmin/templates/dateien/veroeffentlichungen/hauptgutachten/jg2008/wbgu_jg2008.pdf. Accessed 28 June 2016

WBGU (2011) Welt im Wandel – Gesellschaftsvertrag für eine Große Transformation. Wissenschaftlicher Beirat der Bundesregierung Globale Umweltveränderungen. Berlin. http://www.wbgu.de/fileadmin/templates/dateien/veroeffentlichungen/hauptgutachten/jg2011/wbgujg2011.pdf. Accessed 28 June 2016

WBGU (2014) Zivilisatorischer Fortschritt innerhalb planetarischer Leitplanken – Ein Beitrag zur SDG-Debatte. Wissenschaftlicher Beirat der Bundesregierung Globale Umweltveränderungen Politikpapier Nr. 8. WBGU, Berlin

WBGU (2016) Der Umzug der Menschheit: Die transformative Kraft der Städte. Berlin. http://www.wbgu.de/hg2016.pdf. Accessed 27 June 2016

WCED (1987) Report of the world commission on environment and development: our common future. Oxford. http://www.un-documents.net/wced-ocf.htm. Accessed 28 June 2016

Wicke B (2011) Bioenergy production on degraded and marginal land – Assessing its potentials, economic performance and environmental impacts for different settings and geographical scales. PhD Thesis. Utrecht University. https://dspace.library.uu.nl/bitstream/handle/1874/203772/wicke.pdf. Accessed 5 Oct 2016

Wicke B et al (2015) ILUC prevention strategies for sustainable biofuels – synthesis report from the ILUC prevention project. Utrecht University. http://www.uu.nl/faculty/geosciences/EN/research/institutesandgroups/researchinstitutes/copernicusinstitute/research/energyandresources/research/Documents/Synthesis%20report_ILUC%20prevention.pdf

Wolf C et al (2016) ExpRessBio – Methoden. Methoden zur Analyse und Bewertung ausgewählter ökologischer und ökonomischer Wirkungen von Produktsystemen aus land- und forstwirtschaftlichen Rohstoffen. Berichte aus dem TFZ 45. TFZ, Straubing. http://www.tfz.bayern.de/mam/cms08/biokraftstoffe/dateien/tfz_bericht_45_expressbio.pdf. Accessed 3 June 2016

Woods J et al (2015) Land and bioenergy. In: Souza G et al (eds) Bioenergy & sustainability: Bridging the gaps. SCOPE report. Universidade de São Paulo, Sao Paulo. http://bioenfapesp.org/scopebioenergy/images/chapters/bioenergy_sustainability_scope.pdf. Accessed 27 May 2016

WWAP (2014) The United Nations world water development report 2014: water and energy. UNESCO, Paris

Zeddies J et al (2014) Optimierung der Biomassenutzung nach Effizienz in Bereitstellung und Verwendung unter Berücksichtigung von Nachhaltigkeitszielen und Welternährungssicherung. Schlussbericht zum MBEL-Vorhaben FKZ 11NR039. Universität Hohenheim, Hohenheim

Bioeconomy: Key to Unlimited Economic and Consumption Growth?

Armin Grunwald

The load seems to grow incessantly during the mechanical harvest of grass. The silage produced from it is used as animal feed in cattle farming or to generate energy in biogas plants. (© Christine Rösch)

J. Pietzsch (ed.), *Bioeconomy for Beginners*, https://doi.org/10.1007/978-3-662-60390-1_10

10.1 Unlimited Growth in a Limited World?

Consumption shapes the attitude towards life in affluent societies, albeit, historically, not for long (König 2008). For most people in rich countries, being able to buy products and use services is a natural expression of quality of life. Having money available for individual purposes such as clothing, food, entertainment, culture, mobility or holidays, as well as having as few restrictions as possible on consumption, has become an essential prerequisite for satisfaction. Consumption is not simply about satisfying elementary needs. Rather, consumption expresses an attitude towards life in which social status and economic success are also documented, both before oneself and before others. Since the strong economic growth in Germany in the 1950s and 1960s (*Wirtschaftswunder*), this attitude towards life has included the idea that consumption is increasing from year to year. Accordingly, economic growth is politically and economically desired and promoted. Even the growth-critical debate of recent years (Jackson 2009; D'Alisa et al. 2014) has not changed this fundamentally. The central issue in election campaigns continues to be whether and to what extent governments and parties have contributed or intend to contribute to increasing prosperity and consumption.

Unfortunately, this very pleasant development, at least for most people in the industrialised countries and increasingly many people in emerging and developing countries, has its downsides. Prosperity and consumption are largely associated with negative consequences for the environment and for future generations. Products must be manufactured, transported and ultimately disposed of. Energy and resource consumption, emissions and waste are unavoidable consequences of consumption. Shorter useful lives, e.g., of furniture and household appliances, increase material throughput and worsen the environmental balance. The loss of biodiversity, the threat to cultural habitats posed by the dominance of globalised consumption and social problems are also part of the problem spectrum (Dauvergne 2008).

Already, the *Club of Rome*, in its report on the Limits to growth (Meadows et al. 1972), has pointed out that unlimited growth is impossible in a world with limited resources. The recognition of global environmental changes through growing prosperity and consumption (e.g., WBGU 1996) has reinforced this criticism of the emissions side of growth and its accompanying phenomena. In wide circles of Western societies, the position has prevailed that the mere continuation of production and consumption patterns oriented towards the quantitative growth ideal, with their consequences for the consumption of raw materials and emissions, is not compatible with the goal of a sustainable human civilization on planet Earth.

Nevertheless, the consumer spiral of growth continues to turn. In western countries, for example, the amount of clothing purchased has roughly doubled in the past 10 years. Electronic devices from computers to mobile phones to today's ever-smaller mobile devices with Internet access are another area of significant growth that goes hand in hand with ever-shorter product life cycles. But there will also be further waves of growth in traditional areas such as the automotive market. The SUV *(Sport Utility Vehicle)* is a good example of how a new type of product has initiated a new wave of consumption worldwide. Another field of growing consumption with significant environmental impacts is tourism, which has experienced impressive growth rates since the 1960s. Knowledge of negative environmental consequences hardly results in corresponding action – environmental psychologists speak of "cognitive dissonances" here (Leggewie and Welzer 2009). The urgency of tackling environmental problems continues to grow.

10.2 Strategies for Sustainable Development

In order to achieve more environmentally-friendly developments (on further aspects of sustainability, cf. Kopfmüller et al. 2001), the strategic approaches of effectiveness, consistency and sufficiency have been shaping the scientific and social discussion (Huber 1995; Grunwald 2012):

- efficiency strategies are aimed at providing a production or service with the lowest possible use of materials and energy. Increasing material and energy efficiency and resource productivity is intended to achieve a high degree of avoidance of material and energy losses, as well as problematic emissions. New or improved technologies, e.g., in the conversion of energy sources, new production processes and, in some cases, modified products, serve this purpose. There are also more general approaches, such as longevity, multiple use and strategies for more efficient recycling. As classic modernization concepts, these are considered to be largely capable of creating consensus, because they can be easily integrated into existing economic processes. Whether they are sufficient in the long term, however, is controversial. In many fields of application, for example, efficiency increases are used less for the improvement of environmental balance than for increased luxury and comfort, whereby the desired effect is reduced or disappears altogether (rebound effect, see Sorrell 2007).
- consistency strategies focus on a nature-adapted design of material flows and resource use:

> Consistent material flows are therefore those which, on the one hand, are conducted in a largely interference-free manner in the technical natural cycle or, on the other hand, agree with the metabolic processes of the surrounding nature in such a way that they fit into it relatively easily, even in large volumes (Huber 1995, translation A.G.).

A classic example is the bioeconomic idea of at least partially replacing fossil fuels with energy from renewable biomass. Thus, consistency strategies do not only aim at quantitative reductions of environmental pollution in the same manner as efficiency, but also at qualitative changes. The aim is to replace ecologically problematic material flows with more environmentally-friendly ones. This substitution is associated with major challenges for science and technology and requires the research, introduction and implementation of corresponding processes and technologies in production, energy supply and the use of raw materials.

- sufficiency does not focus on the production and technology side of goods and services, but rather on the social demand side, especially among consumers (Schneidewind and Zahrnt 2013). Sufficiency means renouncing material goods and consumption or continuous quantitative growth. For some time now, such considerations have been gaining ground under the label *degrowth*, a concept that continues to demand greater attention (e.g., Jackson 2009; D'Alisa et al. 2014). They aid in the development of a changed understanding of prosperity, in which the orientation towards quantitative growth is replaced by qualitative growth objectives and post-material values such as fulfilment, solidarity, community and a clean environment. The protagonists refer to the results of happiness research, according to which satisfaction no longer rises above a certain level of prosperity. Correspondingly, changed lifestyles and consumption patterns should reduce the pressure of the human economy on the natural environment. However, this model is considered unattractive for large parts of the population and politically difficult to implement:

> » A sufficiency strategy is therefore not connectable and resonant. It remains unrealistic unless extreme external crisis conditions force sufficiency qua normative power of the factual (Huber 1995, translation A.G.).

These strategic approaches are to be understood as ideal-typical in order to structure discussions for the solution of environmental problems. They are not mutually exclusive, but rather can be combined.

10.3 Green Growth or Farewell to Growth?

Bioeconomy can be understood as a combination of efficiency and consistency strategy. The fact that food, production and value chains are largely based on bio-based raw materials, whether renewable biomass or biogenic waste, brings these chains closer to natural cycles than they are in the fossil-based economy. This increase in consistency must be accompanied by efficiency in order to achieve environmentally sound overall balances. As a rule, it requires conversion processes on whose efficiency and raw material productivity the environmental balance decisively depends, e.g., the conversion of microalgae into usable raw materials or energy.

Although sufficiency strategies are not excluded in the context of bioeconomy, they are generally not mentioned. Bioeconomy could be understood as an attempt to go as far as possible with consistency and efficiency in the reduction of environmental pollution in order not to have to demand sufficiency in the first place. From this point of view, a continuation of the classical technology-driven growth paradigm would be possible. The current paradigm of growth, indifferent to environmental problems, would become "green" growth. Since the negative consequences of the human economy for the natural environment are obvious, the attribute "green" is increasingly being assigned to concepts of economic activity. *Green New Deal*, *green economy* and *green growth* are examples of conceptual considerations that focus on environmentally compatible economic restructuring (cf. Grunwald and Kopfmüller 2012). In the United Nations Environment Programme (UNEP), the concept of the *green economy* is an environmentally compatible, growth-promoting and justice-promoting way of doing business. As part of their *Green Economy Initiative* (UNEP 2011), the UN delineates the reforms and political framework conditions with which the transition to a green economy can be realised and, in particular, financed. The *green industry platform* UNEP and the *United Nations Industrial Development Organisation* (UNIDO) aims to bring governments, business and civil society together for joint activities to implement the green economy in the manufacturing sector.

These approaches are based on the conviction that the economic growth paradigm does not necessarily conflict with sustainability and environmental compatibility. Rather, it is possible to achieve *green growth* that is based on the sustainable use of natural resources and that is sustainable in the long term. The opening up of new green markets, the development of eco-innovations (Fussler 1999) and the management of ecosystem services should enable the development of new business fields under adequate framework conditions, and thus simultaneously facilitate growth and environmental protection (OECD 2011).

The bioeconomy seems to be the ideal approach to implementing the basic idea of *green growth*. It thus offers the opportunity to achieve economic growth in harmony with nature conservation and environmental protection (BMBF 2010). In particular, the above-mentioned combination of efficiency and consistency aspects and its strategy, strongly based on technology-based innovation, makes the bioeconomy not only scientifically and economically attractive, but also politically so. The hope here is that the rather politically uncomfortable sufficiency issues could be circumvented by rapid successes in the areas of efficiency and consistency. Bioeconomy, in this sense, promises the reconciliation of economy and environment without a fundamental change in growth-oriented thinking. Although controversies exist in this respect in regard to the bioeconomy, in the following, we will examine these issues through the lens of these political perspectives.

The central question is whether this provides a realistic picture of the potentials of the bioeconomy, but also of its feasibility and the consequences of its realisation. Can the major environmental problems be overcome by means of the bioeconomy? What contributions can the bioeconomy make here, and to what degree? Will sufficiency perhaps be dispensable and the debate on degrowth become meaningless if the bioeconomy is implemented on a large scale? Can we think of unintended side effects of a massive use of the bioeconomy that remain underexposed in view of the noble goals in the debate so far? Would the bioeconomy set in motion a new spiral of exploitation of life by the economy, which, in the long run, might stand in the way of "ways to make peace with nature" (Meyer-Abich 1984)? Is, perhaps, the bioeconomy even seen ideology-critically as only an attempt by powerful actors from politics, economics, science, business and special interests to seize power under the noble mantle of environmental compatibility (Gottwald and Krätzer 2014)? The following substantial and fundamental points of criticism appear to be of particular relevance:

1. Degrowth debate: In the current debate on degrowth (e.g., Jackson 2009; Dietz and O'Neill 2013; D'Alisa et al. 2014), not only is the possibility of further unlimited growth critically questioned, but the overall sense of it is also cast in doubt. The concern is that, with means based solely on efficiency and consistency, and thus on technological progress, such as the bioeconomy, a fundamentally unsustainable economic and social model might be maintained, instead of being changed at its roots (Blühdorn 2007).
2. Respect for life: With the bioeconomy, further steps would be taken on the way to the complete economization of life, which would run counter to the environmentally compatible existence of humans on planet Earth: "The term 'bioeconomy' does not mean an ecologization of the economy, but an economization of the biological, i.e., the living" (Gottwald and Krätzer 2014, translation A.G.). The reevaluation of all living things into the raw material "biomass" is a logical step on a fateful path, which only accelerates the destruction of the basis of human existence (Gottwald and Krätzer 2014).

Both points contain central questions that certainly cannot be conclusively assessed at present. In the following, the first point of criticism will be discussed in more detail. The currently intensively discussed ecomodernistic approach (Manifesto 2015), which, in relation to the alternative of growth or post-growth, is partly – but not completely – associated with the bioeconomy, serves to structure the arguments, so that its consideration allows for a differentiated view.

10.4 Bioeconomy and Ecomodernism

It is instructive to use the current debate on ecomodernism as a background for locating the bioeconomy in the debate on suitable strategies for solving the major environmental problems. The ecomodernist Manifesto (Manifesto 2015) is entirely within the framework of classical-modern ideas of progress, which ultimately go back to David Hume and Francis Bacon and which seek to achieve the most complete possible emancipation of human civilization from nature by technical means. Instead of drawing a conclusion from the global environmental crisis as to the necessity of a "reversal" from the path of classical modernity, its message is that humanity should not stop halfway, or even turn back, but rather should proceed consistently, and even accelerate. As justification, the authors point out that technical progress to date has already led to a considerable reduction in per capita consumption of nature, e.g., the amount of land needed to feed a human being:

> Greater resource productivity associated with modern socio-technological systems has allowed human societies to meet human needs with fewer resource inputs and less impact on the environment (Manifesto 2015).

According to the authors, it would be a mistake to reverse this trend. On the contrary, it must be accelerated in order to achieve more environmentally-friendly development. This can be well illustrated by nutrition: While organic farming, which is usually regarded as more environmentally-friendly, aims at extensifying agriculture, and is thus dependent on a higher land requirement, ecomodernists recommend a highly concentrated, and thus necessarily industrially organised, food industry, which gets by with extremely little land and satisfies the demand for food as completely as possible through technical means. Accordingly, theirs is a vision of a human society that organises itself largely independently of natural resources (Grunwald 2018). A growing world population with increasing prosperity should be made possible on less and less land and using smaller and smaller amounts of raw materials. If we think this through to the end, the planet Earth would be divided into two parts: the smallest possible part of the earth's surface would be used by humans in highly densely populated settlements with highly intensive agriculture or synthetic food production and highly efficient production of goods, while the other and largest possible part would consist of nature, which would largely be left to its own devices and exempted from human use.

The growth paradigm is not called into question. The widespread belief that the finite nature of resources and the limitation of the capacity of the environment to absorb emissions are on an inevitable collision with limitless growth (Meadows et al. 1972) is, in principle, doubted in ecomodernism:

> Despite frequent assertions starting in the 1970s of fundamental 'limits to growth', there is still remarkably little evidence that human population and economic expansion will outstrip the capacity to grow food or procure critical material resources in the foreseeable future. To the degree to which there are fixed physical boundaries to human consumption, they are so theoretical as to be functionally irrelevant (Manifesto 2015).

Therefore, some say, one should not talk about limits to growth, but about the growth of limits (Fücks 2011). As long as the finiteness of resources has not been proven and their quantity cannot be stated beyond doubt, finiteness must not be used as an argument for self-decision. In view of the limited resources in principle, however, ecomodernism ascribes the central role in solving resource and environmental problems to technical progress:

> With proper management, humans are at no risk of lacking sufficient agricultural land for food. Given plentiful land and unlimited energy, substitute for other material inputs to human well-being can easily be found if those inputs become scarce or expensive (Manifesto 2015).

This vision, not to say utopia, is obviously based on a great deal of trust in technological advancement. This is also expressed in the so-called range rule of sustainability in dealing with non-renewable resources (Kopfmüller et al. 2001): According to this, the consumption of non-renewable resources may only be described as sustainable if the temporal range of the resource into the future does not decrease. This seems paradoxical, because each use reduces the available stock – at least if no measures for recycling are planned. The rule can only be complied with if technical progress enables such a considerable increase in the efficiency of the consumption of this resource in the future that the inevitable reduction in the stock as a result of consumption does not have a negative effect on the temporal range of the remaining stock. A certain minimum speed of technical progress is therefore assumed here.

In ecomodernism, beyond the high expectations of technical progress, it is assumed that no relevant unintended side effects will be associated with the efficiency-enhancing technologies and measures, which would counteract the expected positive effects of technical progress and, for example, entail new environmental problems. This is a very delicate premise, since the occurrence of unintended consequences is regarded as a characteristic of modern society, and modern technology in particular (e.g. Grunwald 2019).

It is interesting to observe that the European idea of bioeconomy occupies an alternative position compared to American ecomodernism. Of the three above-mentioned strategies for coping with environmental problems, ecomodernism most relies, and even then only in part, on the idea of efficiency when a reduction in per capita nature demand is specified as the goal, while, at the same time, continuing to grow. However, the reliance on nuclear energy and the resulting unlimited energy supply can even be understood as meaning that efficiency is no longer even important. Efficiency is evidently only an issue in a scarce economy, but not in a world of energy surplus. Even the idea of consistency is alien to ecomodernism, although it is not excluded in principle. If the goal of solving environmental problems is to decouple human civilization from the natural environment, which is made possible by technological progress, it makes no sense to focus on the compatibility of anthropogenic and natural material flows and to make efforts and use resources for this purpose. Consistency is irrelevant in this model.

Bioeconomy, on the other hand, makes consistency a constituent feature. Its programme is not the technically possible decoupling from nature, but the technically possible better adaptation of the human economy to natural cycles, material flows and organisational principles. This also has consequences for a forward-looking ethical assessment (Grunwald 2018).

10.5 Hostility to or Alliance with Nature?

The fact that the possibilities of technical progress for solving environmental problems are to be used within the framework of a responsible strategy that anticipates the potential unintended consequences is unlikely to be doubted ethically. Controversies extend to what precisely "responsible" should mean, which unintended consequences must be expected in which scenarios, how to weigh the environmental relief objectives pursued against possible side effects, and which measures promise the best overall impacts. These are the normal challenges for a technology assessment (Grunwald 2019), as they are processed on the basis of concrete technologies and context-related requirements. At this point, further imperatives for action that go beyond individual technologies and contexts are to be considered and classified:

1. Bioeconomy in the sense of ecomodernism: Technical progress should be accelerated as a key contribution to solving environmental problems.
2. Ecomodernism alone: Environmental problems should be tackled by decoupling human civilisation from nature.
3. Bioeconomy alone: Technical progress should be geared towards consistency with natural material flows and cycles.
4. Efficiency and consistency: Other measures, such as a departure from the growth paradigm or behavioural changes, do not need to be pursued, at least not urgently.

At this conceptual level, bioeconomy and ecomodernism both see technical progress not only as a necessary condition for sustainable development, but also as a sufficient condition. They thus burden technical progress and its possibilities with total responsibility for solving environmental problems. Criticism at this level therefore affects both approaches equally. However, the differences in premises 2 and 3 allow for a differentiated ethical assessment.

Any acceleration of technical progress reduces the chances of learning from experience with new technology, even with unintended consequences or insufficient fulfilment of expectations, for further action. Acceleration increases dependence on technological progress and reduces the chances of being able to think about alternatives or complementary measures at all. It creates factual constraints and undermines consideration of alternatives, which is essential for informed decisions on how to proceed. Ill-considered demands for acceleration also ignore the questions of the risks associated with trust in technological progress and the options that remain if trust in technological progress turns out to be unjustified.

Hans Jonas (1979) warned against risking "the whole thing" on a bet. However, the ecomodernist position does exactly that: it relies completely on technical progress and does not place any demands on technology other than to continually increase efficiency. In doing so, it makes future developments in the Anthropocene dependent on this trust in technical progress being justified. If this hope were not fulfilled, however, "the whole," in the sense of Hans Jonas, would be endangered, since no other option would be available as an alternative. This is nothing more than the position of a moral hazard-maker who bets everything on one card. The ecomodernist position in its purest form is thus ethically unjustifiable (Grunwald 2018). The dominant hope in ecomodernism of solving problems through technical progress must be supplemented. In terms of ethical responsibility, precautions must be taken in the event that its techno-optimistic assumptions are not realised or are only realised in part.

The bioeconomy also relies on technical progress, but has more far-reaching requirements for the environmental compatibility of future technology. By demanding consistency with natural processes, environmental compatibility is to be virtually integrated into the agenda of further technological development. Problem solving is not expected from technical progress itself, but from technical progress that is aligned with the principle of consistency. Here, one can connect to an idea of the highly technicoptimistic Marxist philosopher Ernst Bloch. According to this, technology should no longer be developed and used against nature, an approach that Bloch sees and criticizes as characteristic of traditional modernity (Bloch 1985). Bionics, the orientation of technical solutions towards models from nature, can also be used to refer to the concept of alliance technology by Ernst Bloch (von Gleich et al. 2007). Instead of viewing nature as an enemy and trying to bring it under complete control, as was the goal of the Bacon project (Schäfer 1993), technology should be pursued in alliance with it. This early thought appears compatible with the bioeconomy's demand for strategic consistency of human economic activity with natural material flows. In this way, the bioeconomy appears to be more responsible and ethically sustainable than the ecomodernist position in its purest form.

However, the potential of the bioeconomy is also not guaranteed, nor is the absence of counterproductive side effects of its implementation. Therefore, a one-sided strategy based on bioeconomy would ultimately be problematic in terms of responsible ethics. Even if the prospects of coping with, or at least mitigating, environmental problems appear to be better, in principle, than those in ecomodernism, which is based on an accelerated "continuation as before" approach, due to the higher demands on environmental compatibility, options must also be provided or developed here in the event that expectations are not met. This applies, in particular, to post-growth strategies that use social or sociotechnical innovations to ultimately target other values of action, other lifestyles and behavioural patterns, but also other value chains and social incentive systems (Jackson 2009; Dietz and O'Neill 2013; D'Alisa et al. 2014). Against this background, the expectations in political and economic institutions appear to be one-sided (▶ Sect. 10.3). A discussion between the different positions of the bioeconomy on possible combinations of a technical bioeconomy based on technical efficiency and consistency with social science considerations on adequate lifestyles and value patterns seems overdue.

The bioeconomy is promising, but it alone does not guarantee a more environmentally-friendly technology and economy in the future. This follows solely from the well-known problem of rebound effects, of which there are more than enough examples of disappointed expectations for environmental relief (Sorrell 2007). It should also be remembered that bioeconomic production lines and value chains are by no means environmentally neutral, as the example of energy crop cultivation, with its ecological side effects, shows. In principle, therefore, the overall ecological balance must also be examined in bioeconomic approaches so that it can be regarded as a sustainable bioeconomy (▶ Chap. 9).

10.6 A Learning Process on the Concept of Sustainability

The promises of the bioeconomy are wide-ranging. There is no doubt that it has great potential to contribute to a better environmental balance of human activity. There are also reasons to assume that a more sustainable economic future cannot be achieved without the kinds of technology and value creation that are fundamentally bioeconomically oriented. However, the word "potential" must be taken seriously here: From today's perspective, these are possible contributions to solving environmental problems. It is therefore necessary to ask under which conditions the potentials presented today can be turned into real future solutions.

It goes without saying that there is a more or less high degree of uncertainty about the future arrival of expectations for the bioeconomy. There are no automatisms that turn technical potentials into real contributions to environmentally compatible development. Realising potentials is not a purely technical matter, but rather a political, economic and general social matter of considerable complexity. Since assessments of these hypothetical potentials inevitably guide decisions, e.g., on the political promotion of bioeconomy, not only do the prospects of success and realisation need to be critically examined, but so do the possible negative consequences and risks. They, too, are hypothetical, but this should not stand in the way of an early engagement with them in the sense of an early warning of the technology assessment (Grunwald 2019) in order to mitigate or prevent them.

A future design of the bioeconomy from environmental and sustainability aspects would thus include the following steps:

- Analysis of the (positive and negative) sustainability potentials of the different directions of the bioeconomy and the innovation paths envisaged in each case at the earliest possible stages of development.

- Investigation of the degree and the expression of their prospects of success in problem-solving, as well as their limits.
- Development of prospective innovation paths in view of promoting, as well as inhibiting, factors, e.g., embedded in scenarios of an increasingly bioeconomy-based economy.
- Critical consideration of their premises with regard to the sustainability of quantitative growth and relations to post-growth and sufficiency strategies.
- Analysis of the factors on which it depends as to whether the positive sustainability potential can be realised or whether the sustainability risks can be managed at an early stage, with special consideration given to rebound effects (Sorrell 2007).
- Comparison with the potentials of non-bio-based techniques to achieve greater environmental compatibility and sustainability or analyses of the possible combinations of bio-based and non-bio-based techniques.
- Continuation and concretisation of this multi-stage process in the course of the further development of the bioeconomy.

The design of the bioeconomy should thus form a continuous learning process, oriented towards the normative model of sustainability (Grunwald 2004), in which design goals, implementation options, innovation paths and unintended effects are discussed. In the bioeconomy, it is important to combine technological development and innovation on the one hand with research into and reflection on their hypothetical consequences on the other – in particular, the potential for solving major environmental problems. These learning processes, which take place concretely at the various interfaces between politics, the public, science, industry and technology, incorporate scientific knowledge and ethical orientations. The picture of a more environmentally-friendly economy, including the bioeconomy, is gradually emerging, step by step. Only in this process will it become clear how big the contribution of the bioeconomy will really be in solving the major environmental problems. Future-ethical precautionary arguments therefore call for both a careful opening up of this path in regard to its options and precautions to be taken in the event that expectations cannot be fulfilled, or can only be fulfilled to a small extent, and the choice of other strategies of sustainable development.

References

Bloch E (1985) Das Prinzip Hoffnung. Suhrkamp, Frankfurt

Blühdorn I (2007) Sustaining the unsustainable: symbolic politics and the politics of simulation. Environ Polit 16(2):251–275

BMBF – Federal Ministry of Education and Research (2010) National Research Strategy Bioeconomy 2030. https://www.bmbf.de/pub/Nationale_Forschungsstrategie_Biooekonomie_2030.pdf

D'Alisa G, Demaria F, Kallis G (2014) Degrowth: a vocabulary for a new era. Routledge, San Francisco/London

Dauvergne P (2008) The shadows of consumption. Consequences for the global environment. MIT Press, Cambridge, Mass

Dietz R, O'Neill D (2013) Enough is enough: building a sustainable economy in a world of finite resources. Routledge, San Francisco/London

Fücks R (2011) Das Wachstum der Grenzen. Böll Thema 2(2011):4–6

Fussler C (1999) Die Öko-Innovation. Hirzel, Stuttgart

Gottwald F, Krätzer A (2014) Irrweg Bioökonomie. Kritik an einem totalitären Ansatz. Suhrkamp, Frankfurt

Grunwald A (2004) Strategic knowledge for sustainable development: the need for reflexivity and learning at the interface between science and society. Int J Foresight Innov Policy 1(1/2):150–167

Grunwald A (2012) Sustainability assessment of technologies – an integrative approach. In: Ghenai C (Hrsg) Sustainable development – energy, engineering and technologies – manufacturing and environment. Open access, InTech, pp 35–62

Grunwald A (2018) Diverging pathways to overcoming the environmental crisis: a critique of eco-modernism from a technology assessment perspective. J Clean Prod 197(2018):1854–1862. https://doi.org/10.1016/j.jclepro.2016.07.212

Grunwald A (2019) Technology assessment in practice and theory. Routledge, London

Grunwald A, Kopfmüller J (2012) Nachhaltigkeit. Campus, Frankfurt

Huber J (1995) Nachhaltige Entwicklung. Strategie für eine ökologische und soziale Erdpolitik. Berlin

Jackson T (2009) Prosperity without growth? The transition to a sustainable economy. Sustainable Development Commission, London

Jonas H (1979) The principle of responsibility. Suhrkamp, Frankfurt

König W (2008) Kleine Geschichte der Konsumgesellschaft. Konsum als Lebensform der Moderne, Stuttgart

Kopfmüller J, Brandl V, Jörissen J, Paetau M, Banse G, Coenen R, Grunwald A (2001) Nachhaltige Entwicklung integrativ betrachtet. Konstitutive Elemente, Regeln, Indikatoren. Edition Sigma, Berlin

Leggewie C, Welzer H (2009) Das Ende der Welt, wie wir sie kannten. Frankfurt

Manifesto (2015) The Breakthrough Institute (Ed.) An Ecomodernist Manifesto. www.ecomodernism.org

Meadows D, Meadows D, Randers J, Behrens W (1972) Limits to growth. New American Library, New York

Meyer-Abich K-M (1984) Wege zum Frieden mit der Natur. Praktische Naturphilosophie für die Umweltpolitik. Beck, München

OECD (2011) Towards green growth. OECD, Paris

Schäfer L (1993) Das Bacon-Projekt. Suhrkamp, Frankfurt

Schneidewind U, Zahrnt A (2013) Damit gutes Leben einfacher wird: Perspektiven einer Suffizienzpolitik. oekom, München

Sorrell S (2007) The rebound effect: an assessment of the evidence for economy-wide energy savings from improved energy efficiency. UK Energy Research Centre, London

UNEP – United Nations Environmental Programme (2011) Towards a green economy. Pathways to sustainable development and poverty eradication. A synthesis for policy makers. Nairobi

von Gleich A, Pade C, Petschow U, Pissarskoi E (2007) Bionik: Aktuelle Trends und zukünftige Potentiale. Universität Bremen, Bremen

WBGU – Wissenschaftlicher Beirat Globale Umweltveränderungen (1996) Welt im Wandel. Herausforderungen an die deutsche Wissenschaft. Springer, Berlin

Supplementary Information

J. Pietzsch (ed.), *Bioeconomy for Beginners*, https://doi.org/10.1007/978-3-662-60390-1

Index

MIX
Papier aus verantwortungsvollen Quellen
Paper from responsible sources
FSC® C105338

If you have any concerns about our products,
you can contact us on
ProductSafety@springernature.com

In case Publisher is established outside the EU,
the EU authorized representative is:
Springer Nature Customer Service Center GmbH
Europaplatz 3, 69115 Heidelberg, Germany

Printed by Libri Plureos GmbH
in Hamburg, Germany